AF581664

Mycotoxins Occurence in Feed and Their Influence on Animal Health

Mycotoxins Occurence in Feed and Their Influence on Animal Health

Editors

Maciej Gajęcki
Magdalena Gajęcka
Łukasz Zielonka

MDPI • Basel • Beijing • Wuhan • Barcelona • Belgrade • Manchester • Tokyo • Cluj • Tianjin

Editors
Maciej Gajęcki
University of Warmia and Mazury
Poland

Magdalena Gajęcka
University of Warmia and Mazury
Poland

Łukasz Zielonka
University of Warmia and Mazury
Poland

Editorial Office
MDPI
St. Alban-Anlage 66
4052 Basel, Switzerland

This is a reprint of articles from the Special Issue published online in the open access journal *Toxins* (ISSN 2072-6651) (available at: https://www.mdpi.com/journal/toxins/special_issues/mycotoxins_feed_animal_health).

For citation purposes, cite each article independently as indicated on the article page online and as indicated below:

LastName, A.A.; LastName, B.B.; LastName, C.C. Article Title. *Journal Name* **Year**, *Volume Number*, Page Range.

ISBN 978-3-03943-847-1 (Hbk)
ISBN 978-3-03943-848-8 (PDF)

© 2020 by the authors. Articles in this book are Open Access and distributed under the Creative Commons Attribution (CC BY) license, which allows users to download, copy and build upon published articles, as long as the author and publisher are properly credited, which ensures maximum dissemination and a wider impact of our publications.
The book as a whole is distributed by MDPI under the terms and conditions of the Creative Commons license CC BY-NC-ND.

Contents

About the Editors

Maciej Gajęcki, DVM—Professor of Veterinary Sciences at the Department of Veterinary Prevention and Feed Hygiene, Faculty of Veterinary Medicine, University of Warmia and Mazury in Olsztyn, Poland. He completed a full-time Master's degree program at the Faculty of Veterinary Medicine, University of Agriculture and Technology in Olsztyn, and received a Master of Science degree in Veterinary Medicine in 1973. From May 1973 to September 1975, he was employed at the State-owned Animal Health Care Center and Commercial Pig Farm. Since 1981, he has been employed at the University of Warmia and Mazury in Olsztyn. During his professional career, he has been awarded a doctoral degree in Veterinary Sciences, a postdoctoral degree in Veterinary Sciences and the academic title of Professor of Veterinary Sciences. Over the last 27 years, the research interests of Professor Gajecki and his co-workers have centered around the effects exerted by zearalenone on farm animals (gilts), companion animals (female dogs), and humans (female patients). Professor Gajecki has conducted studies investigating the influence of zearalenone on the health of the reproductive tract and zearalenone mycotoxicosis in female patients, with a particular emphasis on the mycotoxin's effects on the mammary gland. As part of his professional activities, Professor Gajecki provided advisory service to the Ministry of Agriculture and Rural Development. He was President of the Sanitary-Epizootic Council at the General Veterinary Inspectorate (prevention and control of bovine spongiform encephalopathy and Aujeszky's disease). He is also the Director of postgraduate specialization studies in veterinary prevention and feed hygiene at the Polish National Veterinary Chamber. He has been elected three times as the Chairman of Polish delegates participating in annual meetings of the Ad Hoc Intergovernmental Codex Task Force on Animal Feeding (established by the Codex Alimentarius Commission) in Copenhagen.

Magdalena Gajęcka, DVM completed a full-time Master's degree program at the Faculty of Veterinary Medicine, University of Warmia and Mazury in Olsztyn, Poland (including a 10th semester at the Faculty of Veterinary Medicine, Complutense University in Madrid, Spain) and received a Master of Science degree in Veterinary Medicine in 2001. In 2002, she completed postgraduate studies in analytics in environmental protection at the Faculty of Chemistry, Nicolaus Copernicus University in Toruń. In June 2005, she completed postgraduate specialization studies in veterinary prevention and feed hygiene at the Polish National Veterinary Chamber. In 2006, she was awarded a doctoral degree in Veterinary Sciences at the Faculty of Veterinary Medicine, University of Warmia and Mazury in Olsztyn, after successfully defending a doctoral dissertation entitled "The Effect of Experimental Zearalenone Mycotoxicosis on the Female Canine Reproductive Tract". In 2014, she was awarded a postdoctoral degree in Veterinary Sciences at the Faculty of Veterinary Medicine, University of Warmia and Mazury in Olsztyn, after publishing a monothematic series of publications regarded as an outstanding research achievement, entitled "The Effect of Short-Term Zearalenone Mycotoxicosis on Changes in Selected Tissues of Female Dogs". In 2019, she was awarded the academic title of Professor of Veterinary Sciences. Professor Gajecka's research interests focus on feed hygiene, with a particular emphasis on animal waste management, the applicability of natural feed additives and feed materials, distribution and disposal of medicated feeds, safety and quality of animal feedstuffs, mycotoxicosis—implications for animal and human health, zearalenone mycotoxicosis: plants–animals–humans, diagnostic significance of selected Fusarium mycotoxicoses, zearalenone as a destructor of endocrine homeostasis of steroid hormones in female patients, gilts and female dogs.

Łukasz Zielonka is a veterinarian and a professor at the Faculty of Veterinary Medicine, University of Warmia and Mazury in Olsztyn, Poland. He is Head of the Department of Veterinary Prevention and Feed Hygiene. He is an expert on veterinary prevention and feed hygiene, and swine diseases. He has completed postgraduate studies in computer science and analytics in environmental protection. He is the author or co-author of more than 90 original research articles published in peer-reviewed journals, and a reviewer in several scientific journals with a high impact factor. Professor Zielonka is an academic teacher at the Faculty of Veterinary Medicine, University of Warmia and Mazury in Olsztyn. He has delivered lectures at international and national scientific conferences. He is a member of the Polish Society of Veterinary Science, the Polish Society of Toxicology, and the Polish Society of Medical Biology. Professor Zielonka's research interests focus on the biological determinants of mycotoxin contamination of foodstuffs and feedstuffs, the effects of mycotoxins on human and animal health, laboratory analyses of mycotoxins and other biologically active substances, mostly with the use of liquid chromatography-mass spectrometry. He holds a patent and has implemented several research projects. Practical implications of the research conducted by Professor Zielonka include mycotoxin biodegradation with the use of microorganisms (bacteria and yeasts) and insect larvae, biological activities and effects of bovine colostrum, and biologically active plants.

Preface to "Mycotoxins Occurence in Feed and Their Influence on Animal Health"

A large percentage of plant materials and plant-based products used in feed production can be contaminated with mycotoxins (undesirable substances), which pose health risks for humans and different livestock species. The symptoms and adverse health (toxicological) effects associated with exposure to high doses of many mycotoxins are well known. According to the hormesis paradigm, greater attention should be paid to the risks resulting from exposure to low mycotoxin doses that are often encountered in feedstuffs. Dysfunctions in mammals exposed to pure parent compounds, without metabolites or modified mycotoxins, constitute a very interesting research problem.

Most mycotoxins are absorbed in the proximal section of the small intestine, due to considerable physiological variation between intestinal segments. Next, mycotoxins enter the bloodstream, and blood parameters can be used for non-invasive assessment of the animals' health status and for identifying new biomarkers of pathological states. Most targets are cells, where mycotoxins and their metabolites can cause certain pathological states. These mycotoxins not only regulate enzyme metabolism or gene expression, but can also act as ligands that bind to specific receptors on cell membranes or in nuclei, thus participating in signal transduction.

For this reason, the presented studies in the Special Issue of Toxins evaluated selected body systems and functional biomarkers of animals for different doses of mycotoxins causing mycotoxicosis. I hope that the knowledge gained will deepen our knowledge about the effects of mycotoxins on animal health and facilitate decision-making in risk management.

Maciej Gajęcki, Magdalena Gajęcka , Łukasz Zielonka
Editors

Editorial

The Presence of Mycotoxins in Feed and Their Influence on Animal Health

Maciej T. Gajęcki, Magdalena Gajęcka * and Łukasz Zielonka

Department of Veterinary Prevention and Feed Hygiene, Faculty of Veterinary Medicine, University of Warmia and Mazury in Olsztyn, Oczapowskiego 13/29, 10-718 Olsztyn, Poland; gajecki@uwm.edu.pl (M.T.G.); lukaszz@uwm.edu.pl (Ł.Z.)

* Correspondence: mgaja@uwm.edu.pl; Tel.: +48-89-523-32-37; Fax: +48-89-523-36-18

Received: 15 September 2020; Accepted: 14 October 2020; Published: 20 October 2020

Mycotoxins are secondary metabolites of fungi. They are commonly detected in food, feed [1–4], and feed additives [5]. The presence of mycotoxins in food and feed is influenced by numerous factors, including plant species and variety, region, temperature, moisture content, insect damage, storage conditions, and agricultural practices. Reisinger et al. [1] analyzed the prevalence of mycotoxins in maize silage, which is the key component of diets for dairy cattle. In addition to *Fusarium* mycotoxins zearalenone (ZEN), deoxynivalenol (DON), and nivalenol, maize silage in selected European countries also contains emodin, culmorin, enniatin, and beauvericin. The authors found that these mycotoxins compromised the gastrointestinal health of cattle during mixed mycotoxicosis and exerted unknown effects on the composition and function of the ruminal microbiota. Kemboi et al. [4] investigated the contamination of cattle feeds in various regions of Africa. They reported that local farmers made few attempts to minimize feed contamination with mycotoxins, which compromised the health quality of dairy products. African farmers lack the knowledge of how, when, and which detoxicants should be used, and the applied agents are often ineffective. Khoshal et al. [2] analyzed 524 samples of pig diets from around the world for the presence of 800 metabolites. Eighty-eight percent of the samples were contaminated with DON. Feeds also contained mycotoxins produced by fungi of the genera *Fusarium*, *Aspergillus*, *Penicillium*, and *Alternaria*, and their in vitro toxicity was similar to that of DON. The identified compounds were arranged in the following order based on their toxicity: apicidin > enniatin A1 > DON > beauvericin > enniatin B > enniatin B1 > emodin > aurofusarin. An in vivo test also revealed that the presence of additional metabolites did not increase the overall toxicity. Witaszak et al. [3] examined the presence of mycotoxins in dry food for dogs and cats. Their research was motivated by the fact that cereals are a major ingredient in dry food for companion animals. The authors concluded that more effective solutions are needed to reduce the concentrations of mycotoxins in plant materials and animal feeds, and, more importantly, that the proportion of plant materials in the diets of domesticated predators should be decreased. The use of by-products from medicinal fungi in the production of animal feeds also poses a problem [5]. These by-products contain compounds such as adenosine, cordycepin, and pentostatin, as well as substances with cytotoxic and neurotoxic properties. The authors found that *Cordyceps* fungi produced large amounts of unidentified secondary metabolites which were secreted by biosynthetic gene clusters (BGC). These compounds could also be applied in the pharmaceutical industry.

At present, risk evaluations focus mainly on mycotoxins' ability to modify the metabolic profile and their proinflammatory, mutagenic, cytotoxic, and potential carcinogenic effects [6–9]. Rykaczewska et al. [6] reported that pre-pubertal gilts responded differently to ZEN administered at the lowest-observed-adverse-effect level (LOAEL), no-observed-adverse-effect level (NOAEL), and minimal anticipated biological effect level (MABEL) doses. One of the differences was the fact that the proportion of β-ZEL increased in the group of ZEN metabolites. Different responses were observed in female pigs from other age groups. Beta-ZEL exerts varied effects—it induces a minor increase in body weight gain, but also slows down the sexual maturation in gilts. The ZEN levels are initially very low, and its metabolites are not detected in the blood serum (in particular under exposure to the MABEL dose), which confirms that gilts have a high physiological demand for exogenous estrogen-like substances. These substances are readily utilized by pre-pubertal gilts. When administered at higher doses, excess "free" ZEN plays other, not always beneficial, roles, and it can lead to ovarian atrophy or silent heat. The levels of estradiol (E_2) and "free" ZEN increase proportionally to the ZEN dose, which suppresses the concentrations of progesterone (P_4) and testosterone (T). Zielonka et al. [8] analyzed the concentrations of selected steroid hormones (E_2, P_4, and T) in premenopausal pigs administered ZEN at 20 or 40 µg/kg BW for 48 days, and observed that (i) the concentrations of ZEN in the peripheral blood were very low and highly varied on different days of exposure, and their diagnostic value was difficult to determine; (ii) experimentally induced hyperestrogenism or "supraphysiological hormone levels" contributed to a minor increase in the total E_2 levels (which could intensify proliferation processes) and suppressed T concentrations; (iii) the results can be extrapolated to indicate that the analyzed doses of ZEN produced varied responses, where a lower dose probably exerted stimulatory/adaptive effects and a higher dose inhibited physiological processes in the studied animals. Wang et al. [7] confirmed the proinflammatory effects of mycotoxins in vitro. They found that nitric oxide (NO) activity and the relative expression of *iNOS* mRNA increased with a rise in DON dose, and the relative expression of the *COX*-2 gene, which was identical to that of the induced enzyme, also increased in porcine intestinal epithelial cells (IPEC-J2). Intestinal epithelial cells were stimulated by DON to produce an inflammatory response. However, the NF-κB pathway is a potential pathogenic factor which, when activated incorrectly or in excessive amounts, exerts adverse effects on cells. Obremski et al. [9] arrived at similar conclusions in a study investigating the effects of very low ZEN doses on (i) the secretion of proinflammatory cytokines, (ii) the secretion of anti-inflammatory and regulatory cytokines, (iii) oxidative stress markers, and (iv) basic metabolic markers. They found that low ZEN doses induced an inflammatory response. The proinflammatory properties of ZEN and intensified oxidative stress can impair intestinal epithelial function, as demonstrated by oxidative stress markers such as the biochemical changes associated with the metabolism of sugars (intensified glycolysis) and amino acids (proline).

Recent research has demonstrated that mycotoxins exert adverse effects on sensitive structures in the intestines [9,10] and target tissues/organs such as the liver [11]. In a study by Śliżewska et al. [10], ochratoxin A (OTA) present in low concentrations in turkey feed caused body weight loss, depression, and paralysis, leading to a decrease in the mobility and energy levels of birds. The above symptoms were accompanied by catarrhal inflammation of the gastrointestinal mucosa and local ecchymoses on the liver and kidney surface. To prevent the adverse effects of OTA, turkey diets were supplemented with three synbiotic preparations. The tested synbiotics contributed to an increase in the counts of beneficial bacteria and a decrease in the counts of pathogenic gut microbiota. They also increased the activity of α-glucosidase and α-galactosidase while decreasing the activity of fecal enzymes, which exerted a beneficial influence on the health status of turkeys and improved their body weight gains. Skiepko et al. [11] found that low doses of ZEN and DON in feed induced changes in the histology and ultrastructure of the liver in prepubertal gilts, including (i) an increase in the thickness of the perilobular connective tissue and lobe penetration by connective tissue; (ii) an increase in the histology activity index; (iii) the widening of the liver sinusoids; (iv) transient changes in the glycogen content; (v) the increased accumulation of iron in hepatocytes; (vi) changes in the organization of the endoplasmic reticulum in hepatocytes; (vii) changes in the morphology of Kupffer–Browicz cells. The above observations suggest that low doses of mycotoxins administered individually or in combination, even for short periods of time, affected the liver morphology. Another study [12] postulated that mycotoxins could exert adverse effects on the enteric nervous system (ENS) and that their influence should be analyzed in greater detail. The ENS plays a key role in the regulation of most gastrointestinal functions; it participates in adaptive processes and defense mechanisms and acts as one of the first barriers against pathogens and toxins in feed materials. Mycotoxins can exert multidirectional effects depending on their chemical structure, the investigated mammalian species, and the type of nerve fiber bundles and segment of the gastrointestinal tract affected. Mycotoxins can affect the size and morphology of intestinal nerve fibers and the neurochemical characteristics of enteric neurons. These changes are probably induced by adaptive and defense responses that promote homeostasis. The changes observed in the ENS are often the first symptoms of contamination with low mycotoxin doses.

A review of the presented research studies indicates that reference biomarkers should be developed to support quick evaluations of animal health (in particular, the gastrointestinal tract and the accompanying tissues) during ongoing mycotoxicosis, regardless of the dose of the parent compound and its metabolites. The above applies to livestock as well as companion animals that are fed monotonic diets for long periods of time.

Funding: This research received no external funding.

Acknowledgments: The editors are grateful to all the authors who contributed to this Special Issue. Special thanks go to the expert peer reviewers for rigorously evaluating the submitted manuscripts. Lastly, the valuable contributions, organization, and editorial support of the MDPI management team and staff are greatly appreciated.

Conflicts of Interest: The authors declare no conflict of interest.

References

1. Reisinger, N.; Schürer-Waldheim, S.; Mayer, E.; Debevere, S.; Antonissen, G.; Sulyok, M.; Nagl, V. Mycotoxin occurrence in maize silage-a neglected risk for bovine gut health? *Toxins* **2019**, *11*, 577. [CrossRef] [PubMed]
2. Khoshal, A.K.; Novak, B.; Martin, P.G.P.; Jenkins, T.; Neves, M.; Schatzmayr, G.; Oswald, I.P.; Pinton, P. Co-Occurrence of DON and emerging mycotoxins in worldwide finished pig feed and their combined toxicity in intestinal cells. *Toxins* **2019**, *11*, 727. [CrossRef] [PubMed]
3. Witaszak, N.; Waśkiewicz, A.; Bocianowski, J.; Stępień, Ł. Contamination of pet food with mycobiota and *fusarium* mycotoxins-focus on dogs and cats. *Toxins* **2020**, *12*, 130. [CrossRef] [PubMed]
4. Kemboi, D.C.; Antonissen, G.; Ochieng, P.E.; Croubels, S.; Okoth, S.; Kangethe, E.K.; Faas, J.; Lindahl, J.F.; Gathumbi, J.K. A review of the impact of mycotoxins on dairy cattle health: Challenges for food safety and dairy production in Sub-Saharan Africa. *Toxins* **2020**, *12*, 222. [CrossRef] [PubMed]
5. Chen, B.; Sun, Y.; Luo, F.; Wang, C. Bioactive metabolites and potential mycotoxins produced by *cordyceps* fungi: A review of safety. *Toxins* **2020**, *12*, 410. [CrossRef] [PubMed]
6. Rykaczewska, A.; Gajęcka, M.; Onyszek, E.; Cieplińska, K.; Dąbrowski, M.; Lisieska-Żołnierczyk, S.; Bulińska, M.; Babuchowski, A.; Gajęcki, M.T.; Zielonka, Ł. Imbalance in the blood concentrations of selected steroids in pre-pubertal gilts depending on the time of exposure to low doses of zearalenone. *Toxins* **2019**, *11*, 561. [CrossRef] [PubMed]
7. Wang, X.; Zhang, Y.; Zhao, J.; Cao, L.; Zhu, L.; Huang, Y.; Chen, X.; Rahman, S.U.; Feng, S.; Li, Y.; et al. Deoxynivalenol induces inflammatory injury in IPEC-J2 cells via NF-κB signaling pathway. *Toxins* **2019**, *11*, 733. [CrossRef] [PubMed]
8. Zielonka, Ł.; Gajęcka, M.; Lisieska-Żołnierczyk, S.; Dąbrowski, M.; Gajęcki, M.T. The effect of different doses of zearalenone in feed on the bioavailability of zearalenone and alpha-zearalenol, and the concentrations of estradiol and testosterone in the peripheral blood of pre-pubertal gilts. *Toxins* **2020**, *12*, 144. [CrossRef] [PubMed]
9. Obremski, K.; Trybowski, W.; Wojtacha, P.; Gajęcka, M.; Tyburski, J.; Zielonka, Ł. The effect of zearalenone on the cytokine environment, oxidoreductive balance and metabolism in porcine ileal peyer's patches. *Toxins* **2020**, *12*, 350. [CrossRef] [PubMed]
10. Śliżewska, K.; Markowiak-Kopeć, P.; Sip, A.; Lipiński, K.; Mazur-Kuśnirek, M. The effect of using new synbiotics on the turkey performance, the intestinal microbiota and the fecal enzymes activity in turkeys fed ochratoxin a contaminated feed. *Toxins* **2020**, *12*, 578. [CrossRef] [PubMed]

11. Skiepko, N.; Przybylska-Gornowicz, B.; Gajęcka, M.; Gajęcki, M.; Lewczuk, B. Effects of deoxynivalenol and zearalenone on the histology and ultrastructure of pig liver. *Toxins* **2020**, *12*, 463. [CrossRef] [PubMed]
12. Gonkowski, S.; Gajęcka, M.; Makowska, K. Mycotoxins and the enteric nervous system. *Toxins* **2020**, *12*, 461. [CrossRef] [PubMed]

Publisher's Note: MDPI stays neutral with regard to jurisdictional claims in published maps and institutional affiliations.

© 2020 by the authors. Licensee MDPI, Basel, Switzerland. This article is an open access article distributed under the terms and conditions of the Creative Commons Attribution (CC BY) license (http://creativecommons.org/licenses/by/4.0/).

Review

Mycotoxins and the Enteric Nervous System

Sławomir Gonkowski [1], Magdalena Gajęcka [2] and Krystyna Makowska [3,*]

[1] Department of Clinical Physiology, Faculty of Veterinary Medicine, University of Warmia and Mazury in Olsztyn, Oczapowskiego 13, 10-957 Olsztyn, Poland; slawomir.gonkowski@uwm.edu.pl
[2] Department of Veterinary Prevention and Feed Hygiene, Faculty of Veterinary Medicine, University of Warmia and Mazury in Olsztyn, Oczapowskiego Str. 13, 10-718 Olsztyn, Poland; magdalena.gajecka@uwm.edu.pl
[3] Department of Clinical Diagnostics, Faculty of Veterinary Medicine, University of Warmia and Mazury in Olsztyn, Oczapowskiego 14, 10-957 Olsztyn, Poland
* Correspondence: krystyna.makowska@uwm.edu.pl

Received: 5 June 2020; Accepted: 17 July 2020; Published: 19 July 2020

Abstract: Mycotoxins are secondary metabolites produced by various fungal species. They are commonly found in a wide range of agricultural products. Mycotoxins contained in food enter living organisms and may have harmful effects on many internal organs and systems. The gastrointestinal tract, which first comes into contact with mycotoxins present in food, is particularly vulnerable to the harmful effects of these toxins. One of the lesser-known aspects of the impact of mycotoxins on the gastrointestinal tract is the influence of these substances on gastrointestinal innervation. Therefore, the present study is the first review of current knowledge concerning the influence of mycotoxins on the enteric nervous system, which plays an important role, not only in almost all regulatory processes within the gastrointestinal tract, but also in adaptive and protective reactions in response to pathological and toxic factors in food.

Keywords: mycotoxins; enteric nervous system; gastrointestinal tract; mammals; animal pathology; intestines; toxins; feed

Key Contribution: Mycotoxins contained in food affect the living organism, especially the gastrointestinal tract and the enteric nervous system. This impact may be multidirectional and depends not only on the chemical structure of the mycotoxin and mammal species studied, but also on the type of the enteric plexuses and segment of the digestive tract.

---

1. Introduction

Mycotoxins are a group of several biochemicals synthesized as secondary metabolites by various species of fungi [1]. They are commonly found in a wide range of agricultural products, such as cereals (maize, wheat, rye), fresh and dried fruits, grape juice, spices, herbs and many others [2–4]. Moreover, the presence of mycotoxins has also been observed in food products of animal origin and water [3–6]. Previous studies have shown that mycotoxins show multidirectional harmful effects on human and animal health. It is known that mycotoxins may act on many internal organs and systems, including, among others, nervous, reproductive and immunological systems, metabolic processes and endocrine glands [7].

This widespread occurrence of mycotoxins and their adverse effects demonstrate that these substances are a serious health and economic problem of the contemporary world and therefore, mycotoxins are the most widely studied biological toxins [5,6]. However, many aspects of mycotoxin activity on eukaryotic organisms are unknown. One lesser-known issue is the influence of these substances on the enteric nervous system (ENS).

Since mycotoxins are present in food and drinking water, the gastrointestinal (GI) tract is the part of the body that first comes into contact with these toxic factors [8]. A relatively large number of studies have described mycotoxin-induced morphological and functional changes in the GI tract, whose character depends on the type of mycotoxin, mammal species studied, as well as the degree and length of exposure to mycotoxins [9–19]. The most common effects of mycotoxins on the GI tract include inflammatory and necrotic changes, disturbances in secretory activity and metabolism of the enterocytes, damage to the intestinal barrier and dysfunction in intestinal absorption [10,11,16,20]. Unfortunately, the impact of mycotoxins on the ENS has been neglected in toxicological studies for many years. There are a few recent studies published which describe this aspect of mycotoxin activity. These reports have indicated that the ENS plays a crucial role in the regulation of the majority of gastrointestinal functions, takes part in adaptive and protective processes and is one of the first barriers against pathological and toxic factors in food [15–17,21,22] and may also be compromised by the harmful effects of mycotoxins. Therefore, this work is an attempt to summarize the influence of mycotoxins on the ENS. To better understand this influence, a short description of the organization of the ENS is needed.

2. Anatomy of the Enteric Nervous System

The enteric nervous system is a specific part of the autonomic nervous system. It is situated in the wall of the gastrointestinal tract from the esophagus to the rectum and is responsible for the majority of gastrointestinal activities [23]. In terms of the number of nerve cells, the ENS is the second largest (after the brain, and before the spinal cord) nervous structure in mammals, which may contain an estimated 200–500 million neurons [24–26]. For this reason, as well as due to the complicated structure and high autonomy, the ENS is often called the intestinal brain [24].

Millions of neurons comprising the ENS are grouped in the neuronal ganglia, which are interconnected with a dense network of nerve fibers and form ganglionated plexuses. The localization and number of these plexuses depend on the mammal species and the segment of the GI tract. In rodents, the ENS in the esophagus and stomach is built of two types of intramural ganglia. The first type, the myenteric ganglion, is located between longitudinal and circular muscle layers. Myenteric ganglia are interconnected with a dense network of nerve fibers and form the myenteric plexus [27–30]. The second type of intramural ganglia, the submucous ganglion, is located in the submucous layer, near the muscularis mucosae of the mucosal layer. Contrary to muscular ganglia, the nerves interconnected with the submucous ganglia are rather sparse. Therefore, submucous ganglia in the esophagus and stomach do not form plexus [31], although some authors have described submucous plexus in rodent esophagus and stomach [32,33]. However, the situation is different in the small and large intestines in rodents. Both types of enteric ganglia (myenteric and submucous) located in the same places as in the esophagus and stomach are interconnected with a dense network of nerves. Therefore, two kinds of plexuses (the myenteric plexus and submucous plexus) are described in the rodent intestine [34–37].

In large mammals, the organization of the ENS in the esophagus and stomach is similar to rodents [38–40], although some authors have described three types of the enteric plexuses (such as in the intestine—see below) in the porcine stomach [41]. The only exception are ruminants, in which only one type of the enteric ganglia (myenteric ganglia) has been described in the forestomach. These ganglia are located between longitudinal and circular muscular layers, interconnected with the dense nerve fibers and form myenteric plexus [42,43].

In turn, there are three types of the enteric ganglia, which form intramural plexuses in the small and large intestine of large mammal species (for example, in the pig) (Figure 1) [44–46]. The first of them is the myenteric plexus located (similarly to rodents) between the longitudinal and circular muscle layer [45,46]. Moreover, two types of submucous plexuses located in the submucous layer of the intestinal wall have been observed: outer submucous plexus—located in close association with the adjacent circular muscle layer (on its inner side) and the inner submucous plexus—positioned closer to the intestinal lumen, near the muscularis mucosae [47–49]. These plexuses are also often named after their discoverers. The myenteric plexus often called Auerbach's plexus, the outer submucous plexus—Schabadash's plexus, and the inner submucous plexus (in rodents—the submucous plexus)—Meissner's plexus [50,51].

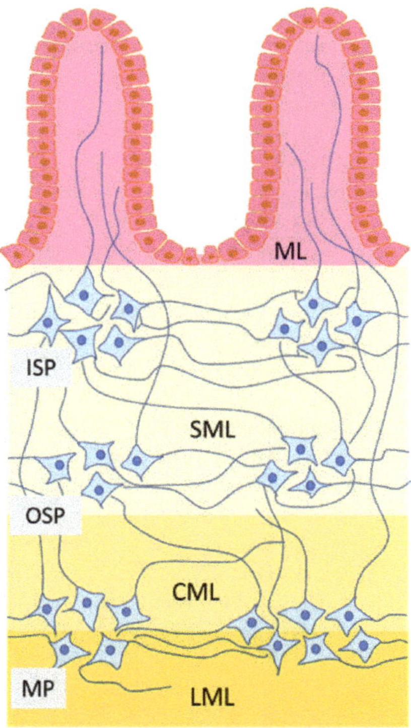

Figure 1. Organization of the enteric nervous system in the intestine of the domestic pig: MP—myenteric plexus, OSP—outer submucous plexus, ISP—inner submucous plexus, LML—longitudinal muscular layer, CML—circular muscular layer, SML—submucosal layer, ML—mucosal—layer.

As regards the organization of the human ENS, the distribution of the nervous structures in the esophagus and stomach is similar to rodents and large mammal species [23,31,52,53]. In the human small and large intestines, the organization of the ENS is not quite clear. Previous publications have described four types of enteric plexuses. In addition to the above-mentioned myenteric, outer submucous and inner submucous plexuses, the presence of an intermediate submucous plexus (IMSP)—a ganglionated plexus located in the submucous layer between the outer and inner submucous plexus has been reported [54]. However, at present, three kinds of plexuses located similarly to the porcine intestine are described in the human small and large intestine. In addition to the myenteric plexus located between the longitudinal and circular muscle layers, they include the plexus submucosus externus (PSE) near the circular muscle layers (on its inner side) and plexus submucosus internus (PSI) located closer to the intestinal lumen [23,31,55–59]. Contrary to the porcine inner submucous plexus, PSI in the human intestine is multi-layered, which means the particular ganglia within this plexus are located at a different depth of the submucous layer [56]. Other publications have shown that submucosal ganglia in the human colon are disseminated throughout the submucosal layer with significant inter-individual differences [60].

In addition to the above-mentioned main types of enteric ganglia, previous studies conducted in various mammal species have also reported the presence of small scattered neuronal ganglia in the mucosal layer (mucosal ganglia) and between the longitudinal muscular layer and serosa (subserosal ganglia), as well as a ganglionated plexus within the muscularis mucosae [31,61].

Enteric neurons are characterized by a high degree of differentiation in terms of morphological, functional and electrophysiological properties [25]. Moreover, the enteric neurons are also highly diverse with regard to their ability to synthesize neuronal active substances. Apart from acetylcholine (the main neuromediator in the ENS), a wide range of other neuronal factors have been described in enteric nervous structures [25,62–65]. The most important neuronal factors include vasoactive intestinal polypeptide (VIP), substance P (SP), galanin (GAL), nitric oxide and calcitonin gene-related peptide (CGRP). These substances may act as neuromediators and/or neuromodulators and participate in many regulatory processes including, among others, intestinal motility, secretion in the GI tract, immunological processes, blood flow, sensory stimuli conduction, intestinal digestion and absorption [16,23,25,62–65]. It should be noted that several active substances have been noted in the enteric neurons. Their exact roles are often still not quite clear. It is also known that the roles of the particular neuronal factors in the regulation of the stomach and intestine activity may depend on the segment of the GI tract and animal species studied. Such a substance is GAL and its participation in the control of the intestinal motility. Previous studies have shown that GAL induces the contraction of the ileal smooth muscles in the rat, guinea-pig rabbit and pig [66], while in the canine ileum and stomach it shows relaxant effects [67]. A similar situation is observed in the case of SP, which strongly stimulates the contraction of intestinal muscles in the rat and dog, while in humans such activity is rather limited [68–70]. Moreover, one substance very often appears to be involved in various GI tract activities. For example, CGRP (which is known as a key factor in sensory and pain stimuli conduction within the GI tract [71,72]) may also participate in the regulation of intestinal motility, mesenteric and intramural blood flow, gastric secretion, absorption of the nutrients in the intestine and protective reactions [73–77]. A detailed discussion of the exact functions played by all neuronal factors located in the enteric neurons is almost impossible because new active substances and their roles in various species are still being discovered. However, the main functions connected with the GI tract of selected neuronal substances occurring in the ENS are presented in Table 1.

Table 1. Functions of selected active substances in the enteric nervous system.

Active Neuronal Substance in the ENS (Alphabetical Order)	Selected Functions	References
Acetylcholine (Ach)	Stimulation of the intestinal motility	[78–80]
	Stimulation of electrolyte, water, enzymes and hormones secretion	[81–84]
	Participation in protective mechanisms	[82,85,86]
	Ant-inflammatory and immunostymulatory effects	[87–89]
	Blood flow regulation	[90]
Cocaine and Amphetamine Regulated Transcript (CART)	Inhibition of gastric acid secretion	[91]
	Regulation of the intestinal motility	[92]
Calcitonin Gene-Related Peptide (CGRP)	Participation in sensory and pain stimuli conduction	[71,72,93–95]
	Regulation of the intestinal motility	[94]
	Blood flow regulation	[96–99]
	Protective roles	[73,99–101]
	The influence on intestinal absorption	[74]

Table 1. *Cont.*

Active Neuronal Substance in the ENS (Alphabetical Order)	Selected Functions	References
Galanin (GAL)	Intestinal motility regulation	[66–70,102]
	Influence on secretory activity	[103–105]
	Participation in inflammatory processes	[103,105,106]
Nitric Oxide (NO)	Inhibition of the intestinal motility	[106–108]
	Participation in inflammatory processes	[109]
	Regulation of intestinal secretion, water and electrolyte transport	[110–113]
	Regulation of blood flow	[114,115]
	Participation in inflammatory processes	[116,117]
Pituitary Adenylate Cyclase-Activating Polypeptide (PACAP)	Inhibition of the intestinal motility	[118,119]
	Stimulation of gastric secretory activity	[120,121]
	Regulation of ion transport and Luminal fluid regulation in the large intestine	[121–123]
	Regulation of blood flow	[124]
Substance P (SP)	Protective roles	[100]
	Sensory stimuli conduction	[93,125]
	Regulation of the intestinal motility	[125–127]
	Regulation of water and electrolytes secretion	[125,128,129]
	Participation in inflammatory processes	[125,130]
Vasoactive Intestinal Polypeptide (VIP)	Neuroprotective functions	[131]
	Regulation of the intestinal motility	[132,133]
	Vasodialtory activity	[132,134]
	Participation in intestinal immunomodulation	[135–137]
	Influences on intestinal secretion	[138–140]

In addition to neurons, the ENS also includes numerous glial cells, which are called enteric glial cells (EGC) [141–143]. Glial cells in the gastrointestinal tract are generally characterized by small size, irregular or stellate shape and numerous processes which are in direct contact with neuronal cell bodies and nerve fibers. Based on previous studies, it is known that ECG may be divided into four major types, and classification of the EGC is similar to that used in the case of glial cells in the central nervous system [144,145]. The first type is "protoplasmic" glial cells (type I glial cells), which are located between neuronal cells in the enteric ganglia and their appearance resembles protoplasmic astrocytes in the brain. The second type of glial cells (type II glial cells) are "fibrous" glial cells, whose processes accompany the nerves connecting the enteric ganglia with each other. These cells are similar to fibrous glial cells located in the central nervous system. Moreover, mucosal glial cells (type III glial cells) located near nerve fibers in the mucosal layer and intramuscular glial cells (type IV glial cells) accompanying the nerve fibers in the muscular layer have been described in the gastrointestinal tract [144,145].

It was initially thought that glial cells are only structural support to neurons, but it is now known that EGC play multidirectional functions in the regulation of various aspects of the ENS and all gastrointestinal tract activities [146,147]. Primarily, they take part in processes connected with the development, protection and nutrition of the enteric neuronal cells [148,149]. They regulate growth, maturation and differentiation of the enteric neurons, and affect synthesis and release of neuromediators and/or neuromodulators, thus constituting a key factor in maintaining intraneuronal homeostasis [145,149–151].

Moreover, EGC (especially mucosal glia) are involved in activities of the intestinal barrier integrity and functions [152]. It is known that EGC synthesize a wide range of several substances, such as glial-derived neurotrophic factor, transforming growth factor-β1 and neurotrophins, and act on the intestinal epithelial cells through paracrine mechanism [152,153]. Experimental studies have also shown that in animals with genetic ablation of EGC, the intestinal epithelial layer loses its integrity and disturbances in vascularization appear and lead to severe inflammatory processes [154].

Enteric glial cells also have important functions during intestinal pathological states. They participate in immune cell modulation in a wide range of the intestinal diseases, including ulcerative colitis, Crohn's disease and colorectal cancer [142]. During inflammatory processes, proliferation of EGC occurs [155]. Glial cells participate in the immune recognition of pathological stimuli and may act as antigen-presenting immune cells [156]. Moreover, an increase in the production of some cytokines, including, among others, interleukins (IL-1β and IL-6) [157–159], as well as nerve growth factor (NGF) [160], glial fibrillary acidic protein (GFAP) [161] and nitric oxide (NO) [162] in glial cells has been noted during inflammatory processes.

It is also known that enteric glial cells play important roles in the pathogenesis of neurodegenerative diseases, including Parkinson's, Alzheimer's and Creutzfeldt-Jakob diseases. They are considered to be a possible trigger point for neurodegenerative processes, which through the gut–brain axis may efficiently affect neurodegenerative processes in the central nervous system [143–163].

An important feature of enteric neurons is the ability to change their morphological, physiological and neurochemical properties under the impact of physiological and pathological factors [15,65,164]. Changes in the ENS have been described during growth and aging, diet changes, as well as various intestinal pathological processes, systemic diseases and the impact of toxic substances [15,65,164,165]. Changes in enteric neurons are a sign of the adaptive and protective reactions and contribute to homeostasis maintenance in the GI tract [164,165]. Moreover, such changes appearing under the impact of disease or toxic substances may be the first signs of subclinical pathological processes or toxicity [166]. Some studies have indicated that mycotoxins may affect the morphology and neurochemical character of the enteric neurons. The following is a short characterization of several mycotoxin-induced changes in the enteric nervous system.

3. Mycotoxins Affecting the Enteric Neurons

3.1. *Deoxynivalenol*

Deoxynivalenol (DON—molecular weight 296.31 g/mol), belongs to the trichothecene family and is a substance produced by Fusarium spp. [9]. It is commonly found in barley, oat, rye, corn and rice [167,168]. The signs of toxicity depend on the dose, mammal species and duration of exposure. The most frequent symptoms of toxicity with DON include loss of appetite, decreased body weight gain, neuroendocrine disorders, vomiting and diarrhea [169].

In the GI tract, toxicity with DON results in a wide range of histopathological changes, such as inflammatory infiltration, necrotic changes in the intestinal villi, edema of lamina propria, a decrease in the number of goblet cells in the jejunum and the ileum, intensification of apoptosis and degeneration of lymphoid cells in the GI tract [10,170]. These changes, together with DON-induced disturbances in the synthesis of many active substances produced by the gastrointestinal mucosa lead to the injury of the intestinal barrier and abnormal nutrient absorption [11].

Within the nervous system, DON-induced changes include abnormal synthesis of neuronal neurotransmitters and/or neuromodulators and in disturbances in neuronal activity [171,172]. Moreover, in neuronal cells, DON induces apoptosis, affects the cerebral lipid peroxidation and influences neuronal calcium homeostasis, and these disturbances in the neuronal cells may lead to anorexic actions [172,173].

During a study performed on male Wistar rats (*Rattus novergicus*) aged 21 days, the influence of relatively low doses of DON on the ENS was described [14]. In that experiment, DON in various doses (from 0.2 mg/kg of chow to 2 mg/kg of chow) was given for 42 days, and the ENS was studied using immunohistochemistry and microscopic analysis. It was shown that this mycotoxin does not affect the myenteric ganglia organization in the jejunum [14]. Between control animals and rats receiving DON there were no differences in the density of glial cells located in the myenteric plexus, or the population density of myenteric neurons. Moreover, DON did not change the density of particular subpopulations of the myenteric neurons, i.e., cholinergic and nitrergic neurons [14]. However, all concentrations of DON studied in the above-mentioned experiment caused a decrease in the area of the general population of the myenteric neuronal cells, as well as cholinergic and nitrergic cell neurons. Moreover, DON also decreased the area of gliocytes located in the myenteric plexus [14] and decreased the myenteric ganglia area. It should be noted that during the cited study, besides changes in the ENS, the animals did not show any other symptoms of toxicity, including a decrease in body weight, diarrhea, loss of appetite or changes in the oxidative status [14]. This indicates that changes in the ENS are the first symptoms of toxicity with low doses of DON.

3.2. *T2 Toxin*

T2 toxin (molecular weight 466.5 g/mol), similar to DON, belongs to the trichothecene family of toxins. It is mainly synthesized by *Fusarium sporotrichioides*, *F. langsethiae*, *F. acuminatum* and *F. poae* and is recognized as the most acutely toxic trichothecene [174]. The impact of T2 toxin on the GI system manifests itself by (among others) histopathological changes in the intestinal mucosal layer (even with low doses), disturbances in the intestinal barrier functionality, influence on the enzymatic activity of enteric cells and inhibition of mucin production [175–178].

T2 toxin also shows neurotoxic activity and exposure to this substance results in a wide range of neurological symptoms, such as ataxia, muscular weakness, anorexia, as well as pathological lesions in the brain with disturbances in the functioning of this organ [179–181]. The main mechanisms underpinning the neurotoxic properties of T2 toxin are connected with reactive oxygen species and oxidative stress, as well as with mitochondrial dysfunction (consisting of the inhibition of the mitochondrial membrane potential and intensification of apoptosis) [182].

The ENS was studied using immunofluorescence in an experiment in vivo performed on juvenile (8-week-old) female domestic pigs of the White Large Polish Breed subjected to oral administration of T2 toxin at the level of 12 μg/kg body weight/day for 42 days [15]. Significant changes in the neurochemical character of the enteric neurons and nerve fibers located in the GI tract wall were described in this study. The character of changes depended on the type of the enteric plexus and the intestinal segment. It was reported that the administration of T-2 toxin increases the number of enteric neurons containing VIP in the porcine stomach and duodenum. These changes concern both myenteric and submucous plexuses and they are more visible in the duodenum, especially in the myenteric and outer submucous plexuses [15]. The same study showed that T-2 toxin also increases the number of nerve fibers containing VIP located in the muscular and mucosal layers of the porcine stomach and duodenum [15]. As previously indicated (Table 1), VIP in one of the potent inhibitory factors in the ENS and causes the hyperpolarization and relaxation of the gastrointestinal muscles and sphincters [132,133]. Moreover, VIP (as a vasodilator) increases blood flow in the wall of the GI tract and mesentery [132,134]. This substance may also affect the secretory activity of the GI tract, and the character of this activity depends on the GI tract segment [138–140]. It is known that VIP inhibits the gastric acid secretion in the stomach, but stimulates the secretion of the intestinal juice. VIP also has neuroprotective properties and increases the survivability of the enteric neurons [131]. Moreover, it is involved in immunological processes and shows anti-inflammatory properties. VIP also inhibits macrophages and inhibits the secretion of pro-inflammatory factors [135–137]. It is assumed that the increase in the number of VIP-positive enteric nervous structures under the impact of T2 toxin is connected with the protective and anti-inflammatory properties of VIP.

The influence of T2 toxin on the number of the enteric neurons containing cocaine and amphetamine-regulated transcript (CART) has been also reported [183]. In this study, T2 toxin was orally administrated to juvenile sows of the Large White Polish breed in a dose of 200 μg/kg of feed (the suggested permissible level of this toxin in the feed for pigs) for 42 days and the immunoreactivity in the ENS was evaluated using immunofluorescence. After the administration of T2 toxin, an increase in the number of CART-positive enteric neurons in all types of enteric plexuses as well as the number of nerve fibers containing CART in the mucosal and muscular layers in the stomach, duodenum and descending colon were described. The most visible changes were noted in the submucous plexus in the stomach and inner submucous plexus in the descending colon, where the number of CART-positive nerves under the impact of T2 toxin more than doubled [183]. It should be underlined that the exact functions of CART in the ENS are not clear [184]. A few studies concerning this issue have shown that CART inhibits the secretion of hydrochloric acid in the stomach and influences colonic motility [91,92]. This activity is probably done via the gut–brain axis because the direct impact of CART on isolated intestinal muscles does not cause changes in intestinal muscle contractility. The regulation of intestinal activity through the gut–brain axis is more likely since CART is known as an important factor regulating the feeding behavior in the central nervous system [185]. Moreover, numerous studies in which an increase in CART levels in the ENS has been observed strongly suggest that this peptide also takes part in protective and adaptive reactions in response to pathological, toxicological and physiological factors [46,166,184].

Another study (also performed with the immunofluorescence technique) concerning the impact of T2 toxin on the ENS was also performed on juvenile female pigs of the Large White Polish breed, which were treated with given T2 toxin orally in the dose of 12 μg/kg body weight/day for 42 days [16]. In this study, it was shown that T2 toxin affects the population of neurons containing calcitonin gene-related peptide located in the enteric plexuses in the porcine descending colon [16]. The administration of T2 toxin caused an increase in the number of CGRP-positive neurons in the myenteric, outer submucous and inner submucous plexuses, as well as an increase in the density of intramucosal nerves immunoreactive to these neuronal factors, without changes in the number of CGRP-positive nerve fibers in the muscular layer [16]. Moreover, it was shown that T2 toxin changes the neurochemical character of CGRP-positive neuronal cells, which were expressed by fluctuations in the degree of co-localization of CGRP with other neuronal factors (including substance P, nitric oxide synthase, galanin, CART peptide and vesicular acetylcholine transporter) in the same enteric nervous structures [16].

CGRP is a substance which primarily occurs in sensory neurons and is involved in sensory and pain stimuli conduction [71,72,93–95]. Moreover, CGRP in the GI tract takes part in the regulation of intestinal motility and increases blood flow in the mesenteric vessels [94,96–99]. It is also known that CGRP inhibits gastric acid secretion in the stomach with simultaneous induction of somatostatin release and regulates the absorption of nutrients from the intestine [186]. Previous studies have also shown that CGRP takes part in inflammatory processes in the intestine [99–101]. The multidirectional functions of CGRP in the ENS appear to be confirmed by a wide range of other neuronal substances present in CGRP-positive enteric neurons (such as substance P, nitric oxide CART peptide and galanin) which also play various roles in the GI tract (Table 1).

These reports of the influence of T2 toxin on the expression of a wide range of neuronal factors responsible for various regulatory processes in the ENS [15,16,183], strongly suggest that even relatively low doses of this mycotoxin may influence various intestinal activities, such as motility, secretion, conduction of sensory stimuli and regulation of the blood flow in the intestinal wall [15,16,183].

3.3. Zearalenon

Zearalenon (ZEN—molecular weight 318.364 g/mol) is synthesized mainly by *Fusarium graminearum, culmorum, crookwellense* and *roseum* and is found in barley, oat, wheat and bread [187]. The toxicity of ZEN is connected with its chemical structure, which allows it to act on the estrogen receptors, which are present in many internal organs [8]. ZEN can cross the blood-brain barrier and may influence neurons in the central nervous system [188,189]. It has been shown that exposure to ZEN leads to the abnormal synthesis of neuronal factors and enzymes in the brain neurons, induces apoptosis of the neuronal cells, increases oxidative stress reactions, influences the development of the nervous system, may cause behavioral aberrations and affects glial cell functions [189–192]. In turn, in the GI system, ZEN (among others) disturbs intestinal homeostasis, changes intestinal microbiome, causes inflammatory cell proliferation and inflammation in the intestinal mucosal layer [11,193–195].

Although the impact of ZEN on the GI tract is relatively well known, studies concerning the influence of this mycotoxin on intestinal innervation are limited to two studies performed on the pigs of the Large White Polish breed (approximately 8 weeks old), in which the nervous structure was evaluated with the immunofluorescence technique [19,21].

These studies have shown that the administration of relatively low doses of ZEN—10 μg/kg body weight/day [19] or 0.1 mg/kg of chow/day [21], administered for 42 days affect the neurochemical coding of nerve fibers in the mucosal and muscular layers of the ileum. For the intramuscular nerves, these changes involved an increase in the number of fibers immunoreactive to CART, substance P, nitric oxide synthase, VIP and pituitary adenylate cyclase-activating peptide and a decrease in the number of fibers containing galanin [19]. In the mucosal layer, ZEN not only caused an increase in the number of nerve fibers containing SP and/or VIP, but also changed the morphology of these nerves [21]. In animals treated with ZEN, nerves immunoreactive to SP and/or VIP become thicker and more visible than in the control animals [21]. It should be underlined that all the above-mentioned neuronal factors play important multidirectional roles in the regulation of the intestinal activity both in physiological conditions as well as during pathological processes, the most important of which are listed in Table 1.

The impact of ZEN on the ENS in the porcine descending colon has also been reported. A study concerning this issue was performed on juvenile (8-week-old) female pigs of the Large White Polish breed, which were treated with a dose of ZEN at the level of 6 μg/kg b.w./day given orally for 42 days [16]. In this study, the ENS evaluation was conducted with the immunofluorescence technique. The impact of ZEN was similar to the influence of T2 toxin. ZEN increased the number of neurons containing CGRP (whose functions in the ENS are described in the subchapter concerning T2-toxin and presented in Table 1) in all types of the enteric plexuses located in the descending colon [16]. Moreover, ZEN-induced changes in the neurochemical character of CGRP-positive enteric neurons were also reported [16]. These changes consisted of an increase in the degree of co-localization of CGRP with other neuronal factors (including substance P, galanin, CART and nitric oxide synthase, which was used as a marker of neuron synthesized nitric oxide) in neurons within all types of the enteric plexuses and intramural nerve fibers [16]. The functions of the above-mentioned neuronal active substances are presented in Table 1.

3.4. Patulin

Patulin (PAT-molecular weight 154.12 g/mol) is produced by various species belonging to *Penicillium, Aspergillus, Paecilomyces* and *Byssochlamys* [196,197] and is present in fruits (especially in apples) and vegetables [196,197]. Previous studies have shown that exposure to patulin causes damage to the intestinal barrier and inflammatory processes in the GI tract and influences the gut microbiota and the production of the mucus by enterocytes [198,199]. The neurotoxic activity of PAT is also known. It causes damage to the DNA in brain neuronal cells, mitochondrial and lysosomal dysfunction, a reduction of ATP levels and intensification of oxidative stress reactions [200,201].

The influence of patulin on the enteric neurons has been the subject of only one study. This study was performed on the cell culture of the enteric neurons prepared from 2–3-month-old C57B6/J OlaHsd mice and included various methods, such as growth and viability testing, a cytotoxicity test, evaluation of calcium signaling, measurement of glucose content, neurite outgrowth measurement and a reactive oxygen species (ROS) test [202]. The enteric neurons were treated with *P coprobium* extract, which decreased their viability with a half-maximal effective concentration (EC_{50}) of 1 ng/μL This study also showed that patulin affects excitability and glucose consumption of the enteric neurons, which results in a patulin-induced reduction of ATP levels and glucose concentration in the enteric neurons. It has been also reported that patulin causes disorders in calcium signaling in the enteric neurons and affects neuronal morphology, which results in a reduction of neurite outgrowth and total neurite mass [202].

3.5. Fumonisins

Fumonisins are synthesized by *Fusarium proliferatum* and *Fusarium verticillioides* and characterized by a high degree of toxicity [203]. Numerous types of these mycotoxins have been described, but the most toxicologically important are fumonisin B1 (molecular weight 721.838 g/mol), fumonisin B2 (molecular weight 705.83 g/mol) and fumonisin B3 (molecular weight 705.8 g/mol), due to their high levels in cereal grains and crop products [12,204]. Among the numerous internal organs and systems which may be affected by fumonisins, the nervous system is one of the most susceptible to the adverse effects of these mycotoxins. It is known that fumonisins may enhance neurodegenerative reactions and impair the developmental processes in neurons located in the central nervous system, and some studies have reported connections between exposure to these mycotoxins and the risk of neurodegenerative diseases, such as multiple sclerosis, Alzheimer's disease and Parkinson's disease [205,206]. Exposure to fumonisins also results in changes in the GI tract, which manifest as disturbances in intestinal absorption, changes in the enterocytes and abnormalities in the intestinal immunological processes leading to increased susceptibility to infections [20].

However, knowledge of the influence of fumonisins on the ENS is extremely limited. One study concerning this issue was performed using the immunohistochemistry method on male Wistar rats (*Rattus novergicus*), which were 21 days old [12]. This study showed that a mixture of fumonisin B1 and B2 added to food in doses of 1 and 3 mg/kg of body weight (i.e., in doses which may be present in "natural" conditions in the food of humans and animals) given for 63 days does not affect the organization of the myenteric plexus in the rat jejunum [12]. Such doses of fumonisins do not result in changes in the general number of myenteric plexus and the number of myenteric neurons causing nitric oxide synthase, which is a marker of structures synthesizing nitric oxide. However, some changes in the myenteric neurons were observed under the impact of the mentioned doses of fumonisins. These changes consisted of a reduction in the size (without changes in their morphology) of neurons located in the myenteric plexus and included both neurons immunoreactive to pan-neuronal marker HuCD and nitric oxide synthase. Suoza et al. (2014) [12] reported that fumonisins not only affect the development and growth of neurons in the central nervous system but may also influence these processes in the ENS.

The influence of fumonisins on the ENS in the rat duodenum and jejunum of adolescent (5-weeks-old) male Wistar rats was also studied by Rudyk et al. (2020), using the immunohistochemistry method and histomorphometric analysis [13]. A mixture of fumonisins B1 and B2 were administered in a dose of 90 mg/kg of body weight for 21 days. That study demonstrated that fumonisins influence the following parameters within myenteric and submucous plexuses: area, perimeter, mean Feret diameter, mean diameter and sphericity [13]. It was also found that the impact of fumonisins on the ENS depends on the segment of the GI tract and the type of the enteric plexus. Fumonisin-induced changes in the duodenum were less visible, concerned only the submucous plexus and consisted of a reduction of area and mean diameter of ganglia, while the other parameters in the submucous plexus and all parameters studied in the myenteric plexus were not subjected to change. In the jejunum, changes were noted in the myenteric and submucous plexuses and consisted of an increase in the sphericity of ganglia and a reduction of other parameters in both types of plexuses. Moreover, the most visible changes were noted in the myenteric plexus.

The mechanisms of the impact of fumonisins on the ENS are unknown, but they probably inhibit ceramide synthase—an enzyme contributing to sphingolipid synthesis [207].

4. Mycotoxin Consumption and Human Gastrointestinal Diseases

The multidirectional adverse effects of mycotoxins on the GI tract (Table 2) cause that exposure to these substances may result in various disturbances of the GI activity in humans. However, the common prevalence of mycotoxins in the human environment and food indicates that participation of these chemicals in the development of intestinal diseases in humans may be an important public health problem all over the world [208].

Table 2. Gastrointestinal signs and effects of mycotoxins on the gastrointestinal tract.

Mycotoxin	Gastrointestinal Signs of Toxicity	References	Influence on the Digestive Tract	References
Doxynivalenol (DON)	Abdominal pain, increased salivation, diarrhea, vomiting, anorexia, decrease body weight gain	[169,209–214]	IPEC-J2 cell line from porcine jejunal epithelium: cytotoxicity, decrease in transepithelial electrical resistance, disruption of epithelial integrity	[176]
			Porcine jejunal explant samples: shortened and coalescent villi, lysis of enterocytes, edema, upregulation of proinflammatory cytokines expression	[215,216]
			Pigs of White Large Polish Breed: increase in the mucosal thickness and the intestinal crypt depth, atrophy of the villi, changes in the number of goblet cells, inflammatory infiltration, intensification of apoptosis, changes in ultrastructure of intestinal cells	[10,11,175, 214,217,218]
			Human Colonic Cell Lines Caco-2, T84, HT-29: decrease in cell proliferation, changes in permeability, genotoxicity, intensification of apoptosis, increase in the expression of proinflammatory cytokines, influence on DNA synthesis	[215,219–221]
			Poultry: decrease in the high of villi	[222,223]
T2 Toxin	Gastrointestinal bleeding, diarrhea, vomiting, decreased feed consumption and weight gain	[224–226]	IPEC-J2 cell line from porcine jejunal epithelium: cytotoxic effects, disruption of intestinal barrier integrity	[176]
			human intestinal Caco-2 cells disturbances in intestinal barrier, enzymatic activity of enteric cells, inhibition of mucin production	[178]

Table 2. *Cont.*

Mycotoxin	Gastrointestinal Signs of Toxicity	References	Influence on the Digestive Tract	References
			Pigs of White Large Polish Breed or crossbred pigs: congestion and hemorrhage of the gastrointestinal mucosal layer, inflammatory infiltration, in high doses—necrotic changes	[175,227–229]
			Sprague-Daw-ley rats: inflammatory and necrotic changes in, lymphocytic necrosis in intestinal Peyer's patches, influence on nutrients absorption, influence on DNA synthesis	[230–232]
Zearalenone (ZEN)	Gastrointestinal symptoms are not typical for ZEN toxicity. Decrease in feed intake and body weight, changes in intestinal microbiome	[195,233]	Pigs of various breeds: increase in the mucosal thickness, increase in the number of goblet cells, increase in lymphocyte number in epithelium, intensification of apoptosis, influence on enzymatic activity of mucosal cells, changes in intestinal microbiome	[10,11,175, 193–195,234, 235]
			Intestinal porcine epithelial cell line (IPEC-1): influence on cell activity by changes in gene expression	[236]
			Poultry: changes in the high of intestinal villi	[237]
Patulin (PAT)	Anorexia, salivation, distended abdomen loss of body weight, bleeding from the digestive tract and diarrhea	[238–243]	Human intestinal Caco-2 cells: the influence on permeability and ion transport in the mucosa, epithelial desquamation and sub mucosal swelling, genotoxicity effects, modulation of tight junctions	[198,199,244]
			Rodents: mucosal layer injury, ulceration, fibrosis in the sub mucosa, necrosis	[238–242]
			Porcine jejunal explant samples: villi atrophy and necrosis, decrease in the number of goblet cells, increase in cell apoptosis	[245]

Table 2. *Cont.*

Mycotoxin	Gastrointestinal Signs of Toxicity	References	Influence on the Digestive Tract	References
Fumonisins (FUM)	reduction of feed consumption and body weight, abdominal pain, diarrhea	[246–249]	Human Colonic Cell Lines Caco-2, HT-29: growth inhibition and apoptosis induction, impact on mitochondrial metabolism, necrosis	[221,250]
			Rodents: inflammatory infiltration increase in the number of mitotic figures in the intestinal crypts, necrotic changes	[251,252]
			Intestinal porcine epithelial cell line (IPEC-1): inhibition of cell proliferation, intestinal barrier dysfunction	[253]

The impact of mycotoxins on the intestinal barrier functions, intestinal immunity, secretory activity and gut microflora, as well as their genotoxic/mutagenic and carcinogenic effects are mainly known from experimental studies (Table 2). Such studies do not always fully reflect the conditions of natural exposure to mycotoxins. The first problem is the dose of mycotoxins, which is very difficult to determine in the human diet [254,255]. The second, more important, problem is the fact that food may contain several or even a dozen mycotoxins at the same time. These mycotoxins may chemically interact with each other, which leads to changes in their toxic properties and bio-availability. In this case, synergistic interactions between mycotoxins is particularly dangerous [255,256]. For example, previous studies have shown that mixtures of ZEN and DON or DON, T2 and ZEN show higher toxicity than these individual mycotoxins [175,257]. Moreover, it is known that human food may also contain other active substances and contaminations, such as bacterial products, pesticides, phytotoxins, chemical contaminations and preservatives, which not only affect mycotoxin activity but may contribute to various disorders in the GI tract [258]. That is why it is so difficult to determine the effective participation of mycotoxins in the development of human gastrointestinal diseases.

A comparison of histopathological changes occurring in the GI tract during human gastrointestinal diseases and changes in the intestine caused by mycotoxins has shown that the negative development in the GI tract in both cases are similar [255]. This may suggest a correlation between a degree of exposure to mycotoxins and the risk of human gastrointestinal diseases, as well as the participation of mycotoxins in the development of various diseases, including inflammatory bowel disease, Crohn's disease, coeliac disease and colorectal cancer [255]. However, only comprehensive epidemiological studies on the relationships between mycotoxin levels in food, blood and urine and the occurrence of particular diseases conducted on a large human population would explain the connection between exposure to mycotoxins and the risk of human gastrointestinal diseases. Unfortunately, such studies are fragmentary and relatively few. These studies have reported that aflatoxins (especially aflatoxin B1) may pose a carcinogenic risk and exposure to these chemicals may increase the risk of gastric and colorectal cancer [259,260]. Other studies suggest a correlation between the exposure to ZEN and colorectal cancer [261], as well as relationships between exposure to aflatoxins and Crohn's Disease, coeliac disease and ulcerative colitis [262]. Despite this, differences in concentration of patulin and citrinin in plasma and urine between healthy people and patients suffering from colorectal cancer have not been observed, which may suggest that these mycotoxins are not key factors leading to this disease [263].

5. Conclusions

Based on previous studies, it is known that mycotoxins affect the enteric nervous system (Table 3). This impact may be multidirectional and depends not only on the chemical structure of the mycotoxin and mammal species studied, but also on the type of the enteric plexuses and segment of the digestive tract. Mycotoxins may act on the size and morphological properties of intestinal nervous structures and the neurochemical character of the enteric neurons. These changes are probably a result of adaptive and protective reactions, which affect homeostasis maintenance. Moreover, mycotoxin-induced changes in the ENS are often the first sign of exposure to low doses of mycotoxins. Understanding the exact mechanisms connected with the influence of mycotoxins on the intestinal innervation may be very important in determining mycotoxin dose limits, which are safe and neutral for the living organism. Unfortunately, the current information about the influence of mycotoxins on the ENS is relatively limited and elucidation of all aspects connected with this issue requires further research.

Table 3. Influence of mycotoxins on the enteric nervous system.

Mycotoxin	Dose Examined	Animal Species or Kind of Tissues	Experimental Method Used in the Study	Character of Changes in the ENS	References
Doxynivalenol	from 0.2 mg/kg of chow to 2 mg/kg of chow	Wistar rats (*Rattus novergicus*)	immunohistochemistry and microscopic analysis	Reduction of the area of general population of the myenteric neurons, glial cells in the myenteric plexus and whole myenteric ganglia.	[14]
T2 Toxin	12 µg/kg body weight/day	domestic pig of the White Large Polish Breed	Immunofluorescence method and microscopic analysis	Increase in the number of VIP-positive enteric neurons and intramucosal and intramuscular nerve fibers containing VIP in the stomach and duodenum.	[15]
	200 µg/kg of feed	domestic pig of the White Large Polish Breed	Immunofluorescence method and microscopic analysis	Increase in the number of CART-positive enteric neurons and intramucosal and intramuscular nerve fibers containing CART in the stomach, duodenum and descending colon.	[55]
	12 µg/kg body weight/day	domestic pig of the White Large Polish Breed	Immunofluorescence method and microscopic analysis	Increase in the number and changes in neurochemical character of CGRP-positive enteric neurons in the descending colon.	[16]
Zearalenon	10 µg/kg body weight/day	domestic pig of the White Large Polish Breed	Immunofluorescence method and microscopic analysis	Increase in the number of nerve fibers immunoreactive to CART, SP, NOS, VIP, PACAP and decrease in the number of GAL-positive nerve fibers in the muscular layer of the ileum.	[19]

Table 3. *Cont.*

Mycotoxin	Dose Examined	Animal Species or Kind of Tissues	Experimental Method Used in the Study	Character of Changes in the ENS	References
	0.1 mg/kg of chow/day	domestic pig of the White Large Polish Breed	Immunofluorescence method and microscopic analysis	Increase in the number of nerve fibers immunoreactive to SP and VIP with changes in their morphology	[21]
	12 µg/kg body weight/day	domestic pig of the White Large Polish Breed	Immunofluorescence method and microscopic analysis	Increase in the number and changes in neurochemical character of neurons immunoreactive to CGRP in the descending colon.	[16]
Patulin	EC50 = 1 ng/µL	culture of the enteric neurons from C57B6/J OlaHsd mice	Growth and viability testing, cytotoxicity test, evaluation of calcium signaling, measurement of glucose content, neurite outgrowth measurement and reactive oxygen species (ROS) test	Reduction of ATP levels and glucose concentration, disorders in calcium signaling in the enteric neurons, changes in their morphology.	[71]
Fumonisins	1 and 3 mg/kg body weight	Wistar rats (*Rattus novergicus*)	immunohistochemistry method	Reduction of the size of neurons in the enteric ganglia.	[12]
	90 mg/kg body weight	Wistar rats (*Rattus novergicus*)	immunohistochemistry method and histomorphometrical analysis	Reduction of area and mean diameter of the submucous plexuses in duodenum. Reduction of area and mean diameter of myenteric and submucous plexuses in the jejunum, increase of sphericity of the enteric ganglia.	[13]

VIP—vasoactive intestinal polypeptide; CART—cocaine- and amphetamine-regulated transcript; CGRP—calcitonin gene related peptide; SP—substance P; NOS—nitric oxide synthase; PACAP—pituitary adenylate cyclase activating peptide; GAL—galanin.

Author Contributions: Conceptualization, S.G., M.G. and K.M.; supervision, S.G.; writing—original draft, S.G., M.G. and K.M.; writing—review and editing, S.G. All authors have read and agreed to the published version of the manuscript.

Funding: Project financially supported by Minister of Science and Higher Education in the range of the program entitled "Regional Initiative of Excellence" for the years 2019–2022: Project No. 010/RID/2018/19, amount of funding 12.000.000 PLN.

Conflicts of Interest: The authors declare that they have no conflict of interest.

References

1. Cimbalo, A.; Alonso-Garrido, M.; Font, G.; Manyes, L. Toxicity of mycotoxins in vivo on vertebrate organisms: A review. *Food Chem. Toxicol.* **2020**, *137*, 111161. [CrossRef] [PubMed]
2. De Ruyck, K.; De Boevre, M.; Huybrechts, I.; De Saeger, S. Dietary mycotoxins, co-exposure, and carcinogenesis in humans: Short review. *Mutat. Res.* **2015**, *766*, 32–41. [CrossRef] [PubMed]

3. González, N.; Marquès, M.; Nadal, M.; Domingo, J.L. Occurrence of environmental pollutants in foodstuffs: A review of organic vs. conventional food. *Food Chem. Toxicol.* **2019**, *125*, 370–375. [CrossRef] [PubMed]
4. Gonkowski, S.; Obremski, K.; Makowska, K.; Rytel, L.; Mwaanga, E.S. Levels of Zearalenone and its metabolites in sun-dried kapenta fish and water of Lake Kariba in Zambi—A preliminary study. *Sci. Total Environ.* **2018**, *637–638*, 1046–1050. [CrossRef] [PubMed]
5. Milićević, D.R.; Skrinjar, M.; Baltić, T. Real and perceived risks for mycotoxin contamination in foods and feeds: Challenges for food safety control. *Toxins* **2010**, *2*, 572–592. [CrossRef] [PubMed]
6. Alshannaq, A.; Yu, J.H. Occurrence, toxicity, and analysis of major mycotoxins in food. *Int. J. Environ. Res. Public Health* **2017**, *14*, 632. [CrossRef] [PubMed]
7. Rykaczewska, A.; Gajęcka, M.; Onyszek, E.; Cieplińska, K.; Dąbrowski, M.; Lisieska-Żołnierczyk, S.; Bulińska, M.; Babuchowski, A.; Gajęcki, M.T.; Zielonka, Ł. Imbalance in the blood concentrations of selected steroids in prepubertal gilts depending on the time of exposure to low doses of zearalenone. *Toxins* **2019**, *11*, 561. [CrossRef]
8. Gajęcka, M.; Zielonka, Ł.; Gajęcki, M. Activity of zearalenone in the porcine intestinal tract. *Molecules* **2017**, *22*, 18. [CrossRef]
9. Khoshal, A.K.; Novak, B.; Martin, P.G.P.; Jenkins, T.; Neves, M.; Schatzmayr, G.; Oswald, I.P.; Pinton, P. Co-Occurrence of DON and emerging mycotoxins in worldwide finished pig feed and their combined toxicity in intestinal cells. *Toxins* **2019**, *11*, 727. [CrossRef]
10. Przybylska-Gornowicz, B.; Tarasiuk, M.; Lewczuk, B.; Prusik, M.; Ziółkowska, N.; Zielonka, Ł.; Gajęcki, M.; Gajęcka, M. The effects of low doses of two Fusarium toxins, zearalenone and deoxynivalenol, on the pig jejunum. A light and electron microscopic study. *Toxins* **2015**, *7*, 4684–4705. [CrossRef]
11. Przybylska-Gornowicz, B.; Lewczuk, B.; Prusik, M.; Hanuszewska, M.; Petrusewicz-Kosińska, M.; Gajęcka, M.; Zielonka, Ł.; Gajęcki, M. The effects of deoxynivalenol and zearalenone on the pig large intestine. A light and electron microscopic study. *Toxins* **2018**, *10*, 148. [CrossRef]
12. Sousa, F.C.; Schamber, C.R.; Amorin, S.S.; Natali, M.R. Effect of fumonisin-containing diet on the myenteric plexus of the jejunum in rats. *Auton. Neurosci.* **2014**, *185*, 93–99. [CrossRef]
13. Rudyk, H.; Tomaszewska, E.; Arciszewski, M.B.; Muszyński, S.; Tomczyk-Warunek, A.; Dobrowolski, P.; Donaldson, J.; Brezvyn, O.; Kotsyumbas, I. Histomorphometrical changes in intestine structure and innervation following experimental fumonisins intoxication in male Wistar rats. *Pol. J. Vet. Sci.* **2020**, *23*, 77–88. [CrossRef] [PubMed]
14. Rissato, D.F.; de Santi Rampazzo, A.P.; Borges, S.C.; Sousa, F.C.; Busso, C.; Buttow, N.C.; Natali, M.R.M. Chronic ingestion of deoxynivalenol-contaminated diet dose-dependently decreases the area of myenteric neurons and gliocytes of rats. *Neurogastroenterol. Motil.* **2020**, *32*, e13770. [CrossRef] [PubMed]
15. Makowska, K.; Obremski, K.; Gonkowski, S. The impact of T-2 toxin on vasoactive intestinal polypeptide-like immunoreactive (VIP-LI) nerve structures in the wall of the porcine stomach and duodenum. *Toxins* **2018**, *10*, 138. [CrossRef] [PubMed]
16. Makowska, K.; Obremski, K.; Zielonka, L.; Gonkowski, S. The influence of low doses of zearalenone and T-2 toxin on calcitonin gene related peptide-like immunoreactive (CGRP-LI) neurons in the ENS of the porcine descending colon. *Toxins* **2017**, *9*, 98. [CrossRef] [PubMed]
17. Alassane-Kpembi, I.; Pinton, P.; Oswald, I.P. Effects of mycotoxins on the intestine. *Toxins* **2019**, *11*, 159. [CrossRef] [PubMed]
18. Liew, W.P.; Mohd-Redzwan, S. Mycotoxin: Its impact on gut health and microbiota. *Front. Cell Infect. Microbiol.* **2018**, *8*, 60. [CrossRef]
19. Gonkowski, S.; Obremski, K.; Calka, J. The influence of low doses of zearalenone on distribution of selected active substances in nerve fibers within the circular muscle layer of porcine ileum. *J. Mol. Neurosci.* **2015**, *56*, 878–886. [CrossRef]
20. Bouhet, S.; Oswald, I. The intestine as a possible target for fumonisin toxicity. *Mol. Nutr. Food Res.* **2007**, *51*, 925–931. [CrossRef]
21. Obremski, K.; Gonkowski, S.; Wojtacha, P. Zearalenone-induced changes in the lymphoid tissue and mucosal nerve fibers in the porcine ileum. *Pol. J. Vet. Sci.* **2015**, *18*, 357–365. [CrossRef] [PubMed]
22. Pinton, P.; Oswald, I.P. Effect of deoxynivalenol and other Type B trichothecenes on the intestine: A review. *Toxins* **2014**, *6*, 1615–1643. [CrossRef] [PubMed]

23. Furness, J.B.; Callaghan, B.P.; Rivera, L.R.; Cho, H.J. The enteric nervous system and gastrointestinal innervation: Integrated local and central control. *Adv. Exp. Med. Biol.* **2014**, *817*, 39–71. [CrossRef] [PubMed]
24. Gershon, M.D. The enteric nervous system: A second brain. *Hosp. Pract.* **1999**, *34*, 31–52. [CrossRef]
25. Furness, J.B. Extrinsic and intrinsic sources of calcitonin gene-related peptide immunoreactivity in the lamb ileum: A morphometric and neurochemical investigation. *Cell Tissue Res.* **2006**, *323*, 183–196.
26. Schneider, S.; Wright, C.M.; Heuckeroth, R.O. Unexpected roles for the second brain: Enteric nervous system as master regulator of bowel function. *Annu. Rev. Physiol.* **2019**, *81*, 235–259. [CrossRef]
27. Morikawa, S.; Komuro, T. Distribution of myenteric NO neurons along the guinea-pig esophagus. *J. Auton. Nerv. Syst.* **1998**, *74*, 91–99. [CrossRef]
28. Reiche, D.; Michel, K.; Pfannkuche, H.; Schemann, M. Projections and neurochemistry of interneurones in the myenteric plexus of the guinea-pig gastric corpus. *Neurosci. Lett.* **2000**, *295*, 109–112. [CrossRef]
29. Zhang, G.Q.; Yang, S.; Li, X.S.; Zhou, D.S. Expression and possible role of IGF-IR in the mouse gastric myenteric plexus and smooth muscles. *Acta Histochem.* **2014**, *116*, 788–794. [CrossRef]
30. Zimmermann, J.; Neuhuber, W.L.; Raab, M. Homer1 (VesL-1) in the rat esophagus: Focus on myenteric plexus and neuromuscular junction. *Histochem. Cell Biol.* **2017**, *148*, 189–206. [CrossRef]
31. Furness, J.B. *The Enteric Nervous System*; Blackwell Publishing: Oxford, UK, 2006; pp. 1–274.
32. Kamikawa, Y.; Shimo, Y. Pharmacological characterization of the opioid receptor in the submucous plexus of the guinea-pig oesophagus. *Br. J. Pharmacol.* **1983**, *78*, 693–699. [CrossRef] [PubMed]
33. Kunisawa, Y.; Komuro, T. Interstitial cells of Cajal associated with the submucosal plexus of the Guinea-pig stomach. *Neurosci. Lett.* **2008**, *434*, 273–276. [CrossRef] [PubMed]
34. Heinicke, E.A.; Kiernan, J.A. An immunohistochemical study of the myenteric plexus of the colon in the rat and mouse. *J. Anat.* **1990**, *170*, 51–62. [PubMed]
35. Sayegh, A.I.; Ritter, R.C. Morphology and distribution of nitric oxide synthase-, neurokinin-1 receptor-, calretinin-, calbindin-, and neurofilament-M-immunoreactive neurons in the myenteric and submucosal plexuses of the rat small intestine. *Anat. Rec. A Discov. Mol. Cell. Evol. Biol.* **2003**, *271*, 209–216. [CrossRef] [PubMed]
36. Monro, R.L.; Bornstein, J.C.; Bertrand, P.P. Synaptic transmission from the submucosal plexus to the myenteric plexus in Guinea-pig ileum. *Neurogastroenterol. Motil.* **2008**, *20*, 1165–1173. [CrossRef]
37. Li, J.P.; Zhang, T.; Gao, C.J.; Kou, Z.Z.; Jiao, X.W.; Zhang, L.X.; Wu, Z.Y.; He, Z.Y.; Li, Y.Q. Neurochemical features of endomorphin-2-containing neurons in the submucosal plexus of the rat colon. *World J. Gastroenterol.* **2015**, *21*, 9936–9944. [CrossRef]
38. Rekawek, W.; Sobiech, P.; Gonkowski, S.; Żarczyńska, K.; Snarska, A.; Waśniewski, T.; Wojtkiewicz, J. Distribution and chemical coding patterns of cocaine- and amphetamine-regulated transcript-like immunoreactive (CART-LI) neurons in the enteric nervous system of the porcine stomach cardia. *Pol. J. Vet. Sci.* **2015**, *18*, 515–522. [CrossRef]
39. Bulc, M.; Palus, K.; Całka, J.; Zielonka, Ł. Changes in immunoreactivity of sensory substances within the enteric nervous system of the porcine stomach during experimentally induced diabetes. *J. Diabetes Res.* **2018**, *2018*, 4735659. [CrossRef]
40. Makowska, K.; Rytel, L.; Lech, P.; Osowski, A.; Kruminis-Kaszkiel, E.; Gonkowski, S. Cocaine- and amphetamine-regulated transcript (CART) peptide in the enteric nervous system of the porcine esophagus. *Comptes Rendus Biol.* **2018**, *341*, 325–333. [CrossRef]
41. Kaleczyc, J.; Klimczuk, M.; Franke-Radowiecka, A.; Sienkiewicz, W.; Majewski, M.; Łakomy, M. The distribution and chemical coding of intramural neurons supplying the porcine stomach—The study on normal pigs and on animals suffering from swine dysentery. *Anat. Histol. Embryol.* **2007**, *36*, 186–193. [CrossRef]
42. Teixeira, A.F.; Wedel, T.; Krammer, H.J.; Kühnel, W. Structural differences of the enteric nervous system in the cattle forestomach revealed by whole mount immunohistochemistry. *Ann. Anat.* **1998**, *180*, 393–400. [CrossRef]
43. Arciszewski, M.B.; Barabasz, S.; Skobowiat, C.; Maksymowicz, W.; Majewski, M. Immunodetection of cocaine- and amphetamine-regulated transcript in the rumen, reticulum, omasum and abomasum of the sheep. *Anat. Histol. Embryol.* **2009**, *38*, 62–67. [CrossRef] [PubMed]

44. Timmermans, J.P.; Barbiers, M.; Scheuermann, D.W.; Stach, W.; Adriaensen, D.; Mayer, B.; De Groodt-Lasseel, M.H. Distribution pattern, neurochemical features and projections of nitrergic neurons in the pig small intestine. *Ann. Anat.* **1994**, *176*, 515–525. [CrossRef]
45. Makowska, K. Chemically induced inflammation and nerve damage affect the distribution of vasoactive intestinal polypeptide-like immunoreactive (VIP-LI) nervous structures in the descending colon of the domestic pig. *Neurogastroenterol. Motil.* **2018**, *30*, e13439. [CrossRef]
46. Makowska, K.; Gonkowski, S. Age and sex-dependent differences in the neurochemical characterization of calcitonin gene-related peptide-like immunoreactive (CGRP-LI) nervous structures in the porcine descending colon. *Int. J. Mol. Sci.* **2019**, *20*, 1024. [CrossRef] [PubMed]
47. Kapp, S.; Schrödl, F.; Neuhuber, W.; Brehmer, A. Chemical coding of submucosal type V neurons in porcine ileum. *Cells Tissues Organs* **2006**, *184*, 31–41. [CrossRef]
48. Gonkowski, S.; Całka, J. Changes in the somatostatin (SOM)-like immunoreactivity within nervous structures of the porcine descending colon under various pathological factors. *Exp. Mol. Pathol.* **2010**, *88*, 416–423. [CrossRef]
49. Gonkowski, S. Substance P as a neuronal factor in the enteric nervous system of the porcine descending colon in physiological conditions and during selected pathogenic processes. *Biofactors* **2013**, *39*, 542–551. [CrossRef]
50. Scheuermann, D.W.; Stach, W. Fluorescence microscopic study of the architecture and structure of an adrenergic network in the plexus myentericus (Auerbach), plexus submucosus externus (Schabadasch) and plexus submucosus internus (Meissner) of the porcine small intestine. *Acta Anat.* **1984**, *119*, 49–59. [CrossRef]
51. Wakabayashi, K.; Takahashi, H.; Ohama, E.; Ikuta, F. Tyrosine hydroxylase-immunoreactive intrinsic neurons in the Auerbach's and Meissner's plexuses of humans. *Neurosci. Lett.* **1989**, *96*, 259–263. [CrossRef]
52. Hwang, S.E.; Hieda, K.; Kim, J.H.; Murakami, G.; Abe, S.; Matsubara, A.; Cho, B.H. Region-specific differences in the human myenteric plexus: An immunohistochemical study using donated elderly cadavers. *Int. J. Colorectal Dis.* **2014**, *29*, 783–791. [CrossRef] [PubMed]
53. Mandić, P.; Filipović, T.; Gasić, M.; Djukić-Macut, N.; Filipović, M.; Bogosavljević, I. Quantitative morphometric analysis of the myenteric nervous plexus ganglion structures along the human digestive tract. *Vojnosanit. Pregl.* **2016**, *73*, 559–565. [CrossRef] [PubMed]
54. Ibba-Manneschi, L.; Martini, M.; Zecchi-Orlandini, S.; Faussone-Pellegrini, M.S. Structural organization of enteric nervous system in human colon. *Histol. Histopathol.* **1995**, *10*, 17–25. [PubMed]
55. Wedel, T.; Roblick, U.; Gleiss, J.; Schiedeck, T.; Bruch, H.P.; Kühnel, W.; Krammer, H.J. Organization of the enteric nervous system in the human colon demonstrated by wholemount immunohistochemistry with special reference to the submucous plexus. *Ann. Anat.* **1999**, *181*, 327–337. [CrossRef]
56. Brehmer, A.; Rupprecht, H.; Neuhuber, W. Two submucosal nerve plexus in human intestines. *Histochem. Cell Biol.* **2010**, *133*, 149–161. [CrossRef]
57. Jabari, S.; de Oliveira, E.C.; Brehmer, A.; da Silveira, A.B. Chagasic megacolon: Enteric neurons and related structures. *Histochem. Cell Biol.* **2014**, *142*, 235–244. [CrossRef]
58. Zetzmann, K.; Strehl, J.; Geppert, C.; Kuerten, S.; Jabari, S.; Brehmer, A. Calbindin D28k-immunoreactivity in human enteric neurons. *Int. J. Mol. Sci.* **2018**, *19*, 194. [CrossRef]
59. Oponowicz, A.; Kozłowska, A.; Gonkowski, S.; Godlewski, J.; Majewski, M. Changes in the distribution of cocaine- and amphetamine-regulated transcript-containing neural structures in the human colon affected by the neoplastic process. *Int. J. Mol. Sci.* **2018**, *19*, 414. [CrossRef]
60. Graham, K.D.; López, S.H.; Sengupta, R.; Shenoy, A.; Schneider, S.; Wright, C.M.; Feldman, M.; Furth, E.; Valdivieso, F.; Lemke, A.; et al. Robust, 3-Dimensional visualization of human colon enteric nervous system without tissue sectioning. *Gastroenterology* **2020**, *158*, 2221–2235.e5. [CrossRef]
61. Crowe, R.; Burnstock, G. The subserosal ganglia of the human taenia. *Neurosci. Lett.* **1990**, *119*, 203–206. [CrossRef]
62. Timmermans, J.P.; Scheuermann, D.W.; Stach, W.; Adriaensen, D.; De Groodt Lesseal, M.H.A. Functional morphology of the enteric nervous system with special reference to large mammals. *Eur. J. Morphol.* **1992**, *30*, 113–122.
63. Timmermans, J.P.; Adriaensen, D.; Cornelissen, W.; Scheuermann, D.W. Structural organization and neuropeptide distribution in the mammalian enteric nervous system, with special attention to those components involved in mucosal reflexes. *Comp. Biochem. Physiol.* **1997**, *118*, 331–340. [CrossRef]

64. Arciszewski, M.B.; Barabasz, S.; Całka, J. Expression of substance P, vasoactive intestinal peptide and galanin in cultured myenteric neurons from the ovine abomasum. *Vet. Med.* **2009**, *3*, 118–124. [CrossRef]
65. Gonkowski, S.; Burliński, P.; Skobowiat, C.; Majewski, M.; Całka, J. Inflammation- and axotomy-induced changes in galanin-like immunoreactive (GAL-LI) nerve structures in the porcine descending colon. *Acta Vet. Hung.* **2010**, *58*, 91–103. [CrossRef] [PubMed]
66. Botella, A.; Delvaux, M.; Frexinos, J.; Bueno, L. Comparative effects of galanin on isolated smooth muscle cells from ileum in five mammalian species. *Life Sci.* **1992**, *50*, 1253–1261. [CrossRef]
67. Fox-Threlkeld, J.E.T.; McDonald, T.J.; Cipris, S.; Woskowska, Z.; Daniel, E.E. Galanin inhibition of vasoactive intestinal polypeptide release and circular muscle motility in the isolated perfused canine ileum. *Gastroenterology* **1991**, *101*, 1471–1476. [CrossRef]
68. Lördal, M.; Johansson, C.; Hellström, P.M. Myoelectric pattern and effects on small bowel transit induced by the tachykinins neurokinin A, neurokinin B, substance P and neuropedtide K in the rat. *Neurogastroenterol. Motil.* **1993**, *5*, 33–40. [CrossRef]
69. Lördal, M.; Theodorsson, E.; Hellström, P.M. Tachykinins influence interdigestive rhythm and contractile strength of human small intestine. *Dig. Dis. Sci.* **1997**, *42*, 1940–1949. [CrossRef]
70. Thor, P.J.; Sendur, R.; Konturek, S.J. Influence of substance P on myoelectric activity of the small bowel. *Am. J. Physiol.* **1982**, *243*, G493–G496. [CrossRef]
71. Roza, C.; Reeh, P.W.; Substance, P. calcitonin gene related peptide and PGE2 co-released from the mouse colon: A new model to study nociceptive and inflammatory responses in viscera, in vitro. *Pain* **2001**, *93*, 213–219. [CrossRef]
72. Wolf, M.; Schrödl, F.; Neuhuber, W.; Brehmer, A. Calcitonin gene-related peptide: A marker for putative primary afferent neurons in the pig small intestinal myenteric plexus? *Anat. Rec.* **2007**, *290*, 1273–1279. [CrossRef] [PubMed]
73. Lambrecht, N.; Burchert, M.; Respondek, M.; Muller, K.M.; Peskar, B.M. Role of calcitonin gene-related peptide and nitric oxide in the gastroprotective effect of capsaicin in the rat. *Gastroenterology* **1993**, *104*, 1371–1380. [CrossRef]
74. Barada, K.A.; Saade, N.E.; Atweh, S.F.; Khoury, C.I.; Nassar, C.F. Calcitonin gene-related peptide regulates amino acid absorption across rat jejunum. *Regul. Pept.* **2000**, *90*, 39–45. [CrossRef]
75. Leung, F.W.; Iwata, F.; Seno, K.; Leung, J.W. Acid-induced mesenteric hyperemia in rats: Role of CGRP, substance P, prostaglandin, adenosine, and histamine. *Dig. Dis. Sci.* **2003**, *48*, 523–532. [CrossRef]
76. De Fontgalland, D.; Wattchow, D.A.; Costa, M.; Brookes, S.J.H. Immunohistochemical characterization of the innervation of human colonic mesenteric and submucosal blood vessels. *Neurogastroenterol. Motil.* **2008**, *20*, 1212–1226. [CrossRef]
77. Kaiser, E.A.; Rea, B.J.; Kuburas, A.; Kovacevich, B.R.; Garcia-Martinez, L.F.; Recober, A.; Russo, A.F. Anti-CGRP antibodies block CGRP-induced diarrhea in mice. *Neuropeptides* **2017**, *64*, 95–99. [CrossRef] [PubMed]
78. Delvalle, N.M.; Fried, D.E.; Rivera-Lopez, G.; Gaudette, L.; Gulbransen, B.D. Cholinergic activation of enteric glia is a physiological mechanism that contributes to the regulation of gastrointestinal motility. *Am. J. Physiol. Gastrointest. Liver Physiol.* **2018**, *315*, G473–G483. [CrossRef] [PubMed]
79. Scheurer, U.; Drack, E.; Halter, F. Cyclooxygenase inhibitors affect Met-enkephalin- and acetylcholine-stimulated motility of the isolated rat colon. *J. Pharmacol. Exp. Ther.* **1985**, *234*, 742–746. [PubMed]
80. Johnson, C.D.; Barlow-Anacker, A.J.; Pierre, J.F.; Touw, K.; Erickson, C.S.; Furness, J.B.; Epstein, M.L.; Gosain, A. Deletion of choline acetyltransferase in enteric neurons results in postnatal intestinal dysmotility and dysbiosis. *FASEB J.* **2018**, *32*, 4744–4752. [CrossRef]
81. Aikawa, N.; Kishibayashi, N.; Karasawa, A.; Ohmori, K. The effect of zaldaride maleate, an antidiarrheal compound, on acetylcholine-induced intestinal electrolyte secretion. *Biol. Pharm. Bull.* **2000**, *23*, 1377–1378. [CrossRef]
82. Ogata, H.; Podolsky, D.K. Trefoil peptide expression and secretion is regulated by neuropeptides and acetylcholine. *Am. J. Physiol.* **1997**, *273*, G348–G354. [CrossRef]
83. Specian, R.D.; Neutra, M.R. Mechanism of rapid mucus secretion in goblet cells stimulated by acetylcholine. *J. Cell Biol.* **1980**, *85*, 626–640. [CrossRef] [PubMed]

84. Hansen, L.; Lampert, S.; Mineo, H.; Holst, J.J. Neural regulation of glucagon-like peptide-1 secretion in pigs. *Am. J. Physiol. Endocrinol. Metab.* **2004**, *287*, E939–E947. [CrossRef]
85. Al-Barazie, R.M.; Bashir, G.H.; Qureshi, M.M.; Mohamed, Y.A.; Al-Sbiei, A.; Tariq, S.; Lammers, W.J.; Al-Ramadi, B.K.; Fernandez-Cabezudo, M.J. Cholinergic activation enhances resistance to oral Salmonella infection by modulating innate immune defense mechanisms at the intestinal barrier. *Front. Immunol.* **2018**, *9*, 551. [CrossRef] [PubMed]
86. Dhawan, S.; Hiemstra, I.H.; Verseijden, C.; Hilbers, F.W.; Te Velde, A.A.; Willemsen, L.E.; Stap, J.; den Haan, J.M.; de Jonge, W.J. Cholinergic receptor activation on epithelia protects against cytokine-induced barrier dysfunction. *Acta Physiol.* **2015**, *213*, 846–859. [CrossRef]
87. Matteoli, G.; Boeckxstaens, G.E. The vagal innervation of the gut and immune homeostasis. *Gut* **2013**, *62*, 1214–1222. [CrossRef] [PubMed]
88. van der Zanden, E.P.; Boeckxstaens, G.E.; de Jonge, W.J. The vagus nerve as a modulator of intestinal inflammation. *Neurogastroenterol. Motil.* **2009**, *21*, 6–17. [CrossRef] [PubMed]
89. Nijhuis, L.E.; Olivier, B.J.; de Jonge, W.J. Neurogenic regulation of dendritic cells in the intestine. *Biochem. Pharmacol.* **2010**, *80*, 2002–2008. [CrossRef] [PubMed]
90. Li, Z.S.; Fox-Threlkeld, J.E.; Furness, J.B. Innervation of intestinal arteries by axons with immunoreactivity for the vesicular acetylcholine transporter (VAChT). *J. Anat.* **1998**, *192*, 107–117. [CrossRef]
91. Okumura, T.; Yamada, H.; Motomura, W.; Kohgo, Y. Cocaine- amphetamine-regulated transcript (CART) acts in the central nervous system to inhibit gastric acid secretion via brain corticotrophin-releasing factor system. *Endocrinology* **2000**, *141*, 2854–2860. [CrossRef]
92. Tebbe, J.J.; Ortmann, E.; Schumacher, K.; Mönnikes, H.; Kobelt, P.; Arnold, R.; Schäffer, M.K.H. Cocaine- and amphetamine-regulated transcript stimulates colonic motility via central CRF receptor activation and peripheral cholinergic pathways in fed conscious rats. *Neurogastroenterol. Motil.* **2004**, *16*, 489–496. [CrossRef] [PubMed]
93. Brehmer, A.; Croner, R.; Dimmler, A.; Papadopoulos, T.; Schrödl, F.; Neuhuber, W. Immunohistochemical characterization of putative primary afferent (sensory) myenteric neurons in human small intestine. *Auton. Neurosci.* **2004**, *112*, 49–59. [CrossRef] [PubMed]
94. Grider, J.R. CGRP as a transmitter in the sensory pathway mediating peristaltic reflex. *Am. J. Physiol.* **1994**, *266*, G1139–G1145. [CrossRef] [PubMed]
95. Kuramoto, H.; Kadowaki, M. Enhancement of CGRP sensory afferent innervation in the gut during the development of food allergy in an experimental murine model. *Biochem. Biophys. Res. Commun.* **2013**, *430*, 895–900.
96. Holzer, P.; Guth, P.H. Neuropeptide control of rat gastric mucosal blood flow. Increase by calcitonin gene-related peptide and vasoactive intestinal polypeptide, but not substance P and neurokinin A. *Circ. Res.* **1991**, *68*, 100–105. [CrossRef]
97. Bulut, K.; Felderbauer, P.; Deters, S.; Hoeck, K.; Schmidt-Choudhury, A.; Schmidt, W.E.; Hoffmann, P. Sensory neuropeptides and epithelial cell restitution: The relevance of SP- and CGRP-stimulated mast cells. *Int. J. Colorectal Dis.* **2008**, *23*, 535–541. [CrossRef]
98. Tam, C.; Brain, S.D. The assessment of vasoactive properties of CGRP and adrenomedullin in the microvasculature: A study using in vivo and in vitro assays in the mouse. *J. Mol. Neurosci.* **2004**, *22*, 117–124. [CrossRef]
99. Pawlik, W.W.; Obuchowicz, R.; Biernat, J.; Sendur, R.; Jaworek, J. Role of calcitonin gene related peptide in the modulation of intestinal circulatory, metabolic, and myoelectric activity during ischemia/reperfusion. *J. Physiol. Pharmacol.* **2000**, *51*, 933–942.
100. Holzer, P. Role of visceral afferent neurons in mucosal inflammation and defense. *Curr. Opin. Pharmacol.* **2007**, *7*, 563–569. [CrossRef]
101. Reinshagen, M.; Flämig, G.; Ernst, S.; Geerling, I.; Wong, H.; Walsh, J.H.; Eysselein, V.E.; Adler, G.J. Calcitonin gene-related peptide mediates the protective effect of sensory nerves in a model of colonic injury. *Pharmacol. Exp. Ther.* **1998**, *286*, 657–661.
102. Bálint, A.; Fehér, E.; Kisfalvi, I.J.; Máté, M.; Zelles, T.; Vizi, E.S.; Varga, G. Functional and immunocytochemical evidence that galanin is a physiological regulator of human jejunal motility. *J. Physiol. Paris* **2001**, *95*, 129–135. [CrossRef]

103. Matkowskyj, K.A.; Nathaniel, R.; Prasad, R.; Weihrauch, D.; Rao, M.; Benya, R.V. Galanin contributes to the excess colonic fluid secretion observed in dextran sulfate sodium murine colitis. *Inflamm. Bowel Dis.* **2004**, *10*, 408–416. [CrossRef]
104. Piqueras, L.; Taché, Y.; Martinez, V. Galanin inhibits gastric acid secretion through a somatostatin-independent mechanism in mice. *Peptides* **2004**, *25*, 1287–1295. [CrossRef]
105. Matkowskyj, K.; Royan, S.V.; Blunier, A.; Hecht, G.; Rao, M.; Benya, R.V. Age-dependent differences in galanin-dependent colonic fluid secretion after infection with Salmonella typhimurium. *Gut* **2009**, *58*, 1201–1206. [CrossRef] [PubMed]
106. Daniel, E.E.; Haugh, C.; Woskowska, Z.; Cipris, S.; Jury, J.; Fox-Threlkeld, J.E.T. Role of nitric oxide—Related inhibition in intestinal function: Relation to vasoactive intestinal polypeptide. *Am. J. Physiol.* **1994**, *266*, 31–39. [CrossRef] [PubMed]
107. Grider, J.R. Interplay of VIP and nitric oxide in regulation of the descending relaxation phase of peristalsis. *Am. J. Physiol.* **1993**, *264*, G334–G340. [CrossRef]
108. Groneberg, D.; Voussen, B.; Friebe, A. Integrative control of gastrointestinal motility by nitric oxide. *Curr. Med. Chem.* **2016**, *23*, 2715–2735. [CrossRef]
109. Walker, M.Y.; Pratap, S.; Southerland, J.H.; Farmer-Dixon, C.M.; Lakshmyya, K.; Gangula, P.R. Role of oral and gut microbiome in nitric oxide-mediated colon motility. *Nitric Oxide* **2018**, *73*, 81–88. [CrossRef]
110. Mourad, F.H.; O'Donnell, L.J.; Andre, E.A.; Bearcroft, C.P.; Owen, R.A.; Clark, M.L.; Farthing, M.J. L-Arginine, nitric oxide, and intestinal secretion: Studies in rat jejunum in vivo. *Gut* **1996**, *39*, 539–544. [CrossRef]
111. Izzo, A.A.; Mascolo, N.; Capasso, F. Nitric oxide as a modulator of intestinal water and electrolyte transport. *Dig. Dis. Sci.* **1998**, *43*, 1605–1620. [CrossRef]
112. Mourad, F.H.; Barada, K.A.; Abdel-Malak, N.; Bou Rached, N.A.; Khoury, C.I.; Saade, N.E.; Nassar, C.F. Interplay between nitric oxide and vasoactive intestinal polypeptide in inducing fluid secretion in rat jejunum. *J. Physiol.* **2003**, *550*, 863–871. [CrossRef] [PubMed]
113. Barry, M.K.; Alois, J.D.; Pickering, S.P.; Yeo, C.J. Nitric oxide modulates water and electrolyte transport in the ileum. *Ann. Surg.* **1994**, *219*, 382–388. [CrossRef]
114. Fan, W.Q.; Smolich, J.J.; Wild, J.; Yu, V.Y.; Walker, A.M. Nitric oxide modulates regional blood flow differences in the fetal gastrointestinal tract. *Am. J. Physiol.* **1996**, *271*, G598–G604. [CrossRef]
115. Jansson, L.; Carlsson, P.O.; Bodin, B.; Andersson, A.; Källskog, O. Neuronal nitric oxide synthase and splanchnic blood flow in anaesthetized rats. *Acta. Physiol. Scand.* **2005**, *183*, 257–262. [CrossRef] [PubMed]
116. Roediger, W.E. Nitric oxide damage to colonocytes in colitis-by-association: Remote transfer of nitric oxide to the colon. *Digestion* **2002**, *65*, 191–195. [CrossRef]
117. Varga, S.; Juhász, L.; Gál, P.; Bogáts, G.; Boro, M.; Palásthy, Z.; Szabó, A.; Kaszaki, J. Neuronal nitric oxide mediates the anti-inflammatory effects of intestinal ischemic preconditioning. *J. Surg. Res.* **2019**, *244*, 241–250. [CrossRef] [PubMed]
118. Katsoulis, S.; Schmidt, W.E. Role of PACAP in the regulation of gastrointestinal motility. *Ann. N. Y. Acad. Sci.* **1996**, *805*, 364–378. [CrossRef] [PubMed]
119. Ozawa, M.; Aono, M.; Moriga, M. Central effects of pituitary adenylate cyclase activating polypeptide (PACAP) on gastric motility and emptying in rats. *Dig. Dis. Sci.* **1999**, *44*, 735–743. [CrossRef] [PubMed]
120. Felley, C.P.; Qian, J.M.; Mantey, S.; Pradhan, T.; Jensen, R.T. Chief cells possess a receptor with high affinity for PACAP and VIP that stimulates pepsinogen release. *Am. J. Physiol.* **1992**, *263*, G901–G907. [CrossRef]
121. Läuff, J.M.; Modlin, I.M.; Tang, L.H. Biological relevance of pituitary adenylate cyclase-activating polypeptide (PACAP) in the gastrointestinal tract. *Regul. Pept.* **1999**, *84*, 1–12. [CrossRef]
122. Kuwahara, A.; Kuwahara, Y.; Mochizuki, T.; Yanaihara, N. Action of pituitary adenylate cyclase-activating polypeptide on ion transport in guinea pig distal colon. *Am. J. Physiol.* **1993**, *264*, G433–G441. [CrossRef] [PubMed]
123. Fuchs, M.; Adermann, K.; Raab, H.R.; Forssmann, W.G.; Kuhn, M. Pituitary adenylate cyclase-activating polypeptide: A potent activator of human intestinal ion transport. *Ann. N. Y. Acad. Sci.* **1996**, *805*, 640–647. [CrossRef] [PubMed]
124. Wei, M.X.; Hu, P.; Wang, P.; Naruse, S.; Nokihara, K.; Wray, V.; Ozaki, T. Possible key residues that determine left gastric artery blood flow response to PACAP in dogs. *World. J. Gastroenterol.* **2010**, *16*, 4865–4870. [CrossRef] [PubMed]

125. Shimizu, Y.; Matsuyama, H.; Shiina, T.; Takewaki, T.; Furness, J.B. Tachykinins and their functions in the gastrointestinal tract. *Cell. Mol. Life Sci.* **2008**, *65*, 295–311. [CrossRef]
126. Shibata, C.; Sasaki, I.; Naito, H.; Ohtani, N.; Matsuno, S.; Mizumoto, A.; Iwanaga, Y.; Itoh, Z.; Tohoku, J. Effects of substance P on gastric motility differ depending on the sites and vagal innervation in conscious dogs. *Exp. Med.* **1994**, *174*, 119–128. [CrossRef]
127. Donnerer, J.; Barthó, L.; Holzer, P.; Lembeck, F. Intestinal peristalsis associated with release of immunoreactive substance P. *Neuroscience* **1984**, *11*, 913–918. [CrossRef]
128. Greenwood, B.; Doolittle, T.; See, N.A.; Koch, T.R.; Dodds, W.J.; Davison, J.S. Effects of substance P and vasoactive intestinal polypeptide on contractile activity and epithelial transport in the ferret jejunum. *Gastroenterology* **1990**, *98*, 1509–1517. [CrossRef]
129. Perdue, M.H.; Galbraith, R.; Davison, J.S. Evidence for substance P as a functional neurotransmitter inguinea pig small intestinal mucosa. *Regul. Pept.* **1987**, *18*, 63–74. [CrossRef]
130. Pothoulakis, C.; Castagliuolo, I.; LaMont, J.T.; Jaffer, A.; OKeane, J.C.; Snider, R.M.; Leeman, S.E. CP-96,345, a substance P antagonist, inhibits rat intestinal responses to Clostridium difficile toxin A but not choleratoxin. *Proc. Natl. Acad. Sci. USA* **1994**, *91*, 947–951. [CrossRef]
131. Arciszewski, M.B.; Ekblad, E. Effects of vasoactive intestinal peptide and galanin on survival of cultured porcine myenteric neurons. *Regul. Pept.* **2005**, *125*, 185–192. [CrossRef]
132. Eklund, S.; Jodal, M.; Lundgren, O.; Sjöqvist, A. Effects of vasoactive intestinal polypeptide on blood flow, motility and fluid transport in the gastrointestinal tract of the cat. *Acta Physiol. Scand.* **1979**, *105*, 461–468. [CrossRef] [PubMed]
133. Krantis, A.; Mattar, K.; Glasgow, I. Rat gastroduodenal motility in vivo: Interaction of GABA and VIP in control of spontaneous relaxations. *Am. J. Physiol.* **1998**, *275*, G897–G903. [CrossRef] [PubMed]
134. Iwasaki, M.; Akiba, Y.; Kaunitz, J.D. Recent advances in vasoactive intestinal peptide physiology and pathophysiology: Focus on the gastrointestinal system. *F1000Resrarch* **2019**, *8*, F1000. [CrossRef] [PubMed]
135. Talbot, J.; Hahn, P.; Kroehling, L.; Nguyen, H.; Li, D.; Littman, D.R. Feeding-dependent VIP neuron-ILC3 circuit regulates the intestinal barrier. *Nature* **2020**, *579*, 575–580. [CrossRef]
136. Kovsca Janjatovic, A.; Valpotic, H.; Kezic, D.; Lacković, G.; Gregorovic, G.; Sladoljev, S.; Mršić, G.; Popovic, M.; Valpotic, I. Secretion of immunomodulating neuropeptides (VIP, SP) and nitric oxide synthase in porcine small intestine during postnatal development. *Eur. J. Histochem.* **2012**, *56*, e30. [CrossRef]
137. Ottaway, C.A. Neuroimmunomodulation in the intestinal mucosa. *Gastroenterol. Clin. N. Am.* **1991**, *20*, 511–529.
138. Burleigh, D.E.; Banks, M.R. Stimulation of intestinal secretion by vasoactive intestinal peptide and cholera toxin. *Auton. Neurosci. Basic Clin.* **2007**, *133*, 64–75. [CrossRef]
139. Nassar, C.F.; Abdallah, L.E.; Barada, K.A.; Atweh, S.F.; Saadé, N.F. Effects of intravenous vasoactive intestinal peptide injection on jejunal alanine absorption and gastric acid secretion in rats. *Regul. Pept.* **1995**, *55*, 261–267. [CrossRef]
140. Mourad, F.H.; Nassar, C.F. Effect of vasoactive intestinal polypeptide (VIP) antagonism on rat jejunal fluid and electrolyte secretion induced by cholera and Escherichia coli enterotoxins. *Gut* **2000**, *47*, 382–386. [CrossRef]
141. Jessen, K.R.; Mirsky, R. Glial cells in the enteric nervous system contain glial fibrillary acidic protein. *Nature* **1980**, *286*, 736–737. [CrossRef]
142. Pochard, C.; Coquenlorge, S.; Freyssinet, M.; Naveilhan, P.; Bourreille, A.; Neunlist, M.; Rolli-Derkinderen, M. The multiple faces of inflammatory enteric glial cells: Is Crohn's disease a gliopathy? *Am. J. Physiol. Gastrointest. Liver Physiol.* **2018**, *315*, G1–G11. [CrossRef] [PubMed]
143. Seguella, L.; Capuano, R.; Sarnelli, G.; Esposito, G. Play in advance against neurodegeneration: Exploring enteric glial cells in gut-brain axis during neurodegenerative diseases. *Expert Rev. Clin. Pharmacol.* **2019**, *12*, 555–564. [CrossRef] [PubMed]
144. Gulbransen, B.D.; Sharkey, K.A. Novel functional roles for enteric glia in the gastrointestinal tract. *Nat. Rev. Gastroenterol. Hepatol.* **2012**, *9*, 625–632. [CrossRef]
145. de Mattos Coelho-Aguiar, J.; Bon-Frauches, A.C.; Gomes, A.L.; Veríssimo, C.P.; Aguiar, D.P.; Matias, D.; Thomasi, B.B.; Gomes, A.S.; Brito, G.A.; Moura-Neto, V. The enteric glia: Identity and functions. *Glia* **2015**, *63*, 921–935. [CrossRef] [PubMed]
146. Jessen, K.R. Glial cells. *Int. J. Biochem. Cell Biol.* **2004**, *36*, 1861–1867. [CrossRef] [PubMed]

147. Grubišić, V.; Gulbransen, B.D. Enteric glia: The most alimentary of all glia. *J. Physiol.* **2017**, *595*, 557–570. [CrossRef] [PubMed]
148. Abdo, H.; Derkinderen, P.; Gomes, P.; Chevalier, J.; Aubert, P.; Masson, D.; Galmiche, J.P.; Vanden Berghe, P.; Neunlist, M.; Lardeux, B. Enteric glial cells protect neurons from oxidative stress in part via reduced glutathione. *FASEB J.* **2010**, *24*, 1082–1094. [CrossRef]
149. Neunlist, M.; Rolli-Derkinderen, M.; Latorre, R.; Van Landeghem, L.; Coron, E.; Derkinderen, P.; De Giorgio, R. Enteric glial cells: Recent developments and future directions. *Gastroenterology* **2014**, *147*, 1230–1237. [CrossRef]
150. Ruhl, A. Glial cells in the gut. *Neurogastroenterol. Motil.* **2005**, *17*, 777–790. [CrossRef]
151. Boesmans, W.; Cirill, C.; Van den Abbeel, V.; Van den Haute, C.; Depoortere, I.; Tack, J.; Vanden Berghe, P. Neuro-transmitters involved in fast excitatory neurotransmission directly activate enteric glial cells. *Neurogastroenterol. Motil.* **2013**, *25*, e151–e160. [CrossRef]
152. Vergnolle, N.; Cirillo, C. Neurons and glia in the enteric nervous system and epithelial barrier function. *Physiology* **2018**, *33*, 269–280. [CrossRef] [PubMed]
153. Bauman, B.D.; Meng, J.; Zhang, L.; Louiselle, A.; Zheng, E.; Banerjee, S.; Roy, S.; Segura, B.J. Enteric glial-mediated enhancement of intestinal barrier integrity is compromised by morphine. *J. Surg. Res.* **2017**, *219*, 214–221. [CrossRef] [PubMed]
154. Bush, T.G.; Savidge, T.C.; Freeman, T.C.; Cox, H.J.; Campbell, E.A.; Mucke, L.; Johnson, M.H.; Sofroniew, M.V. Fulminant jejuno-ileitis following ablation of enteric glia in adult transgenic mice. *Cell* **1998**, *93*, 189–201. [CrossRef]
155. Joseph, N.M.; He, S.; Quintana, E.; Kim, Y.G.; Núñez, G.; Morrison, S.J. Enteric glia are multipotent in culture but primarily form glia in the adult rodent gut. *J. Clin. Investig.* **2011**, *121*, 3398–3411. [CrossRef]
156. Da Silveira, A.B.; de Oliveira, E.C.; Neto, S.G.; Luquetti, A.O.; Fujiwara, R.T.; Oliveira, T.C.; Brehmer, A. Enteroglial cells act as antigen-presenting cells in chagasic megacolon. *Hum. Pathol.* **2011**, *42*, 522–553. [CrossRef]
157. Rühl, A.; Franzke, S.; Collins, S.M.; Stremmel, W. Interleukin-6 expression and regulation in rat enteric glial cells. *Am. J. Physiol. Gastrointest. Liver Physiol.* **2001**, *280*, G1163–G1171. [CrossRef]
158. Murakami, M.; Ohta, T.; Ito, S. Lipopolysaccharides enhance the action of bradykinin in enteric neurons via secretion of interleukin-1beta from enteric glial cells. *J. Neurosci. Res.* **2009**, *87*, 2095–2104. [CrossRef]
159. Kermarrec, L.; Durand, T.; Gonzales, J.; Pabois, J.; Hulin, P.; Neunlist, M.; Neveu, I.; Naveilhan, P. Rat enteric glial cells express novel isoforms of Interleukine-7 regulated during inflammation. *Neurogastroenterol. Motil.* **2019**, *31*, e13467. [CrossRef]
160. von Boyen, G.B.; Steinkamp, M.; Reinshagen, M.; Schäfer, K.H.; Adler, G.; Kirsch, J. Nerve growth factor secretion in cultured enteric glia cells is modulated by proinflammatory cytokines. *J. Neuroendocrinol.* **2006**, *18*, 820–825. [CrossRef]
161. Rosenbaum, C.; Schick, M.A.; Wollborn, J.; Heider, A.; Scholz, C.S.; Cecil, A.; Niesler, B.; Hirrlinger, J.; Walles, H.; Metzger, M. Activation of myenteric glia during acute inflammation in vitro and in vivo. *PLoS ONE* **2016**, *11*, e0151335. [CrossRef]
162. Cirillo, C.; Sarnelli, G.; Esposito, G.; Grosso, M.; Petruzzelli, R.; Izzo, G.; D'Armiento, C.F.P.; Rocco, A.; Nardone, G.; Iuvone, T.; et al. Increased mucosal nitric oxide production in ulcerative colitis is mediated in part by the enteroglial-derived S100B protein. *Neurogastroenterol. Motil.* **2009**, *21*, 1209-e112. [CrossRef]
163. Caputi, V.; Giron, M.C. Microbiome-gut-brain axis and toll-like receptors in Parkinson's disease. *Int. J. Mol. Sci.* **2018**, *19*, 1689. [CrossRef]
164. Vasina, V.; Barbara, G.; Talamonti, L.; Stanghellini, V.; Corinaldesi, R.; Tonini, M.; De Ponti, F.; De Georgio, R. Enteric neuroplasticy evoked by inflammation. *Auton. Neurosci.* **2006**, *126–127*, 264–272. [CrossRef] [PubMed]
165. Obata, Y.; Pachnis, V. The Effect of microbiota and the immune system on the development and organization of the enteric nervous system. *Gastroenterology* **2016**, *151*, 836–844. [CrossRef] [PubMed]
166. Szymanska, K.; Gonkowski, S. Neurochemical characterization of the enteric neurons within the porcine jejunum in physiological conditions and under the influence of bisphenol A (BPA). *Neurogastroenterol. Motil.* **2019**, *31*, e13580. [CrossRef] [PubMed]
167. Wang, L.; Shao, H.; Luo, X.; Wang, R.; Li, Y.; Li, Y.; Luo, Y.; Chen, Z. Effect of ozone treatment on deoxynivalenol and wheat quality. *PLoS ONE* **2016**, *11*, e0147613. [CrossRef] [PubMed]

168. Habrowska-Górczyńska, D.E.; Kowalska, K.; Urbanek, K.A.; Domińska, K.; Sakowicz, A.; Piastowska-Ciesielska, A.W. Deoxynivalenol modulates the viability, ROS production and apoptosis in prostate cancer cells. *Toxins* **2019**, *11*, 265. [CrossRef]
169. Payros, D.; Alassane-Kpembi, I.; Pierron, A.; Loiseau, N.; Pinton, P.; Oswald, I.P. Toxicology of deoxynivalenol and its acetylated and modified forms. *Arch. Toxicol.* **2016**, *90*, 2931–2957. [CrossRef] [PubMed]
170. Matejova, I.; Modra, H.; Blahova, J.; Franc, A.; Fictum, P.; Sevcikova, M.; Svobodova, Z. The effect of mycotoxin deoxynivalenol on haematological and biochemical indicators and histopathological changes in rainbow trout (Oncorhynchus Mykiss). *BioMed Res. Int.* **2014**, *2014*, 310680. [CrossRef]
171. Faeste, C.K.; Pierre, F.; Ivanova, L.; Sayyari, A.; Massotte, D. Behavioural and metabolomic changes from chronic dietary exposure to low-level deoxynivalenol reveal impact on mouse well-being. *Arch. Toxicol.* **2019**, *93*, 2087–2102. [CrossRef]
172. Girardet, C.; Bonnet, M.S.; Jdir, R.; Sadoud, M.; Thirion, S.; Tardivel, C.; Roux, J.; Lebrun, B.; Mounien, L.; Trouslard, J.; et al. Central inflammation and sickness-like behavior induced by the food contaminant deoxynivalenol: A PGE2-independent mechanism. *Toxicol. Sci.* **2011**, *124*, 179–191. [CrossRef] [PubMed]
173. Tominaga, M.; Momonaka, Y.; Yokose, C.; Tadaishi, M.; Shimizu, M.; Yamane, T.; Oishi, Y.; Kobayashi-Hattori, K. Anorexic action of deoxynivalenol in hypothalamus and intestine. *Toxicon* **2016**, *118*, 54–60. [CrossRef]
174. Hussein, H.S.; Brasel, J.M. Toxicity, metabolism, and impact of mycotoxins on humans and animals. *Toxicology* **2001**, *167*, 101–134. [CrossRef]
175. Obremski, K.; Zielonka, L.; Gajecka, M.; Jakimiuk, E.; Bakuła, T.; Baranowski, M.; Gajecki, M. Histological estimation of the small intestine wall after administration of feed containing deoxynivalenol, T-2 toxin and zearalenone in the pig. *Pol. J. Vet. Sci.* **2008**, *11*, 339–345.
176. Goossens, J.; Pasmans, F.; Verbrugghe, E.; Vandenbroucke, V.; De Baere, S.; Meyer, E.; Haesebrouck, F.; De Backer, P.; Croubels, S. Porcine intestinal epithelial barrier disruption by the Fusarium mycotoxins deoxynivalenol and T-2 toxin promotes transepithelial passage of doxycycline and paromomycin. *BMC Vet. Res.* **2012**, *8*, 245. [CrossRef] [PubMed]
177. Osselaere, A.; Li, S.J.; De Bock, L.; Devreese, M.; Goossens, J.; Vandenbroucke, V.; Van Bocxlaer, J.; Boussery, K.; Pasmans, F.; Martel, A.; et al. Toxic effects of dietary exposure to T-2 toxin on intestinal and hepatic biotransformation enzymes and drug transporter systems in broiler chickens. *Food Chem. Toxicol.* **2013**, *55*, 150–155. [CrossRef]
178. Lin, R.; Sun, Y.; Ye, W.; Zheng, T.; Wen, J.; Deng, Y. T-2 toxin inhibits the production of mucin via activating the IRE1/XBP1 pathway. *Toxicology* **2019**, *424*, 152230. [CrossRef]
179. Sheng, K.; Lu, X.; Yue, J.; Gu, W.; Gu, C.; Zhang, H.; Wu, W. Role of neurotransmitters 5-hydroxy-tryptamine and substance P in anorexia induction following oral exposure to the trichothecene T-2 toxin. *Food Chem. Toxicol.* **2019**, *123*, 1–8. [CrossRef]
180. Guo, P.; Liu, A.; Huang, D.; Wu, Q.; Fatima, Z.; Tao, Y.; Cheng, G.; Wang, X.; Yuan, Z. Brain damage and neurological symptoms induced by T-2 toxin in rat brain. *Toxicol. Lett.* **2018**, *286*, 96–107. [CrossRef]
181. Chaudhary, M.; Rao, P.V. Brain oxidative stress after dermal and subcutaneous exposure of T-2 toxin in mice. *Food Chem. Toxicol.* **2010**, *48*, 3436–3442. [CrossRef]
182. Dai, C.; Xiao, X.; Sun, F.; Zhang, Y.; Hoyer, D.; Shen, J.; Tang, S.; Velkov, T. T-2 toxin neurotoxicity: Role of oxidative stress and mitochondrial dysfunction. *Arch. Toxicol.* **2019**, *93*, 3041–3056. [CrossRef]
183. Makowska, K.; Gonkowski, S.; Zielonka, L.; Dabrowski, M.; Calka, J. T2 toxin-induced changes in cocaine- and amphetamine-regulated transcript (CART)-like immunoreactivity in the enteric nervous system within selected fragments of the porcine digestive tract. *Neurotox. Res.* **2017**, *31*, 136–147. [CrossRef] [PubMed]
184. Makowska, K.; Gonkowski, S. Cocaine- and amphetamine-regulated transcript (CART) peptide in mammals gastrointestinal system—A review. *Ann. Anim. Sci.* **2017**, *1*, 2–21. [CrossRef]
185. Stanley, S.A.; Small, C.J.; Murphy, K.G.; Rayes, E.; Abbott, C.R.; Seal, L.J.; Morgan, D.G.; Sunter, D.; Dakin, C.L.; Kim, M.S.; et al. Actions of cocaine- and amphetamine-regulated transcript (CART) peptide on regulation of appetite and hypothalamo-pituitary axes in vitro and in vivo in male rats. *Brain Res.* **2001**, *893*, 186–194. [CrossRef]
186. Helton, W.S.; Mulholland, M.M.; Bunnett, N.W.; Debas, H.T. Inhibition of gastric and pancreatic secretion in dogs by CGRP: Role of somatostatin. *Am. J. Physiol.* **1989**, *256*, G715–G720. [CrossRef] [PubMed]

187. Knutsen, H.-K.; Alexander, J.; Barregård, L.; Bignami, M.; Brüschweiler, B.; Ceccatelli, S.; Cottrill, B.; Dinovi, M.; Edler, L.; Grasl-Kraupp, B.; et al. Risks for animal health related to the presence of zearalenone and its modified forms in feed. *EFSA J.* **2017**, *15*, 4851.
188. Lephart, E.D.; Thompson, J.M.; Setchell, K.D.; Adlercreutz, H.; Weber, K.S. Phytoestrogens decrease brain calcium-binding proteins but do not alter hypothalamic androgen metabolizing enzymes in adult male rats. *Brain Res.* **2000**, *859*, 123–131. [CrossRef]
189. Venkataramana, M.; Chandra Nayaka, S.; Anand, T.; Rajesh, R.; Aiyaz, M.; Divakara, S.T.; Murali, H.S.; Prakash, H.S.; Lakshmana Rao, P.V. Zearalenone induced toxicity in SHSY-5Y cells: The role of oxidative stress evidenced by N-acetyl cysteine. *Food Chem. Toxicol.* **2014**, *65*, 335–342. [CrossRef] [PubMed]
190. Ren, Z.H.; Deng, H.D.; Deng, Y.T.; Deng, J.L.; Zuo, Z.C.; Yu, S.M.; Hu, Y.C. Effect of the Fusarium toxins, zearalenone and deoxynivalenol, on the mouse brain. *Environ. Toxicol. Pharmacol.* **2016**, *46*, 62–70. [CrossRef] [PubMed]
191. Kiss, D.S.; Ioja, E.; Toth, I.; Barany, Z.; Jocsak, G.; Bartha, T.; Horvath, T.L.; Zsarnovszky, A. Comparative analysis of zearalenone effects on thyroid receptor alpha (TRα) and beta (TRβ) expression in rat primary cerebellar cell cultures. *Int. J. Mol. Sci.* **2018**, *19*, 1440. [CrossRef]
192. Khezri, A.; Herranz-Jusdado, J.G.; Ropstad, E.; Fraser, T.W. Mycotoxins induce developmental toxicity and behavioural aberrations in zebrafish larvae. *Environ. Pollut.* **2018**, *242*, 500–506. [CrossRef] [PubMed]
193. Gajęcka, M.; Stopa, E.; Tarasiuk, M.; Zielonka, Ł.; Gajęcki, M. The expression of type-1 and type-2 nitric oxide synthase in selected tissues of the gastrointestinal tract during mixed mycotoxicosis. *Toxins* **2013**, *5*, 2281–2292. [CrossRef] [PubMed]
194. Lahjouji, T.; Bertaccini, A.; Neves, M.; Puel, S.; Oswald, I.P.; Soler, L. Acute exposure to zearalenone disturbs intestinal homeostasis by modulating the Wnt/β-Catenin signaling pathway. *Toxins* **2020**, *12*, 113. [CrossRef] [PubMed]
195. Cieplińska, K.; Gajęcka, M.; Dąbrowski, M.; Rykaczewska, A.; Lisieska-Żołnierczyk, S.; Bulińska, M.; Zielonka, Ł.; Gajęcki, M.T. Time-dependent changes in the intestinal microbiome of gilts exposed to low zearalenone doses. *Toxins* **2019**, *11*, 296. [CrossRef]
196. Artigot, M.P.; Loiseau, N.; Laffitte, J.; Mas-Reguieg, L.; Tadrist, S.; Oswald, I.P.; Puel, O. Molecular cloning and functional characterization of two CYP619 cytochrome P450s involved in biosynthesis of patulin in Aspergillus clavatus. *Microbiology* **2009**, *155*, 1738–1747. [CrossRef]
197. Biango-Daniels, M.N.; Hodge, K.T. Paecilomyces rot: A New apple disease. *Plant Dis.* **2018**, *102*, 1581–1587. [CrossRef]
198. Assunção, R.; Pinhão, M.; Loureiro, S.; Alvito, P.; João Silva, M. A Multi-endpoint approach to the combined toxic effects of patulin and ochratoxin a in human intestinal cells. *Toxicol. Lett.* **2019**, *313*, 120–129. [CrossRef]
199. Mohan, H.M.; Collins, D.; Maher, S.; Walsh, E.G.; Winter, D.C.; O'Brien, P.J.; Brayden, D.J.; Baird, A.W. The mycotoxin patulin increases colonic epithelial permeability in vitro. *Food. Chem. Toxicol.* **2012**, *50*, 4097–4102. [CrossRef]
200. Malekinejad, H.; Aghazadeh-Attari, J.; Rezabakhsh, A.; Sattari, M.; Ghasemsoltani-Momtaz, B. Neurotoxicity of mycotoxins produced in vitro by Penicillium roqueforti isolated from maize and grass silage. *Hum. Exp. Toxicol.* **2015**, *34*, 997–1005. [CrossRef]
201. Vidal, A.; Ouhibi, S.; Gali, R.; Hedhili, A.; De Saeger, S.; De Boevre, M. The mycotoxin patulin: An updated short review on occurrence, toxicity and analytical challenges. *Food. Chem. Toxicol.* **2019**, *129*, 249–256. [CrossRef]
202. Brand, B.; Stoye, N.M.; Guilherme, M.D.S.; Nguyen, V.T.T.; Baumgaertner, J.C.; Schüffler, A.; Thines, E.; Endres, K. Identification of patulin from Penicillium coprobium as a toxin for enteric neurons. *Molecules* **2019**, *24*, 2776. [CrossRef] [PubMed]
203. Peltomaa, R.; Vaghini, S.; Patiño, B.; Benito-Peña, E.; Moreno-Bondi, M.C. Species-specific optical genosensors for the detection of mycotoxigenic Fusarium fungi in food samples. *Anal. Chim. Acta* **2016**, *935*, 231–238. [CrossRef] [PubMed]
204. Palencia, E.R.; Mitchell, T.R.; Snook, M.E.; Glenn, A.E.; Gold, S.; Hinton, D.M.; Riley, R.T.; Bacon, C.W. Analyses of black Aspergillus species of peanut and maize for ochratoxins and fumonisins. *J. Food Prot.* **2014**, *77*, 805–813. [CrossRef]

205. Waes, J.G.; Starr, L.; Maddox, J.; Aleman, F.; Voss, K.A.; Wilberding, J.; Riley, R.T. Maternal fumonisin exposure and risk for neural tube defects: Mechanisms in an in vivo mouse model. *Birth Defects Res. A Clin. Mol. Teratol.* **2005**, *73*, 487–497. [CrossRef] [PubMed]
206. Purzycki, C.B.; Shain, D.H. Fungal toxins and multiple sclerosis: A compelling connection. *Brain Res. Bull.* **2010**, *82*, 4–6. [CrossRef] [PubMed]
207. Jenkins, G.R.; Tolleson, W.H.; Newkirk, D.K.; Roberts, D.W.; Rowland, K.L.; Saheki, T.; Kobayashi, K.; Howard, P.C.; Melchior, W.B., Jr. Identification of fumonisin B1 as an inhibitor of argininosuccinate synthetase using fumonisin affinity chromatography and in vitro kinetic studies. *J. Biochem. Mol. Toxicol.* **2000**, *14*, 320–328. [CrossRef]
208. Kolf-Clauw, M.; Castellote, J.; Joly, B.; Bourges-Abella, N.; Raymond-Letron, I.; Pinton, P.; Oswald, I.P. Development of a pig jejunal explant culture for studying the gastrointestinal toxicity of the mycotoxin deoxynivalenol: Histopathological analysis. *Toxicol. In Vitro* **2009**, *23*, 1580–1584. [CrossRef]
209. Pestka, J.J. Deoxynivalenol: Mechanisms of action, human exposure, and toxicological relevance. *Arch. Toxicol.* **2010**, *84*, 663–679. [CrossRef]
210. Pestka, J.J.; Smolinski, A.T. Deoxynivalenol: Toxicology and potential effects on humans. *J. Toxicol. Environ. Health B Crit. Rev.* **2005**, *8*, 39–69. [CrossRef]
211. Pestka, J.J.; Lin, W.S.; Miller, E.R. Emetic activity of the trichothecene 15-acetyldeoxynivalenol in swine. *Food Chem. Toxicol.* **1987**, *25*, 855–858. [CrossRef]
212. Young, L.G.; McGirr, L.; Valli, V.E.; Lumsden, J.H.; Lun, A. Vomitoxin in corn fed to young pigs. *J. Anim. Sci.* **1983**, *57*, 655–664. [CrossRef] [PubMed]
213. Prelusky, D.B.; Trenholm, H.L. The efficacy of various classes of anti-emetics in preventing deoxynivalenol-induced vomiting in swine. *Nat. Toxins* **1983**, *1*, 296–302. [CrossRef]
214. Lee, H.J.; Ryu, D. Worldwide occurrence of mycotoxins in cereals and cereal-derived food products: Public health perspectives of their co-occurrence. *J. Agric. Food Chem.* **2017**, *65*, 7034–7051. [CrossRef]
215. Pierron, A.; Mimoun, S.; Murate, L.S.; Loiseau, N.; Lippi, Y.; Bracarense, A.P.; Liaubet, L.; Schatzmayr, G.; Berthiller, F.; Moll, W.D.; et al. Intestinal toxicity of the masked mycotoxin deoxynivalenol-3-β-D-glucoside. *Arch. Toxicol.* **2016**, *90*, 2037–2046. [CrossRef] [PubMed]
216. Alizadeh, A.; Braber, S.; Akbari, P.; Garssen, J.; Fink-Gremmels, J. Deoxynivalenol impairs weight gain and affects markers of gut health after low-dose, short-term exposure of growing pigs. *Toxins* **2015**, *7*, 2071–2095. [CrossRef]
217. Bracarense, A.P.; Lucioli, J.; Grenier, B.; Pacheco, G.D.; Moll, W.D.; Schatzmayr, G.; Oswald, I.P. Chronic ingestion of deoxynivalenol and fumonisin, alone or in interaction, induces morphological and immunological changes in the intestine of piglets. *Br. J. Nutr.* **2012**, *107*, 1776–1786. [CrossRef] [PubMed]
218. Gerez, J.R.; Pinton, P.; Callu, P.; Grosjean, F.; Oswald, I.P.; Bracarense, A.P. Deoxynivalenol alone or in combination with nivalenol and zearalenone induce systemic histological changes in pigs. *Exp. Toxicol. Pathol.* **2015**, *67*, 89–98. [CrossRef]
219. Kasuga, F.; Hara-Kudo, Y.; Saito, N.; Kumagai, S.; Sugita-Konishi, Y. In vitro effect of deoxynivalenol on the differentiation of human colonic cell lines Caco-2 and T84. *Mycopathologia* **1998**, *142*, 161–167. [CrossRef]
220. Bensassi, F.; El Golli-Bennour, E.; Abid-Essefi, S.; Bouaziz, C.; Hajlaoui, M.R.; Bacha, H. Pathway of deoxynivalenol-induced apoptosis in human colon carcinoma cells. *Toxicology* **2009**, *264*, 104–109. [CrossRef] [PubMed]
221. Kouadio, J.H.; Mobio, T.A.; Baudrimont, I.; Moukha, S.; Dano, S.D.; Creppy, E. Comparative study of cytotoxicity and oxidative stress induced by deoxynivalenol, zearalenone or fumonisin B1 in human intestinal cell line Caco-2. *Toxicology* **2005**, *213*, 56–65. [CrossRef]
222. Awad, W.A.; Böhm, J.; Razzazi-Fazeli, E.; Zentek, J. Effects of feeding deoxynivalenol contaminated wheat on growth performance, organ weights and histological parameters of the intestine of broiler chickens. *J. Anim. Physiol. Anim. Nutr.* **2006**, *90*, 32–37. [CrossRef]
223. Girish, C.K.; Smith, T.K. Effects of feeding blends of grains naturally contaminated with Fusarium mycotoxins on small intestinal morphology of turkeys. *Poult. Sci.* **2008**, *87*, 1075–1082. [CrossRef] [PubMed]
224. Tremel, H.; Strugala, G.; Forth, W.; Fichtl, B. Dexamethasone decreases lethality of rats in acute poisoning with T-2 toxin. *Arch. Toxicol.* **1985**, *57*, 74–75. [CrossRef]
225. Chi, M.S.; Robison, T.S.; Mirocha, C.J.; Reddy, K.R. Acute toxicity of 12,13-epoxytrichothecenes in one-day-old broiler chicks. *Appl. Environ. Microbiol.* **1978**, *35*, 636–640. [CrossRef] [PubMed]

226. Richard, J.L. Some major mycotoxins and their mycotoxicoses—An overview. *Int. J. Food Microbiol.* **2007**, *119*, 3–10. [CrossRef]
227. Pang, V.F.; Lorenzana, R.M.; Beasley, V.R.; Buck, W.B.; Haschek, W.M. Experimental T-2 toxicosis in swine. III. Morphologic changes following intravascular administration of T-2 toxin. *Fundam. Appl. Toxicol.* **1987**, *8*, 298–309. [CrossRef]
228. Beasley, V.R.; Lundeen, G.R.; Poppenga, R.H.; Buck, W.B. Distribution of blood flow to the gastrointestinal tract of swine during T-2 toxin-induced shock. *Fundam. Appl. Toxicol.* **1987**, *9*, 588–594. [CrossRef]
229. Ványi, A.; Glávits, R.; Gajdács, E.; Sándor, G.; Kovács, F. Changes induced in newborn piglets by the trichothecene toxin T-2. *Acta. Vet. Hung.* **1991**, *39*, 29–37. [PubMed]
230. Bratich, P.M.; Buck, W.B.; Haschek, W.M. Prevention of T-2 toxin-induced morphologic effects in the rat by highly activated charcoal. *Arch. Toxicol.* **1990**, *64*, 251–253. [CrossRef] [PubMed]
231. Kumagai, S.; Shimizu, T. Effects of fusarenon-X and T-2 toxin on intestinal absorption of monosaccharide in rats. *Arch. Toxicol.* **1988**, *61*, 489–495. [CrossRef] [PubMed]
232. Thompson, W.L.; Wannemacher, R.W., Jr. In vivo effects of T-2 mycotoxin on synthesis of proteins and DNA in rat tissues. *Toxicol. Appl. Pharmacol.* **1990**, *105*, 483–491. [CrossRef]
233. Gao, X.; Sun, L.; Zhang, N.; Li, C.; Zhang, J.; Xiao, Z.; Qi, D. Gestational zearalenone exposure causes reproductive and developmental toxicity in pregnant rats and female offspring. *Toxins* **2017**, *9*, 21. [CrossRef] [PubMed]
234. Obremski, K.; Gajęcka, M.; Zielonka, Ł.; Jakimiuk, E.; Gajęcki, M. Morphology and ultrastructure of small intestine mucosa in gilts with zearalenone mycotoxicosis. *Pol. J. Vet. Sci.* **2005**, *8*, 301–307. [PubMed]
235. Obremski, K.; Poniatowska-Broniek, G. Zearalenone induces apoptosis and inhibits proliferation in porcine ileal Peyer's patch lymphocytes. *Pol. J. Vet. Sci.* **2015**, *18*, 153–161. [CrossRef] [PubMed]
236. Taranu, I.; Braicu, C.; Marin, D.E.; Pistol, G.C.; Motiu, M.; Balacescu, L.; Beridan Neagoe, I.; Burlacu, R. Exposure to zearalenone mycotoxin alters in vitro porcine intestinal epithelial cells by differential gene expression. *Toxicol. Lett.* **2015**, *232*, 310–325. [CrossRef] [PubMed]
237. Girgis, G.N.; Barta, J.R.; Brash, M.; Smith, T.K. Morphologic changes in the intestine of broiler breeder pullets fed diets naturally contaminated with Fusarium mycotoxins with or without coccidial challenge. *Avian Dis.* **2010**, *54*, 67–73. [CrossRef]
238. McKinley, E.R.; Carlton, W.W.; Boon, G.D. Patulin mycotoxicosis in the rat: Toxicology, pathology and clinical pathology. *Food Chem. Toxicol.* **1982**, *20*, 289–300. [CrossRef]
239. Escoula, L.; More, J.; Baradat, C. The toxins by Byssochlamys nivea Westling. I. Acute toxicity of patulin in adult rats and mice. *Ann. Rech. Vet.* **1977**, *8*, 41–49.
240. Speijers, G.J.; Franken, M.A.; van Leeuwen, F.X. Subacute toxicity study of patulin in the rat: Effects on the kidney and the gastro-intestinal tract. *Food Chem. Toxicol.* **1988**, *26*, 23–30. [CrossRef]
241. McKinley, E.R.; Carlton, W.W. Patulin mycotoxicosis in Swiss ICR mice. *Food Cosmet. Toxicol.* **1980**, *18*, 181–187. [CrossRef]
242. McKinley, E.R.; Carlton, W.W. Patulin mycotoxicosis in the Syrian hamster. *Food Cosmet. Toxicol.* **1980**, *18*, 173–179. [CrossRef]
243. Dailey, R.E.; Brouwer, E.; Blaschka, A.M.; Reynaldo, E.F.; Green, S.; Monlux, W.S.; Ruggles, D.I. Intermediate-duration toxicity study of patulin in rats. *J. Toxicol. Environ. Health* **1977**, *2*, 713–725. [CrossRef] [PubMed]
244. McLaughlin, J.; Lambert, D.; Padfield, P.J.; Burt, J.P.; O'Neill, C.A. The mycotoxin patulin, modulates tight junctions in caco-2 cells. *Toxicol. In Vitro* **2009**, *23*, 83–89. [CrossRef] [PubMed]
245. Maidana, L.; Gerez, J.R.; El Khoury, R.; Pinho, F.; Puel, O.; Oswald, I.P.; Bracarense, A.P. Effects of patulin and ascladiol on porcine intestinal mucosa: An ex vivo approach. *Food Chem. Toxicol.* **2016**, *98*, 189–194. [CrossRef]
246. Rotter, B.A.; Thompson, B.K.; Prelusky, D.B.; Trenholm, H.L.; Stewart, B.; Miller, J.D.; Savard, M.E. Response of growing swine to dietary exposure to pure fumonisin B1 during an eight-week period: Growth and clinical parameters. *Nat. Toxins* **1996**, *4*, 42–50. [CrossRef]
247. Dilkin, P.; Zorzete, P.; Mallmann, C.A.; Gomes, J.D.; Utiyama, C.E.; Oetting, L.L.; Corrêa, B. Toxicological effects of chronic low doses of aflatoxin B(1) and fumonisin B(1)-containing Fusarium moniliforme culture material in weaned piglets. *Food Chem. Toxicol.* **2003**, *41*, 1345–1353. [CrossRef]

248. Ledoux, D.R.; Brown, T.P.; Weibking, T.S.; Rottinghaus, G.E. Fumonisin toxicity in broiler chicks. *J. Vet. Diagn. Investig.* **1992**, *4*, 330–333. [CrossRef]
249. Bhat, R.V.; Shetty, P.H.; Rao, P.A.; Rao, V.S. A foodborne disease outbreak due to the consumption of moldy sorghum and maize containing fumonisin mycotoxins. *J. Toxicol. Clin. Toxicol.* **1997**, *35*, 249–255. [CrossRef]
250. Schmelz, E.M.; Dombrink-Kurtzman, M.A.; Roberts, P.C.; Kozutsumi, Y.; Kawasaki, T.; Merrill, A.H., Jr. Induction of apoptosis by fumonisin B1in HT29 cells is mediated by the accumulation of endogenous free sphingoid bases. *Toxicol. Appl. Pharmacol.* **1998**, *148*, 252–260. [CrossRef]
251. Theumer, M.G.; Lopez, A.G.; Masih, D.T.; Chulze, S.N.; Rubinstein, H.R. Immunobiological effects of fumonisin B1in experimental subchronic mycotoxicoses in rats. *Clin. Diagn. Lab. Immunol.* **2002**, *9*, 149–155.
252. Casado, J.M.; Theumer, M.; Masih, D.T.; Chulze, S.; Rubinstein, H.R. Experimental subchronic mycotoxicoses in mice: Individual and combined effects of dietary exposure to fumonisins and aflatoxin B1. *Food Chem. Toxicol.* **2001**, *39*, 579–586. [CrossRef]
253. Bouhet, S.; Hourcade, E.; Loiseau, N.; Fikry, A.; Martinez, S.; Roselli, M.; Galtier, P.; Mengheri, E.; Oswald, I.P. The mycotoxin, fumonisin B1alters the proliferation and the barrier function of porcine intestinal epithelial cells. *Toxicol. Sci.* **2004**, *77*, 165–171. [CrossRef]
254. Carballo, D.; Tolosa, J.; Ferrer, E.; Berrada, H. Dietary exposure assessment to mycotoxins through total diet studies. A review. *Food Chem. Toxicol.* **2019**, *128*, 8–20. [CrossRef] [PubMed]
255. Maresca, M.; Fantini, J. Some food-associated mycotoxins as potential risk factors in humans predisposed to chronic intestinal inflammatory diseases. *Toxicon* **2010**, *56*, 282–294. [CrossRef] [PubMed]
256. Capriotti, A.L.; Caruso, G.; Cavaliere, C.; Foglia, P.; Samperi, R.; Laganà, A. Multiclass mycotoxin analysis in food, environmental and biological matrices with chromatography/mass spectrometry. *Mass Spectrom. Rev.* **2012**, *31*, 466–503. [CrossRef] [PubMed]
257. Chen, F.; Ma, Y.; Xue, C.; Ma, J.; Xie, Q.; Wang, G.; Bi, Y.; Cao, Y. The combination of deoxynivalenol and zearalenone at permitted feed concentrations causes serious physiological effects in young pigs. *J. Vet. Sci.* **2008**, *9*, 39–44. [CrossRef]
258. Sergent, T.; Ribonnet, L.; Kolosova, A.; Garsou, S.; Schaut, A.; De Saeger, S.; Van Peteghem, C.; Larondelle, Y.; Pussemier, L.; Schneider, Y.J. Molecular and cellular effects of food contaminants and secondary plant components and their plausible interactions at the intestinal level. *Food Chem. Toxicol.* **2008**, *46*, 813–841. [CrossRef]
259. Harrison, J.C.; Carvajal, M.; Garner, R.C. Does aflatoxin exposure in the United Kingdom constitute a cancer risk? *Environ. Health Perspect.* **1993**, *99*, 99–105. [CrossRef]
260. Eom, S.Y.; Yim, D.H.; Zhang, Y.; Yun, J.K.; Moon, S.I.; Yun, H.Y.; Song, Y.J.; Youn, S.J.; Hyun, T.; Park, J.S.; et al. Dietary aflatoxin B1 intake, genetic polymorphisms of CYP1A2, CYP2E1, EPHX1, GSTM1, and GSTT1, and gastric cancer risk in Korean. *Cancer Causes Control* **2013**, *24*, 1963–1972. [CrossRef]
261. Przybyłowicz, K.E.; Arłukowicz, T.; Danielewicz, A.; Morze, J.; Gajęcka, M.; Zielonka, Ł.; Fotschki, B.; Sawicki, T. Association between mycotoxin exposure and dietary habits in colorectal cancer development among a Polish population: A study protocol. *Int. J. Environ. Res. Public Health* **2020**, *17*, 698. [CrossRef]
262. Roy, R.N.; Russell, R.I. Crohn's Disease and Aflatoxins. *J. R. Soc. Health* **1992**, *112*, 277–279. [CrossRef] [PubMed]
263. Ouhibi, S.; Vidal, A.; Martins, C.; Gali, R.; Hedhili, A.; De Saeger, S.; De Boevre, M. LC-MS/MS methodology for simultaneous determination of patulin and citrinin in urine and plasma applied to a pilot study in colorectal cancer patients. *Food Chem. Toxicol.* **2020**, *136*, 110994. [CrossRef] [PubMed]

© 2020 by the authors. Licensee MDPI, Basel, Switzerland. This article is an open access article distributed under the terms and conditions of the Creative Commons Attribution (CC BY) license (http://creativecommons.org/licenses/by/4.0/).

Article

Co-Occurrence of DON and Emerging Mycotoxins in Worldwide Finished Pig Feed and Their Combined Toxicity in Intestinal Cells

Abdullah Khan Khoshal [1], Barbara Novak [2], Pascal G. P. Martin [1], Timothy Jenkins [2], Manon Neves [1], Gerd Schatzmayr [2], Isabelle P. Oswald [1,*] and Philippe Pinton [1,*]

1 Toxalim (Research Center in Food Toxicology), Université de Toulouse, INRAE, ENVT, INP-Purpan, UPS, 180 chemin de tournefeuille, Cedex 3, F-31027 Toulouse, France; abdullah.khoshal@inra.fr (A.K.K.); pascal.martin@inra.fr (P.G.P.M.); manon.neves@inra.fr (M.N.)

2 BIOMIN Research Center, Technopark 1, 3430 Tulln, Austria; barbara.novak@biomin.net (B.N.); timothy.jenkins@biomin.net (T.J.); gerd.schatzmayr@biomin.net (G.S.)

* Correspondence: isabelle.oswald@inra.fr (I.P.O.); philippe.pinton@inra.fr (P.P.)

Received: 8 November 2019; Accepted: 6 December 2019; Published: 11 December 2019

Abstract: Food and feed can be naturally contaminated by several mycotoxins, and concern about the hazard of exposure to mycotoxin mixtures is increasing. In this study, more than 800 metabolites were analyzed in 524 finished pig feed samples collected worldwide. Eighty-eight percent of the samples were co-contaminated with deoxynivalenol (DON) and other regulated/emerging mycotoxins. The Top 60 emerging/regulated mycotoxins co-occurring with DON in pig feed shows that 48%, 13%, 8% and 12% are produced by *Fusarium*, *Aspergillus*, *Penicillium* and *Alternaria* species, respectively. Then, the individual and combined toxicity of DON and the 10 most prevalent emerging mycotoxins (brevianamide F, cyclo-(L-Pro-L-Tyr), tryptophol, enniatins A1, B, B1, emodin, aurofusarin, beauvericin and apicidin) was measured at three ratios corresponding to pig feed contamination. Toxicity was assessed by measuring the viability of intestinal porcine epithelial cells, IPEC-1, at 48-h. BRV-F, Cyclo and TRPT did not alter cell viability. The other metabolites were ranked in the following order of toxicity: apicidin > enniatin A1 > DON > beauvericin > enniatin B > enniatin B1 > emodin > aurofusarin. In most of the mixtures, combined toxicity was similar to the toxicity of DON alone. In terms of pig health, these results demonstrate that the co-occurrence of emerging mycotoxins that we tested with DON does not exacerbate toxicity.

Keywords: global survey; finished pig feed; co-occurrence; emerging mycotoxins; DON; toxicity; combined toxicity; IPEC-1

Key Contribution: A worldwide survey of finished pig feed demonstrates the co-occurrence of DON and emerging mycotoxins. Assessment of their combined toxicity with DON at realistic ratios revealed that their toxicity was similar to that of DON alone.

---

1. Introduction

Mycotoxins are low molecular weight fungal secondary metabolites that trigger a detrimental response when ingested by humans and animals. They are mainly produced by filamentous fungi belonging to *Aspergillus*, *Fusarium* and *Penicillium* species [1]. Mycotoxin contamination can occur all along the food chain from field to storage, including the food process. This depends upon the requirements of fungi, and *Fusarium* mostly occurs in the field, whereas *Aspergillus* and *Penicillium* mostly occurs during storage.

Because of their toxicity and occurrence, deoxynivalenol (DON), zearalenone, aflatoxins, ochratoxin A, patulin, fumonisins and T-2/HT-2 toxins are regulated in Europe. For example, the maximum recommendations limits that are set up for complete piglet feed are 0.9, 0.1, 0.05, 5 and 0.25 mg/kg feed for DON, zearalenone, ochratoxin, fumonisins and T2 + HT2, respectively [2,3]. However, in addition to regulated mycotoxins, many other fungal secondary metabolites are being identified in food and feed [4,5]. Metabolites that are neither routinely determined, nor legislatively regulated, have been defined as 'emerging mycotoxins' [6], while the derivatives of regulated mycotoxins that are undetectable using conventional analytical techniques due to their modified structure, are defined as 'modified/masked mycotoxins' [4,7]. Recent findings showed that more than 70% of the world's cereal grains are contaminated by mycotoxins [8,9], often in a mixture [10].

Among regulated mycotoxins, DON very frequently contaminates cereals (wheat, barley, oats, rye and maize, and less frequently rice, sorghum and triticale) and cereal-based food and feed. DON belongs to the group of B-trichothecenes, and is one of the most widely distributed contaminants in human food and animal feed. In a total of more than 25,000 samples collected from 28 European countries between 2007 and 2014, DON was found in 47% of 4000 feed samples and 45% of 1621 unprocessed grains with no defined end use, respectively [11]. Even though DON is considered as a non-carcinogenic compound [12], the maximum level of this toxin in food and feed have been set up in different countries. For example, in complete piglet feed, the maximum limits are 0.9, 1 and 5 mg/kg feed in Europe, Canada and the USA, respectively [2,13]. Exposure to high concentrations of DON is associated with diarrhea, vomiting (emesis), leukocytosis and gastrointestinal bleeding. Chronic exposure affects growth, immunity and intestinal barrier function in animals [14–16]. This toxin interacts with the peptidyl transferase region of the 60S ribosomal subunit, inducing 'ribotoxic stress', resulting in the activation of mitogen-activated protein kinases (MAPKs) and their downstream pathways [14,17].

Among emerging mycotoxins, those that occur most frequently are enniatins (ENNs), beauvericin (BEA), apicidin (API), aurofusarin (AFN), culmorin, butenolide, fusaric acid, moniliformin, fusaproliferin and emodin (EMO). They are produced by *Fusarium* species except EMO, which is produced by *Aspergillus* species [6,18]. ENNs and BEA were detected in food (63% and 80%), feed (32% and 79%), and unprocessed grains (24% and 46%) collected between 2010 and 2014 in 12 European countries [19]. AFN, API, brevianamide-F (BRV-F), EMO and tryptophol (TRPT) were also found in pig feed (80%, 52%, 65%, 63% and 75%) [20], Egyptian animal feed (73%, 17%, 86%, 98% and 90%) [21] and feed raw materials (84%, 55%, 5%, 74%, 59%) [22].

Multiple mycotoxins are frequently present in food and feed [10]. The co-occurrence of DON, aflatoxins, fumonisins, zearalenone and other fungal secondary metabolites in maize seeds and grains, as well as in animal feed, has been reported [21–23]. The presence of different fungi on the same raw material, the ability of fungal species to produce several toxins, as well as the various commodities present in completed feed, can explain this multiple contamination [24,25]. Compound feed is particularly prone to multiple contaminations, as it typically contains a mixture of several raw materials.

The co-occurrence of mycotoxins is challenging for at least two reasons: (i) The toxicity of mycotoxins when present together cannot always be predicted based upon their individual toxicity and (ii) the risk assessment is performed on a chemical-by-chemical basis [24,25]. Scientific interest in the toxicity of these mixtures of mycotoxins is currently increasing rapidly [26–29]. Several studies have investigated the combined toxicity of regulated mycotoxins on the intestine [30–33], but the combined effect of regulated and other mycotoxins is poorly documented [34,35].

Among farm animals, pig is one of the most sensitive species to mycotoxins [36]. As feed raw materials are potentially contaminated by several fungi at a time, and completed feed is made from various commodities, pig can be exposed, through its rich cereal diet, to high concentrations of mixtures of mycotoxins [10,37]. The sanitary and economic losses due to mycotoxin contamination are important in the pig industry, even if they are hard to estimate precisely [38].

The aims of this study were thus to (i) determine the prevalence and concentration of mycotoxins present in finished pig feed and (ii) assess the intestinal combined toxicity of DON in mixture with the 10 most prevalent emerging mycotoxins present in pig feed using realistic ratios.

2. Results

2.1. Occcurrence and Abundance of Emerging Mycotoxins and DON in Finished Pig Feed

A total of 524 finished pig feed samples collected worldwide were analyzed, and more than 235 different metabolites were detected, including regulated mycotoxins, emerging mycotoxins and modified/masked mycotoxins. Table 1 lists the 60 most prevalent fungal metabolites that contaminated more than 20% of the finished pig feed samples. Among regulated mycotoxins, DON was detected in 463 samples (88%), mostly in the Northern Hemisphere and in relatively similar concentrations in samples from all countries (median concentration 206 μg/kg) (Figure 1A,B).

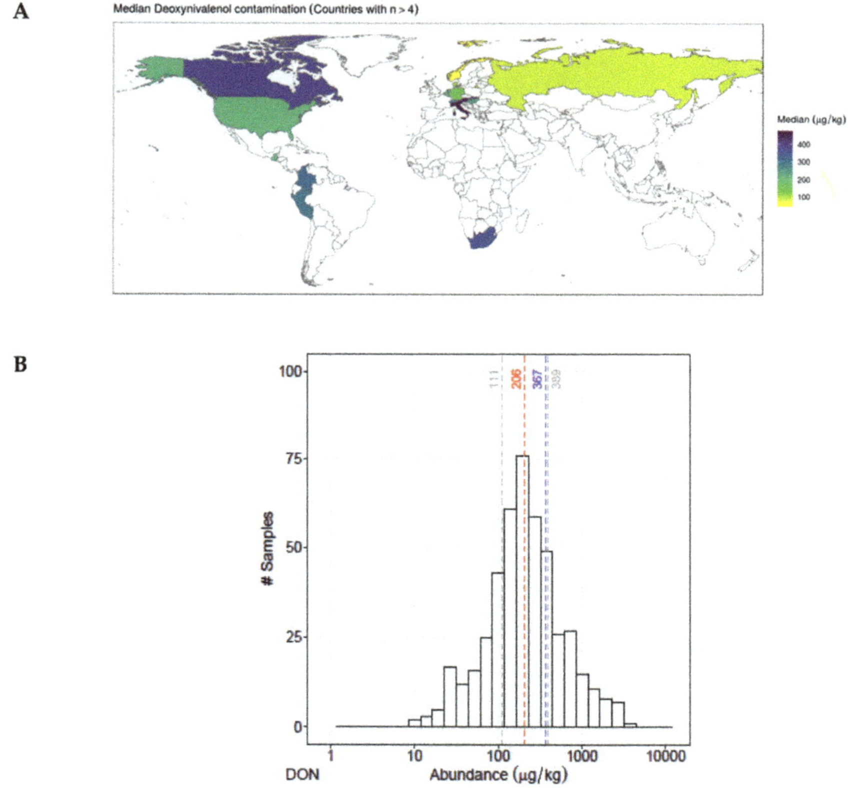

Figure 1. (**A**) Worldwide contamination of deoxynivalenol (DON). Concentration of DON is highlighted by colors, where yellow indicates > 100 μg/kg, green > 200 μg/kg, dark green > 300 and dark blue indicates > 400 μg/kg. (**B**) Abundance of DON in pig finished feed (P25, median, mean, P75). X axis represents the distribution of the concentration, Y axis describes the number of contaminated samples.

All DON-contaminated samples were co-contaminated by other mycotoxins. The distribution of the samples was checked. Table 1 gives the descriptive statistics for DON and the most abundant metabolites that co-occur with DON.

Table 1. Top 60 emerging and regulated mycotoxins co-occurring with DON in finished pig feed.

	Metabolites	Occurrence *n* (%)	Co-occurrence with DON *n* (%)	Contamination Level (µg/kg Feed)			Correlation (DON and Other Mycotoxins)	
				P25	**P50**	**P75**	**Coefficient**	***p*-Value**
1	Deoxynivalenol	463 (88%)	463 (100%)	111	206	389	1.00	NA
2	Culmorin	492 (94%)	458 (99%)	38	107	247	0.50	0.00
3	Zearalenone	502 (96%)	449 (97%)	9	18	46	0.64	0.00
4	Brevianamide F	500 (95%)	446 (96%)	17	28	45	0.17	0.00
5	Cyclo-(L-Pro-L-Tyr)	494 (94%)	434 (94%)	117	196	371	0.14	0.00
6	Enniatin B	479 (91%)	430 (93%)	9	32	83	0.02	0.61
7	Enniatin B1	481 (92%)	430 (93%)	10	37	87	0.04	0.39
8	Asperglaucide	470 (90%)	419 (90%)	18	38	94	-0.11	0.02
9	Emodin	472 (90%)	418 (90%)	3	5	10	-0.01	0.91
10	Aurofusarin	464 (89%)	417 (90%)	87	211	548	0.59	0.00
11	Moniliformin	469 (90%)	416 (90%)	7	17	45	0.11	0.03
12	Beauvericin	464 (89%)	411 (89%)	4	7	13	0.32	0.00
13	Enniatin A1	459 (88%)	411 (89%)	5	16	33	0.09	0.08
14	3-Nitropropion acid	455 (87%)	407 (88%)	3	6	10	-0.03	0.57
15	Tryptophol	454 (87%)	407 (88%)	119	197	319	0.10	0.04
16	15-Hydroxyculmorin	429 (82%)	391 (84%)	76	142	277	0.79	0.00
17	Equisetin	424 (81%)	386 (83%)	5	10	23	0.01	0.80
18	Infectopyron	409 (78%)	366 (79%)	108	263	449	-0.16	0.00
19	DON-3 Glucoside	380 (73%)	362 (78%)	10	21	47	0.79	0.00
20	Neoechinulin A	407 (78%)	357 (77%)	10	19	42	0.06	0.22
21	Tenuazonic-acid	384 (73%)	347 (75%)	53	90	182	0.04	0.49
22	Alternariol	366 (70%)	333 (72%)	2	4	9	0.01	0.79
23	Rugulusovin	373 (71%)	332 (72%)	4	7	14	0.15	0.01
24	Tentoxin	342 (65%)	319 (69%)	2	3	6	-0.03	0.56
25	Apicidin	341 (65%)	310 (67%)	3	7	11	-0.13	0.02
26	Fumonisin B1	332 (63%)	304 (66%)	26	70	254	0.14	0.02
27	Nivalenol	315 (60%)	296 (64%)	10	24	57	0.14	0.02
28	Cyclo-(L-Pro-L-Val)	337 (64%)	286 (62%)	85	137	246	0.15	0.01
29	Epiequisetin	307 (59%)	285 (62%)	2	4	7	0.05	0.42
30	Citreorosein	317 (60%)	283 (61%)	3	5	8	0.12	0.05
31	Enniatin A	306 (58%)	282 (61%)	2	3	5	-0.01	0.81
32	Alternariolmethylether	307 (59%)	275 (59%)	2	3	5	0.09	0.14
33	Altersetin	301 (57%)	275 (59%)	13	29	76	0.18	0.00
34	Asperphenamate	311 (59%)	269 (58%)	5	11	27	-0.02	0.75
35	Lotaustralin	288 (55%)	257 (56%)	15	30	85	-0.10	0.13
36	Butenolid	253 (48%)	242 (52%)	22	37	70	0.32	0.00
37	Kojic acid	262 (50%)	241 (52%)	43	74	148	-0.06	0.39
38	Enniatin B2	258 (49%)	238 (51%)	2	3	5	-0.01	0.84
39	Fumonisin B2	258 (49%)	237 (51%)	19	50	143	0.16	0.01
40	Zearalenone Sulfate	236 (45%)	222 (48%)	10	25	53	0.25	0.00
41	Antibiotic Y	233 (44%)	215 (46%)	40	111	402	-0.02	0.75
42	T2 Toxin	235 (45%)	209 (45%)	2	4	9	0.12	0.07
43	Macrosporin	219 (42%)	202 (44%)	2	3	8	0.02	0.76
44	N-Benzoyl-Phenylalanine	220 (42%)	191 (41%)	3	5	11	-0.02	0.82
45	Flavoglaucin	206 (39%)	175 (38%)	7	16	34	0.05	0.51
46	Curvularin	196 (37%)	171 (37%)	2	4	8	-0.09	0.27
47	Questiomycin A	178 (34%)	162 (35%)	4	10	20	0.22	0.01
48	Rubellin D	179 (34%)	161 (35%)	4	8	18	0.10	0.21
50	Bikaverin	171 (33%)	153 (33%)	10	25	56	0.27	0.00
50	Fusarinolic-acid	157 (30%)	153 (33%)	47	130	320	0.3	0.00
51	Fumonisin B4	165 (31%)	149 (32%)	11	23	68	0.2	0.03
52	Cytochalasin J	170 (32%)	146 (32%)	13	29	63	0.1	0.46
53	Ergometrine	152 (29%)	145 (31%)	6	11	24	0.0	0.57
54	Ergocristine	151 (29%)	143 (31%)	2	5	13	0.2	0.02
55	Fumonisin B3	154 (29%)	136 (29%)	24	48	103	0.1	0.15
56	HT2-toxin	149 (28%)	134 (29%)	13	20	30	0.2	0.01
57	Monocerin	144 (27%)	133 (29%)	1	2	3	0.2	0.02
58	Chrysogin	136 (26%)	126 (27%)	7	12	17	0.4	0.00
59	Ergosin	128 (24%)	123 (27%)	3	6	13	-0.1	0.39
60	5-Hydroxyculmorin	121 (23%)	117 (25%)	107	170	304	0.7	0.00

The 60 mycotoxins found in more than 20% of the 524 samples of finished pig feed. Their concentrations in the three quartiles (P25, P50 and P75) are expressed in µg/kg of feed. The correlation of their concentration with DON and the associated P-value was calculated using Spearman's rank correlation coefficient.

From this list, the most prevalent emerging mycotoxins co-occurring with DON and which were commercially available were selected for the toxicological studies. Because of its high toxicity [39], apicidin (API) was included in the list. The worldwide distributions of these compounds are presented in Supplementary Figure S1. Except for API, which was detected in only 67% of DON-contaminated samples, the selected emerging mycotoxins were present in more than 87% of them (Table 1). Three compounds API, emodin (EMO) and beauvericin (BEA) were detected in a median concentration range of 5 to 10 µg/kg feed. The median concentration of four metabolites, enniatins A1, B, B1 (ENN-A1, B, B1) and brevianamide F (BRV-F), was in the range of 15-40 µg/kg feed, and the last three compounds aurofusarin (AFN), cyclo-(L-Pro-L-Tyr) (Cyclo) and tryptophol (TRPT) had median concentrations close to 200 µg/kg feed, like DON (Supplementary Figure S2). Despite their high co-occurrence with DON, with the exception of AFN, and to a lesser extent BEA, the concentration of these mycotoxins showed limited correlation with DON concentration (Table 1 and Supplementary Figure S3).

2.2. *Intestinal Toxicity of Emerging Mycotoxins Found in Pig Feed, alone or Combined with DON*

2.2.1. Individual Toxicity of DON and Emerging Mycotoxins

The individual intestinal toxicity of 10 selected emerging mycotoxins, as well as that of DON, was first analyzed at a wide range of concentrations (Supplementary Figure S4). As shown in Supplementary Figure S4, all tested metabolites exhibited dose-dependent toxicity toward intestinal epithelial cells, except BRV-F, Cyclo and TRPT, that were not toxic for this porcine cell line. AFN, EMO and ENN-B1 reduced the viability of IPEC-1, but their toxicity was less than that of DON. The toxicity of BEA and ENN-B was close to that of DON, whereas API and ENN-A1 were more toxic than DON. As shown in Figure 2, low doses of API (0.01–0.3 µM) significantly stimulated the proliferation of IPEC-1. When the doses leading to a 50% reduction in the cell viability (IC_{50}) of these emerging mycotoxins were compared with the dose of DON, then BRV-F, Cyclo and TRPT were classified as non-toxic metabolites, while EMO, AFN and ENN-B1 were given as moderately toxic metabolites, and finally API, ENN-A1, ENN-B and BEA as highly toxic metabolites (Table 2).

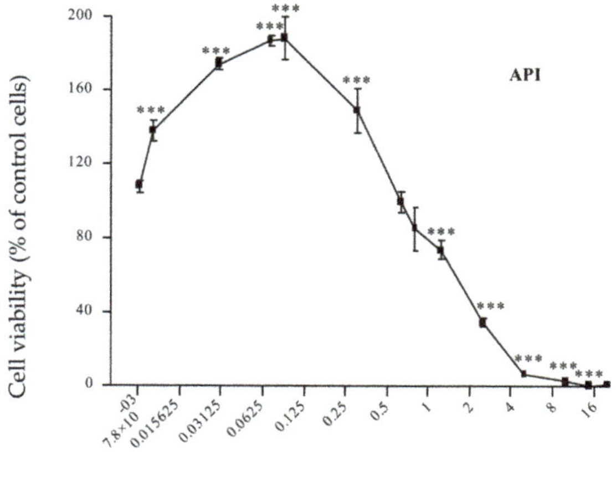

Figure 2. Dose effect curve of individual toxicity of apicidin (API). Data are mean ± SEM of three biological replicates. Bonferroni multiple comparison test. Significant difference between control and different doses of API *** $p < 0.001$.

Table 2. IC_{50} values of the selected emerging mycotoxins on IPEC-1 cells.

Metabolites	Abbreviation	IC_{50} (μM)	Toxicity
Brevianamide F	BRV-F	Non-Toxic	Non-toxic
Cyclo-(L-Pro-L-Tyr)	Cyclo	Non-Toxic	
Tryptophol	TRPT	Non-Toxic	
Aurofusarin	AFN	19.1 ± 3.4	Moderately toxic
Emodin	EMO	19.0 ± 0.7	
Enniatin B1	ENN-B1	13.5 ± 2.5	
Enniatin B	ENN-B	4.4 ± 0.9	Highly toxic
Beauvericin	BEA	4.3 ± 1.8	
Deoxynivalenol	DON	3.2 ± 0.7	
Enniatin A1	ENN-A1	1.6 ± 0.3	
Apicidin	API	1.5 ± 0.5	

Data are the mean ± the standard error of the mean (SEM) of three biological replicates.

2.2.2. Combined Toxicity of DON and Emerging Mycotoxins

Next, the combined toxicity of DON and the selected emerging mycotoxins was assessed. As mentioned above, the concentration of these secondary metabolites was not correlated with the concentration of DON (Table 1 and Supplementary Figure S3). Thus, to account for the different situations to which pigs can be exposed, three ratios were tested (Supplementary Table S1). Ratio 1 was calculated using the P25 (1st quartile) concentration of the emerging mycotoxin and P75 (3rd quartile) concentration of DON. Ratio 2 was calculated using the median (2nd quartile) concentration of DON and each emerging mycotoxin. Ratio 3 was calculated using the P75 concentration of the emerging mycotoxin and P25 concentration of DON. For each ratio, serial dilutions were tested to obtain a dose-effect curve that encompassed the realistic concentrations of the mixture of DON and the tested metabolites.

2.2.3. Combined Toxicity of DON and the Non-Toxic Secondary Metabolites (BRV-F, Cyclo and TRPT)

First, the combined toxicity of DON and the 'non-toxic' secondary metabolites BRV-B, Cyclo and TRPT were analyzed. As shown in Figure 3, whatever the ratios tested, the toxicity of the combination of DON and the compound being tested was similar to the toxicity of DON alone.

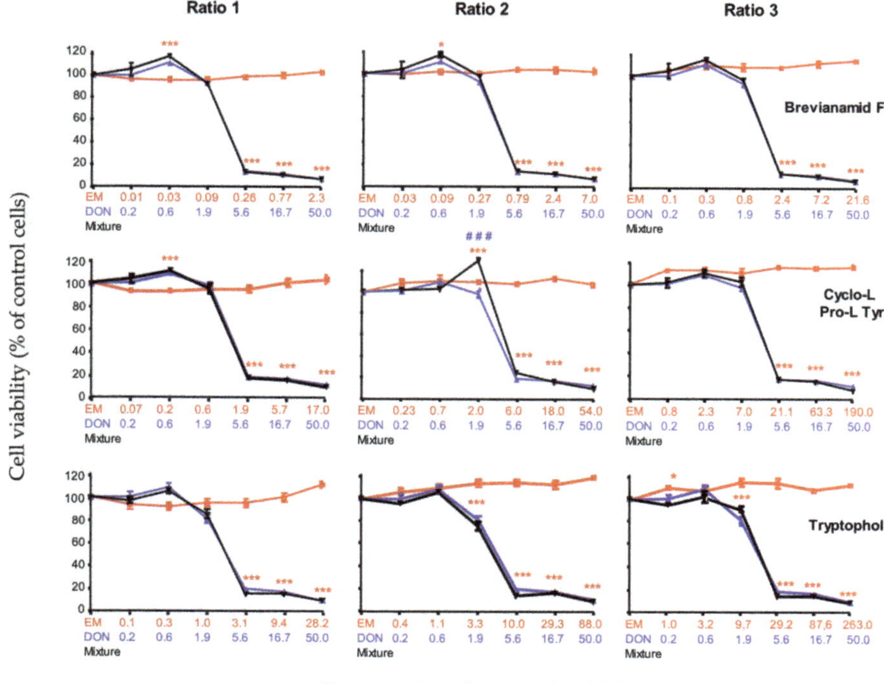

Figure 3. Dose-effect curves of deoxynivalenol (DON) (blue lines and symbols), emerging mycotoxins (brevianamide-F (BRV-F), cyclo-(L-Pro-L-Tyr) (Cyclo) and tryptophol (TRPT)) alone (red lines and symbols), or in combination with DON (black lines and symbols) at different ratios: ratio 1 was calculated from the P25 concentration of emerging mycotoxin and P75 concentration of DON; ratio 2 was calculated from the median concentration of emerging mycotoxin and DON; ratio 3 was calculated from the P75 concentration of emerging mycotoxin and the P25 concentration of DON. Six serial dilutions of each ratio were tested (Emerging mycotoxin alone, DON alone, mixture). Data are mean ± SEM of three biological replicates. Bonferroni multiple comparison test. Significant difference between emerging mycotoxins alone and the mixtures *** $p < 0.001$. Significant difference between DON alone and the mixtures # # # $p < 0.001$.

2.2.4. Combined Toxicity of DON and the Moderately Toxic Secondary Metabolites (AFN, EMO and ENN-B1)

Next, the combined toxicity of DON and the moderately toxic metabolites AFN, EMO and ENN-B1 was assessed (Figure 4). The toxicity of AFN, EMO and ENN-B1 was minimal when used at ratio 1. In these conditions, the toxicity of the combination of DON and emerging toxins was similar to the toxicity of DON alone. Ratio 2 reached toxic concentrations of AFN and ENN-B1, but the toxicity of the mixture was similar to the toxicity of DON alone, except at the highest concentration of ENN-B1 (4.1 μM), where the toxicity of the mixture (ENN-B1 4.1 μM + DON 50 μM) was higher than the toxicity of ENN-B1 alone, but lower than the toxicity of DON alone.

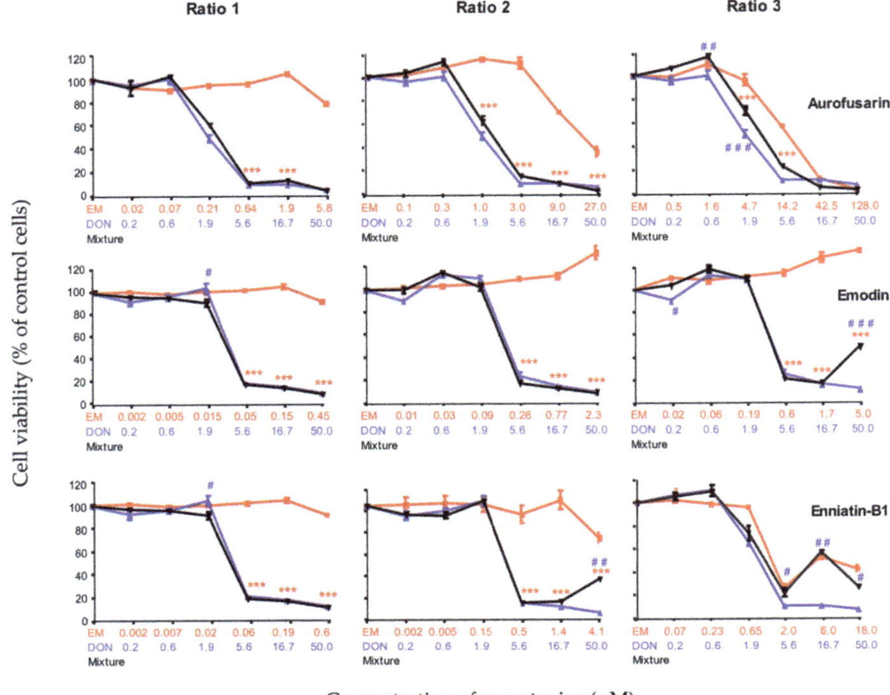

Figure 4. Dose-effect curves of DON (blue lines and symbols), emerging mycotoxins (AFN, EMO and ENN-B1) alone (red lines and symbols) or in combination with DON (black lines and symbols) at different ratios: ratio 1 was calculated from the P25 concentration of emerging mycotoxin and P75 concentration of DON; ratio 2 was calculated from the median concentration of emerging mycotoxin and DON; ratio 3 was calculated from the P75 concentration of emerging mycotoxin and P25 concentration of DON. Six serial dilutions of each ratio were tested (Emerging mycotoxin alone, DON alone, mixture). Data are mean ± SEM of three biological replicates. Bonferroni multiple comparison test. Significant difference between emerging mycotoxins alone and the mixtures *** $p < 0.001$. Significant difference between DON alone and the mixtures # # # $p < 0.001$.

When cytotoxicity was tested at ratio 3, the toxicity of AFN + DON was identical to the one of DON alone except at the concentrations DON 0.6 µM + AFN 1.6 µM and DON 1.9 µM + AFN 4.7 µM, when the toxicity of the mixture was slightly lower than the toxicity of DON alone. At this ratio, EMO alone was still not toxic, and even induced proliferation (up to 130% of treated cells).

The combined toxicity of DON and EMO was similar to the toxicity of DON alone except at the highest concentration of EMO, when the mixture of DON (50 µM) and EMO (5 µM) was still less toxic than DON alone. The combined toxicity of ENN-B1 + DON was the same as the toxicity of DON alone, except at the highest concentrations of ENN-B1 (6 and 18 µM), when the toxicity of the mixture was the same as the toxicity of ENN-B1 alone, but lower than the toxicity of DON alone.

In conclusion, our data showed that the toxicity of the combination of DON and emerging mycotoxins such as AFN, EMO and ENN-B1 was similar to or lower than the toxicity of DON alone, whatever the ratio used.

2.2.5. Combined Toxicity of DON and The Highly Toxic Secondary Metabolites (ENN-B, BEA, ENN-A1and API)

The combined toxicity of DON and highly toxic compounds (ENN-B, BEA, ENN-A1 and API) was also analyzed at different ratios (Figure 5). At the first tested ratio (ratio 1), ENN-B, BEA, ENN-A1 and API were not toxic, and their combined toxicity in the presence of DON was similar to the toxicity of DON alone. Ratio 2 reached toxic concentrations of ENN-B, ENN-A1 and API.

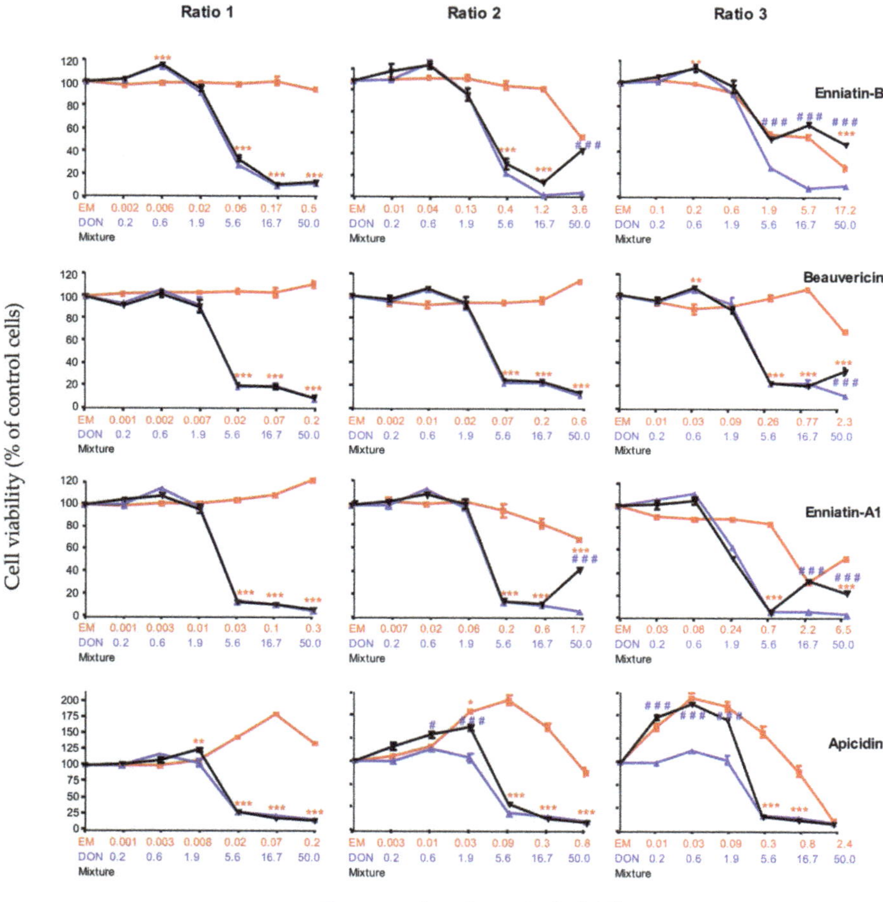

Figure 5. Dose-effect curves of DON (blue lines and symbols), emerging mycotoxins (ENN-B, BEA, ENN-A1 and API) alone (red lines and symbols) or in combination with DON (black lines and symbols) at different ratios: ratio 1 was calculated from the P25 concentration of the emerging mycotoxin and the P75 concentration of DON; ratio 2 was calculated from the median concentration of emerging mycotoxin and DON; ratio 3 was calculated from the P75 concentration of emerging mycotoxin and the P25 concentration of DON. Six serial dilutions of each ratio were tested (Emerging mycotoxin alone, DON alone, mixture). Data are mean ± SEM of three biological replicates. Bonferroni multiple comparison test. Significant difference between emerging mycotoxins alone and the mixtures *** $p < 0.001$. Significant difference between DON alone and the mixtures # # # $p < 0.001$.

For ENN-B and ENN-A1, the toxicity of the mixture was similar to the toxicity of DON alone, except at the highest concentrations of ENN-B (3.6 µM) and ENN-A1 (1.7 µM). The combined toxicity of DON (50 μM) + ENN-B (3.6 μM) was similar to the toxicity of ENN-B alone, but lower than the toxicity of DON alone, whereas the combined toxicity of DON (50 μM) + ENN-A1 (1.7 μM) was higher than the toxicity of ENN-A1 alone, but still lower than the toxicity of DON alone. The combined toxicity of API and DON was similar to the toxicity of DON only for the higher doses. For the doses lower than DON (5.6 μM) + API (0.09 μM) the combined toxicity was lower than the toxicity of DON alone, but higher than API alone.

At ratio 3, the combined toxicity of DON and ENN-B displayed a different characteristic. At a high concentration, the combined toxicity of DON (50 μM) + ENN-B (17.2 μM) was lower than the toxicity of either DON alone or ENN-B alone. At this ratio, the toxicity of the mixture of DON and BEA was similar to the toxicity of DON alone, except at the highest concentration of BEA (2.3 μM) when mixed with DON (50 μM), where the toxicity was higher than that of BEA alone, but lower than that of DON alone. The combined toxicity of DON (16.7μM) + ENN-A1 (2.2 μM) at ratio 3 was similar to the toxicity of ENN-A1 alone, but lower than the toxicity of DON alone; the toxicity of DON (50 μM) + ENN-A1 (6.5 μM) was higher than the toxicity of ENN-A1 alone, but lower than the toxicity of DON alone. Finally, the combined toxicity of DON and API, at concentrations lower than DON (5.6 μM) + API (0.3 μM), was always similar to API alone, but lower than DON alone, whereas at higher concentrations, it was similar to the one of DON alone and higher than the one of API alone. As mentioned above, a proliferation of IPEC-1 cells was observed at low concentrations of API.

In conclusion, our data showed that the toxicity of the combination of DON and highly toxic emerging mycotoxins such as ENN-B, BEA, ENN-A1 and API, whatever the ratio used, was similar to or lower than the toxicity of DON alone.

3. Discussion

Progress in analytical methods enabled the discovery of numerous fungal secondary metabolites that are the subject of increasing attention today due to their prevalence in human food and animal feed [19,20]. In the present study, 524 samples of finished pig feed were analyzed. In addition to regulated mycotoxins such as DON, zearalenone and fumonisin B1, less known secondary metabolites were detected.

As already described in other surveys [20–22], BRV-F, Cyclo, TRPT, ENNs, EMO, BEA and AFN, culmorin, and moniliformin were highly prevalent emerging mycotoxins detected in more than 85% of pig feed samples. The diversity of the metabolites detected is very likely related to the wide range of fungal species that contaminate the raw materials used to make pig feed. Indeed, *Fusarium* species produce ENNs (A1, B and B1), BEA, AFN, API, culmorin and moniliformin, while *Penicillium* species produce BRV-F and Cyclo, *Aspergillus* EMO and *Acremonium* TRPT [22,40].

The toxicity of these new poorly documented metabolites was also investigated in the present study. The results of our analyses showed that, even at high concentrations of up to 300 μM, BRV, Cyclo and TRPT are not toxic to intestinal cells. Similar results were recently obtained using another porcine intestinal cell line, IPEC-J2, and a different readout, cellular protein content [20]. Interestingly, at a much higher concentration (2 mM), TRPT induced DNA damage in HepG2, A549 and THP-1 cells [41]. AFN and EMO were identified as moderately toxic compounds at relative IC_{50} values of 19.1 μM and 19 μM, respectively. These emerging mycotoxins were found to be more toxic for IPEC-J2 with relative IC_{50} of 9.3 μM and 13.1 μM, respectively [20]. On the other hand, human multiple myeloma blood cells were less sensitive to EMO (IC_{50} 38 μM) [42]. According to their IC_{50}, ENNs were ranked in the following order of toxic potency ENN-A1 > ENN-B > ENN-B1. Similar ranking was reported for HT-29 [43] and IPEC-J2 [44]. ENN-A1 is also more toxic than ENN-B1 for Caco-2 and HepG2, but ENN-B displayed no toxicity at all [43]. The mechanism of toxicity of ENNs is related to their ionophoric properties [19] that facilitate the transport of mono- or divalent cations such as K^+ or Ca^{2+} across membranes, but the relative sensitivity of the different cell lines to the different ENNs is still not understood. The high toxicity of BEA has already been reported in other cell lines of human and porcine origin, including Caco-2, HT-29 and IPEC-J2 [20,43]. BEA also has ionophoric properties and increase ions permeability (Na^+ K^+ and Ca^{2+}) in biological membrane, this mechanism of action participates to the toxicity of this mycotoxin [6,19,44].

Data on the toxicity of API are very limited. We confirmed the toxicity of this metabolite when used at high concentration (> 0.5 μM) [20,39]; we also observed that low concentrations of API (<0.1 μM) stimulate the proliferation of porcine intestinal cells. The proliferative effect of low doses of some mycotoxins has already been described. For example, up to 200% proliferation has been observed in lymphocytes and splenocytes exposed to low doses (> 0.1μM) of DON, nivalenol, aflatoxin B1 or fumonisin B1 [45]. Because of its high prevalence and its low IC_{50}, it would be of great interest to deepen our knowledge of API. The toxicity of other very prevalent metabolites, such as culmorin and moniliformin, could not be evaluated in this study because they were not commercially available in the quantities required for cellular experiments.

Different studies have described the co-occurrence of mycotoxins in cereals and other raw materials used for animal feed in several regions of the globe [21,46,47]. In the present study, we investigated the co-occurrence of mycotoxins in pig feed. We identified 60 metabolites that co-occur with DON in more than 20% of samples, confirming the prevalence of co-contamination, and as a result, the exposure of animals to a mixture of mycotoxins. Our results are in accordance with those of previous surveys of raw materials and finished feed [20,46]. Although the exact composition of the analyzed pig feed is not known, the main component of the majority of the samples is maize. The extraction efficiencies are between 85%-95% determined in seven different matrices [48]. However, the focus of our study is to provide a general picture of the worldwide contamination of pig feed and to test realistic ratios of fungal compounds in vitro.

Despite the very high frequency of co-contamination by DON and emerging mycotoxins, the correlation between the concentrations of DON and emerging mycotoxins was very low. The correlation between DON and emerging mycotoxins is poorly documented. In winter wheat, Blandino and co-workers [49] addressed the correlation between DON and other mycotoxins produced by *Fusarium graminearum* and *F. culmorum*, and showed that correlations between DON and either culmorin or moniliformin were significant (0.94 and 0.42, respectively). By contrast, the correlations between DON and AFN or BEA were not significant (0.2 and -0.14, respectively). Correlations between the concentration of DON and its modified forms enabled EFSA to calculate ratios between these different toxins [11].

In the present study, the absence of correlation between the concentrations of the emerging mycotoxins and DON could be explained by the diversity of fungi that produce the various metabolites via different biosynthetic pathways. Furthermore, pigs feed is made of different raw materials, which also explains the lack of correlation between the amounts of the different fungal metabolites involved.

The main objective of the present study was to assess the combined toxicity of DON and emerging mycotoxins. As no correlation was found, to encompass the situations to which animals may be exposed, different plausible ratios were tested. These ratios were based on the P25, the median and the P75 concentrations of DON and emerging mycotoxins observed in pig feed, because these summary statistics are robust to extreme outlier values. In most cases, we observed that when the non-toxic metabolites (BRV, Cyclo and TRPT) were present in mixture with DON, whatever the doses or the ratio, the effect was driven by DON. We observed a similar trend for the combined toxicity of DON with moderate and highly toxic metabolites ENNs, BEA, API, AFN and EMO. The effect of the mixtures was mostly similar to the effect of DON alone. The only exception was when very high concentrations of DON were used, in which cases surprisingly, the toxicity of the mixture was lower than the toxicity of DON alone.

To the best of our knowledge, this is the first study to investigate the interactions of different mycotoxins using plausible ratios. Most published studies used toxins of equal toxicity, regardless of their concentration in food or feed. For example, exposure of Caco-2 or IPEC-1 cells to low doses of DON, combined with isotoxic concentrations of nivalenol and/or their acetylated derivatives, led to a synergistic effect [31,32]. Similarly, mixtures of ENNs as well as mixtures of DON, ENNs and alternariol managed to induce synergistic cytotoxicity in Caco-2 cells [50,51]. A synergistic inflammatory effect was also observed in porcine intestinal explants co-exposed to DON and nivalenol [33].

In the future, these original data on the intestinal toxicity of realistic mixtures of DON and emerging mycotoxins should be completed by toxicity studies on other types of cells to account for all target organs. In vivo experiments are also needed to confirm the in vitro data. More widely, the effects of other mixtures of mycotoxins or of mycotoxins with other food contaminants should be investigated at realistic doses. As the pig is a good model for human toxicity studies of food contaminants [52], the results would be useful to estimate the effects of similar mixtures of toxins on human health. Better knowledge of the occurrence and toxicity of the real mixtures present in food is a precondition for the assessment of health risk [29].

4. Conclusions

This global survey of finished pig feed confirmed that such feed is co-contaminated by DON and emerging mycotoxins. However, despite the high percentages of co-occurrence, no correlation was found between the concentration of DON and most of these emerging mycotoxins. Using ratios based on the concentration of DON and emerging mycotoxins in feed, we observed that the toxicity of most of the mixtures was similar to the toxicity of DON alone. This demonstrates that, when these emerging mycotoxins are present together with DON, the toxicity of the mixture is not exacerbated.

5. Materials and Methods

5.1. Extraction and Analysis of Metabolites

A total of 524 finished pig feed samples were collected from 2014 to 2018 on the world market, but most in Europe (76.5%) and North America (15.8%) and fewer in Asia (3.2%), South Africa (1.5%), Australia (2 samples) and some of unknown origin (2.5%) (Figure 1 and Supplementary Figure S1), and more than 800 analytes, including fungal and bacterial secondary metabolites, were sought. Samples were provided by the BIOMIN Mycotoxin Survey, and the analyses were performed using the Liquid chromatography–mass spectrometry/mass spectrometry (LC–MS/MS) multi-mycotoxin method described by Malachova [5].

A QTrap5500 LC-MS/MS System (Applied Biosystems, Foster City, CA, USA) equipped with a TurboIonSpray electrospray ionization (ESI) source and a 1290 Series ultra high performance liquid chromatography (UHPLC) System (Agilent Technologies, Waldbronn, Germany) was used for detection and quantification of the analytes (Supplementary Figures S5 and S6). Chromatographic separation was performed on a Gemini® C18-column (150 × 4.6 mm i.d., 5 μm particle size) equipped with a C18 security guard cartridge (4 × 3 mm i.d.) (all from Phenomenex, Torrance, CA, USA) at 25 °C. Elution was carried out in binary gradient mode. Both mobile phases contained 5 mm ammonium acetate ($NH_4CH_3CO_2$) and were composed of methanol/water/acetic acid in a ratio of 10:89:1 (*v*/*v*/*v*; eluent A) or 97:2:1 (*v*/*v*/*v*; eluent B). After an initial time of 2 min at 100% A, the proportion of B was increased linearly to 50% within 3 min. Further linear increase of B to 100% within 9 min was followed by a hold-time of 4 min at 100% B and 2.5 min column re-equilibration at 100% A. The flow rate was 1000 μL/min. ESI-MS/MS was performed in the scheduled multiple reaction monitoring (sMRM) mode both in positive and negative polarities in two separate chromatographic runs. The sMRM detection window of each analyte was set to the respective retention time ± 27 s and ± 42 s in positive and in negative mode, respectively. The target scan time was set to 1 s. Confirmation of positive analyte identification is obtained by the acquisition of two sMRMs per analyte (with the exception of moniliformin and 3-nitropropionic acid, that each exhibit only one fragment ion), which yields 4.0 identification points according to commission decision 2002/657/EC (EU, 2002).

A total of 235 mycotoxins and other fungal secondary metabolites were detected and quantified in the samples of finished pig feed analyzed. The threshold of relevant concentrations was set at > 1.0 μg/kg or the limit of detection, whichever was higher. Samples were collected only by trained staff, or after the instruction of untrained staff according to a protocol. A minimum of 500 g homogenized sample was sent to the laboratory of the Institute of Bioanalytics and Agro-Metabolomics at the University of Natural Resources and Life Sciences Vienna (BOKU) in Tulln, Austria. Samples were milled and extracted with a mixture of acetonitrile, water and acetic acid (79:20:1, per volume) on a shaker for 90 min. The solution was centrifuged, after which the supernatant was diluted and injected into an LC-MS/MS system (electrospray ionization and mass spectrometric detection using a quadrupole mass filter). Quantification was done by comparing an external calibration using a multi-analyte stock solution. During the period 2014 to 2018, the number of substances measured using this method increased each year, and more substances were included in the survey. Nevertheless, the list of compounds investigated in this manuscript were those measured in 2014. All concentration data were collected in a single file, and sample information such as sampling year and month, country and region of origin and sample matrix were added for subsetting. Data were imported and analyzed in R v 3.5.1 mainly using tidy-verse packages (www.tidyverse.org). Spearman correlation coefficients and associated *p*-values were calculated with the corr.test function from the psych R package. Data were plotted (including maps) using ggplot2.

5.2. Toxins

For the cytotoxicity test, toxins were purchased from Sigma (St Quentin Fallavier, France): deoxynivalenol (DON) (purity > 98%), tryptophol (TRPT) (purity > 97%), apicidin (API) (purity > 98%) and emodin (EMO) (purity > 90%). Enniatins (ENNs) (A1, B, B1, purity > 99%), cyclo-(L-Pro-L-Tyr) (Cyclo) (purity > 98%), and brevianamide-F (BRV-F) (purity > 95%) were purchased from BioAustralis (Smithfield, Australia), beauvericin (BEA) (purity > 95%) and aurofusarin (AFN), (purity > 97%) were purchased from Santa Cruz Biotechnology (Dallas, TX, USA). All mycotoxins were dissolved in dimethyl sulfoxide (DMSO) (Sigma) to prepare stock solutions stored at −20 °C. Working dilutions were freshly prepared in cell culture medium for each experiment.

To convert the concentration in the pig feed into the concentration to which intestinal cells might be exposed, we assumed, as already done in previous studies [53], that mycotoxins were ingested in one meal, diluted in 1 L of gastrointestinal fluid and were entirely bioaccessible. Next, the ratio of DON to emerging mycotoxins was calculated based on three plausible scenarios according to the concentration of DON and emerging mycotoxins in the feed (Supplementary Table S1).

Several 3-fold dilutions of each individual toxin and mixtures at different ratios were performed to account for the concentrations present in feed.

5.3. Cell Culture and Cytotxicity Assay

IPEC-1, derived from the small intestine of a newborn unsuckled piglet were maintained in complete medium (Dulbecco's modified Eagle's medium (DMEM)/Ham's F12 medium (Sigma) plus 1% Penicillin Streptomycin (Eurobio, Courtaboeuf, France), 5% fetal bovine serum (FBS) (Eurobio), 1% L-glutamine (Eurobio), 5 μg/L epidermal growth factor (EGF) (Becton–Dickinson, Le Pont de Claix, France), and 1% ITS solution (insulin (5 μg/mL), transferrin (5 μg/mL), selenium (5 ng/mL), (Sigma Aldrich)) at 39 °C under 5% CO_2, as previously described [54].

For the cytotoxicity experiments, cells were seeded in 96-white-well flat-bottom cell culture plates (Costar, Cambridge, MA, USA) at a rate of 10^4 cells per well in 100 μL culture medium. After 24 h, the medium was replaced by complete medium without FBS containing the mycotoxins and incubated for a further 48 h. Toxicity was then assessed using the CellTiter-Glo® Luminescent Cell Viability Assay (Promega, Charbonnières-les-Bains, France) that determine the number of viable cells based on the quantitation of adenosine triphosphate (ATP). The luminescent signal produced by the luciferase reaction, reflecting the presence of metabolically active cells, was read using a multiplate reader (TECAN, Lyon, France). The results were obtained by calculating the percentage of viability obtained by calculating the ratio of the luminescence in treated samples and the luminescence in non-treated samples.

5.4. Statistical Analysis

The reported values on viability are expressed as the means ± the standard error of the mean (SEM) of at least three biological replicates, with duplicate wells for each dose. The IC_{50} value, the dose of each toxin leading to 50% viability, was determined using CompuSyn statistical software (CompuSyn Version-1 Inc. Paramus, NJ, USA). Significant differences between groups were analyzed using the Bonferroni multiple comparison test in GraphPad (GraphPad Prism 4 La Jolla, CA, USA).

Supplementary Materials: The following are available online at http://www.mdpi.com/2072-6651/11/12/727/s1, Figure S1: Worldwide contamination of emerging mycotoxins. Figure S2: Abundance of emerging mycotoxins in finished pig feed. Figure S3: Scatter plot of the correlation between DON and 10 selected emerging mycotoxins. Figure S4: Dose-effect curves of the individual toxicity of DON and 9 selected fungal emerging mycotoxins Table S1: Concentration of the metabolites in 3 quartiles (P25, P50 and P75) and ratios between the concentration of DON and 10 selected emerging mycotoxins. Figure S5: LC-ESI(+)-MRM chromatograms of one pig feed sample. Figure S6: Overlay of the extracted ion chromatograms of both quantifier as well as qualifier of the 11 investigated compounds in one pig feed sample.

Author Contributions: A.K.K., I.P.O. and P.P. designed the experiment, A.K.K. and M.N. performed the experiments, B.N. and G.S. contributed to provide raw data on finished pig feed, T.J. analyzed raw data on finished pig feed, P.G.P.M. performed statistical analysis on correlation, A.K.K. analyzed the cytotoxicity results, A.K.K., P.P. and I.P.O. wrote the paper, B.N. and P.G.P.M. reviewed the paper.

Funding: This research was supported by the ANR grants "EmergingMyco" (ANR-18-CE34-0014) and "Newmyco" (ANR-15-CE21-0010) and a grant from BIOMIN. A.K.K was partially supported by the regional contract 15066653.

Acknowledgments: We would like to thank Terciolo for her critical review of the paper and Daphne Goodfellow for English editing.

Conflicts of Interest: B.N., T.J. and G.S. are employed by BIOMIN. However, this circumstance did not influence the design of the experimental studies or bias the presentation and interpretation of results. The other authors, A.K.K., P.G.P.P., M.N., P.P. and I.P.O. declare no conflict of interest.

References

1. Bennett, J.W.; Klich, M. Mycotoxins. *Clin. Microbiol. Rev.* **2003**, *16*, 497–516. [CrossRef] [PubMed]
2. European Commission. Commission Recommendation of 17 August 2006 on the presence of deoxynivalenol, zearalenone, ochratoxin A, T-2 and HT-2 and fumonisins in products intended for animal feeding. *Off. J. Eur. Union* **2006**, *L229*, 7–9.
3. European Commission. Commission Recommendation of 27 March 2013 on the presence of T-2 and HT-2 toxin in cereals and cereal productsText with EEA relevance. *Off. J. Eur. Union* **2013**, *L91*, 12–15.
4. Berthiller, F.; Crews, C.; Dall'Asta, C.; Saeger, S.D.; Haesaert, G.; Karlovsky, P.; Oswald, I.P.; Seefelder, W.; Speijers, G.; Stroka, J. Masked mycotoxins: A review. *Mol. Nutr. Food Res.* **2013**, *57*, 165–186. [CrossRef] [PubMed]
5. Malachová, A.; Sulyok, M.; Beltrán, E.; Berthiller, F.; Krska, R. Optimization and validation of a quantitative liquid chromatography–tandem mass spectrometric method covering 295 bacterial and fungal metabolites including all regulated mycotoxins in four model food matrices. *J. Chromatogr. A* **2014**, *1362*, 145–156. [CrossRef]
6. Gruber-Dorninger, C.; Novak, B.; Nagl, V.; Berthiller, F. Emerging Mycotoxins: Beyond Traditionally Determined Food Contaminants. *J. Agric. Food Chem.* **2016**, *65*, 7052–7070. [CrossRef]
7. Rychlik, M.; Humpf, H.-U.; Marko, D.; Dänicke, S.; Mally, A.; Berthiller, F.; Klaffke, H.; Lorenz, N. Proposal of a comprehensive definition of modified and other forms of mycotoxins including "masked" mycotoxins. *Mycotoxin Res.* **2014**, *30*, 197–205. [CrossRef]
8. Eskola, M.; Kos, G.; Elliott, C.T.; Hajšlová, J.; Mayar, S.; Krska, R. Worldwide contamination of food-crops with mycotoxins: Validity of the widely cited 'FAO estimate' of 25%. *Crit. Rev. Food Sci. Nutr.*. in press. [CrossRef]
9. Schatzmayr, G.; Streit, E. Global occurrence of mycotoxins in the food and feed chain: Facts and figures. *World Mycotoxin J.* **2013**, *6*, 213–222. [CrossRef]
10. Streit, E.; Schatzmayr, G.; Tassis, P.; Tzika, E.; Marin, D.; Taranu, I.; Tabuc, C.; Nicolau, A.; Aprodu, I.; Puel, O.; et al. Current Situation of Mycotoxin Contamination and Co-occurrence in Animal Feed—Focus on Europe. *Toxins* **2012**, *4*, 788–809. [CrossRef]

11. Knutsen, H.K.; Alexander, J.; Barregard, L.; Bignami, M.; Bruschweiler, B.; Ceccatelli, S.; Cottrill, B.; Dinovi, M.; Grasl-Kraupp, B.; Hogstrand, C.; et al. Risks to human and animal health related to the presence of deoxynivalenol and its acetylated and modified forms in food and feed. *EFSA J.* **2017**, *15*, 4718.
12. IARC. Toxins derived from *Fusarium* graminearum, F. culmorum and F. crookwellense: Zearalenone, deoxynivalenol, nivalenol and fusarenone X. *IARC Monogr. Eval. Carcinog. Risks Hum.* **1993**, *56*, 397–444.
13. Charmley, L.; Trenholm, H. RG-8 Regulatory Guidance:Contaminants in Feed (Formerly RG-1, Chapter 7). Available online: https://www.inspection.gc.ca/animals/feeds/regulatory-guidance/rg-8/eng/1347383943203/1347384015909?chap=1#s1c1 (accessed on 21 November 2019).
14. Pestka, J.J. Deoxynivalenol: Mechanisms of action, human exposure, and toxicological relevance. *Arch. Toxicol.* **2010**, *84*, 663–679. [CrossRef] [PubMed]
15. Pinton, P.; Oswald, I. Effect of Deoxynivalenol and Other Type B Trichothecenes on the Intestine: A Review. *Toxins* **2014**, *6*, 1615–1643. [CrossRef] [PubMed]
16. Payros, D.; Alassane-Kpembi, I.; Pierron, A.; Loiseau, N.; Pinton, P.; Oswald, I.P. Toxicology of deoxynivalenol and its acetylated and modified forms. *Arch. Toxicol.* **2016**, *90*, 2931–2957. [CrossRef] [PubMed]
17. Pierron, A.; Mimoun, S.; Murate, L.S.; Loiseau, N.; Lippi, Y.; Bracarense, A.-P.F.L.; Schatzmayr, G.; He, J.W.; Zhou, T.; Moll, W.-D.; et al. Microbial biotransformation of DON: Molecular basis for reduced toxicity. *Sci. Rep.* **2016**, *6*, 29105. [CrossRef]
18. Jajić, I.; Dudaš, T.; Krstović, S.; Krska, R.; Sulyok, M.; Bagi, F.; Savić, Z.; Guljaš, D.; Stankov, A. Emerging *Fusarium* Mycotoxins Fusaproliferin, Beauvericin, Enniatins, and Moniliformin in Serbian Maize. *Toxins* **2019**, *11*, 357. [CrossRef]
19. EFSA. Scientific Opinion on the risks to human and animal health related to the presence of beauvericin and enniatins in food and feed: Beauvericin and enniatins in food and feed. *EFSA J.* **2014**, *12*, 3802. [CrossRef]
20. Novak, B.; Rainer, V.; Sulyok, M.; Haltrich, D.; Schatzmayr, G.; Mayer, E. Twenty-Eight Fungal Secondary Metabolites Detected in Pig Feed Samples: Their Occurrence, Relevance and Cytotoxic Effects In Vitro. *Toxins* **2019**, *11*, 537. [CrossRef]
21. Abdallah, M.F.; Girgin, G.; Baydar, T.; Krska, R.; Sulyok, M. Occurrence of multiple mycotoxins and other fungal metabolites in animal feed and maize samples from Egypt using LC-MS/MS: Toxic fungal and bacterial metabolites in feed and maize from Egypt. *J. Sci. Food Agric.* **2017**, *97*, 4419–4428. [CrossRef]
22. Streit, E.; Schwab, C.; Sulyok, M.; Naehrer, K.; Krska, R.; Schatzmayr, G. Multi-Mycotoxin Screening Reveals the Occurrence of 139 Different Secondary Metabolites in Feed and Feed Ingredients. *Toxins* **2013**, *5*, 504–523. [CrossRef] [PubMed]
23. Anjorin, S.T.; Fapohunda, S.; Sulyok, M.; Krska, R. Natural Co-occurrence of Emerging and Minor Mycotoxins on Maize Grains from Abuja, Nigeria. *Ann. Agric. Environ. Sci.* **2016**, *1*, 21–29.
24. Alassane-Kpembi, I.; Schatzmayr, G.; Taranu, I.; Marin, D.; Puel, O.; Oswald, I.P. Mycotoxins co-contamination: Methodological aspects and biological relevance of combined toxicity studies. *Crit. Rev. Food Sci. Nutr.* **2017**, *57*, 3489–3507. [CrossRef]
25. Fremy, J.-M.; Alassane-Kpembi, I.; Oswald, I.P.; Cottrill, B.; Van Egmond, H.P. A review on combined effects of moniliformin and co-occurring *Fusarium* toxins in farm animals. *World Mycotoxin J.* **2019**, *12*, 281–291. [CrossRef]
26. Assunção, R.; Silva, M.J.; Alvito, P. Challenges in risk assessment of multiple mycotoxins in food. *World Mycotoxin J.* **2016**, *9*, 791–811. [CrossRef]
27. More, S.J.; Bampidis, V.; Benford, D.; Bennekou, S.H.; Bragard, C.; Halldorsson, T.I.; Herna, A.F.; Koutsoumanis, K.; Naegeli, H.; Schlatter, J.R.; et al. Guidance on harmonised methodologies for human health, animal health and ecological risk assessment of combined exposure to multiple chemicals. *EFSA J.* **2019**, *17*, 5634.
28. Meek, B.; Roobis, A.R.; Crofton, K.M.; Heinemeyer, G.; Raaij, M.V.; Vickers, C. Risk assessment of combined exposure to multiple chemicals: A WHO/IPCS framework. *Regul. Toxicol. Pharmacol.* **2011**, *60*, S1–S14.
29. Kortenkamp, A.; Faust, M. Regulate to reduce chemical mixture risk. *Science* **2018**, *361*, 224–226. [CrossRef] [PubMed]
30. Ghareeb, K.; Awad, W.A.; Böhm, J.; Zebeli, Q. Impacts of the feed contaminant deoxynivalenol on the intestine of monogastric animals: Poultry and swine. *J. Appl. Toxicol.* **2015**, *35*, 327–337. [CrossRef]

31. Alassane-Kpembi, I.; Kolf-Clauw, M.; Gauthier, T.; Abrami, R.; Abiola, F.A.; Oswald, I.P.; Puel, O. New insights into mycotoxin mixtures: The toxicity of low doses of Type B trichothecenes on intestinal epithelial cells is synergistic. *Toxicol. Appl. Pharmacol.* **2013**, *272*, 191–198. [CrossRef]
32. Alassane-Kpembi, I.; Puel, O.; Oswald, I.P. Toxicological interactions between the mycotoxins deoxynivalenol, nivalenol and their acetylated derivatives in intestinal epithelial cells. *Arch. Toxicol.* **2015**, *89*, 1337–1346. [CrossRef]
33. Alassane-Kpembi, I.; Puel, O.; Pinton, P.; Cossalter, A.-M.; Chou, T.-C.; Oswald, I. Co-exposure to low doses of the food contaminants deoxynivalenol and nivalenol has a synergistic inflammatory effect on intestinal explants. *Arch. Toxicol.* **2017**, *91*, 2677–2687. [CrossRef] [PubMed]
34. Kolf-Clauw, M.; Sassahara, M.; Lucioli, J.; Rubira-Gerez, J.; Alassane-Kpembi, I.; Lyazhri, F.; Borin, C.; Oswald, I.P. The emerging mycotoxin, enniatin B1, down-modulates the gastrointestinal toxicity of T-2 toxin in vitro on intestinal epithelial cells and ex vivo on intestinal explants. *Arch. Toxicol.* **2013**, *87*, 2233–2241. [CrossRef] [PubMed]
35. Escrivá, L.; Jennen, D.; Caiment, F.; Manyes, L. Transcriptomic study of the toxic mechanism triggered by beauvericin in Jurkat cells. *Toxicol. Lett.* **2018**, *284*, 213–221. [CrossRef] [PubMed]
36. Pierron, A.; Alassane-Kpembi, I.; Oswald, I.P. Impact of mycotoxin on immune response and consequences for pig health. *Anim. Nutr.* **2016**, *2*, 63–68. [CrossRef] [PubMed]
37. Pinton, P.; Braicu, C.; Nougayrede, J.-P.; Laffitte, J.; Taranu, I.; Oswald, I.P. Deoxynivalenol Impairs Porcine Intestinal Barrier Function and Decreases the Protein Expression of Claudin-4 through a Mitogen-Activated Protein Kinase-Dependent Mechanism. *J. Nutr.* **2010**, *140*, 1956–1962. [CrossRef] [PubMed]
38. Council for Agricultural Science and Technology. *Mycotoxins: Risks in Plant, Animal, and Human Systems;* Task Force report; Council for Agricultural Science and Technology: Ames, IA, USA, 2003; ISBN 978-1-887383-22-6.
39. Bauden, M.; Tassidis, H.; Ansari, D. In vitro cytotoxicity evaluation of HDAC inhibitor Apicidin in pancreatic carcinoma cells subsequent time and dose dependent treatment. *Toxicol. Lett.* **2015**, *236*, 8–15. [CrossRef]
40. Glenn, A.E.; Bacon, C.W.; Price, R.; Hanlin, R.T. Molecular phylogeny of *Acremonium* and its taxonomic implications. *Mycologia* **1996**, *88*, 369–383. [CrossRef]
41. Kosalec, I.; Ramić, S.; Jelić, D.; Antolović, R.; Pepeljnjak, S.; Kopjar, N. Assessment of Tryptophol Genotoxicity in Four Cell Lines In Vitro: A Pilot Study with Alkaline Comet Assay. *Arch. Ind. Hyg. Toxicol.* **2011**, *62*, 41–49. [CrossRef]
42. Muto, A.; Hori, M.; Sasaki, Y.; Saitoh, A.; Yasuda, I.; Maekawa, T.; Uchida, T.; Asakura, K.; Nakazato, T.; Kaneda, T.; et al. Emodin has a cytotoxic activity against human multiple myeloma as a Janus-activated kinase 2 inhibitor. *Mol. Cancer Ther.* **2007**, *6*, 987–994. [CrossRef]
43. Meca, G.; Font, G.; Ruiz, M.J. Comparative cytotoxicity study of enniatins A, A1, A2, B, B1, B4 and J3 on Caco-2 cells, Hep-G2 and HT-29. *Food Chem. Toxicol.* **2011**, *49*, 2464–2469. [CrossRef] [PubMed]
44. Fraeyman, S.; Meyer, E.; Devreese, M.; Antonissen, G.; Demeyere, K.; Haesebrouck, F.; Croubels, S. Comparative in vitro cytotoxicity of the emerging *Fusarium* mycotoxins beauvericin and enniatins to porcine intestinal epithelial cells. *Food Chem. Toxicol.* **2018**, *121*, 566–572. [CrossRef] [PubMed]
45. Taranu, I.; Marin, D.E.; Burlacu, R.; Pinton, P.; Damian, V.; Oswald, I.P. Comparative aspects of *in vitro* proliferation of human and porcine lymphocytes exposed to mycotoxins. *Arch. Anim. Nutr.* **2010**, *64*, 383–393. [CrossRef] [PubMed]
46. Gruber-Dorninger, C.; Jenkins, T.; Schatzmayr, G. Global Mycotoxin Occurrence in Feed: A Ten-Year Survey. *Toxins* **2019**, *11*, 375. [CrossRef]
47. Alkadri, D.; Rubert, J.; Prodi, A.; Pisi, A.; Mañes, J.; Soler, C. Natural co-occurrence of mycotoxins in wheat grains from Italy and Syria. *Food Chem.* **2014**, *157*, 111–118. [CrossRef]
48. Sulyok, M.; Stadler, D.; Steiner, D.; Krska, R. Validation of an LC-MS/MS based dilute-and-shoot approach for the quantification of >500 mycotoxins and other secondary metabolites in food crops: Challenges and solutions. submitted.
49. Blandino, M.; Scarpino, V.; Sulyok, M.; Krska, R.; Reyneri, A. Effect of agronomic programmes with different susceptibility to deoxynivalenol risk on emerging contamination in winter wheat. *Eur. J. Agron.* **2017**, *85*, 12–24. [CrossRef]
50. Prosperini, A.; Font, G.; Ruiz, M.J. Interaction effects of *Fusarium* enniatins (A, A1, B and B1) combinations on in vitro cytotoxicity of Caco-2 cells. *Toxicol. Vitr.* **2014**, *28*, 88–94. [CrossRef]

51. Fernández-Blanco, C.; Font, G.; Ruiz, M.-J. Interaction effects of enniatin B, deoxinivalenol and alternariol in Caco-2 cells. *Toxicol. Lett.* **2016**, *241*, 38–48. [CrossRef]
52. Ireland, J.J.; Roberts, R.M.; Palmer, G.H.; Bauman, D.E.; Bazer, F.W. A commentary on domestic animals as dual-purpose models that benefit agricultural and biomedical research. *Am. Soc. Anim. Sci.* **2008**, *86*, 2797–2805.
53. García, G.R.; Payros, D.; Pinton, P.; Dogi, C.A.; Laffitte, J.; Neves, M.; González Pereyra, M.L.; Cavaglieri, L.R.; Oswald, I.P. Intestinal toxicity of deoxynivalenol is limited by Lactobacillus rhamnosus RC007 in pig jejunum explants. *Arch. Toxicol.* **2018**, *92*, 983–993. [CrossRef] [PubMed]
54. Pinton, P.; Nougayrède, J.-P.; Del Rio, J.-C.; Moreno, C.; Marin, D.E.; Ferrier, L.; Bracarense, A.-P.; Kolf-Clauw, M.; Oswald, I.P. The food contaminant deoxynivalenol, decreases intestinal barrier permeability and reduces claudin expression. *Toxicol. Appl. Pharmacol.* **2009**, *237*, 41–48. [CrossRef] [PubMed]

© 2019 by the authors. Licensee MDPI, Basel, Switzerland. This article is an open access article distributed under the terms and conditions of the Creative Commons Attribution (CC BY) license (http://creativecommons.org/licenses/by/4.0/).

Article

Deoxynivalenol Induces Inflammatory Injury in IPEC-J2 Cells via NF-κB Signaling Pathway

Xichun Wang, Yafei Zhang, Jie Zhao, Li Cao, Lei Zhu, Yingying Huang, Xiaofang Chen, Sajid Ur Rahman, Shibin Feng, Yu Li * and Jinjie Wu *

Department of Veterinary Medicine, College of Animal Science and Technology, Anhui Agricultural University, Hefei 230036, China; wangxichun@ahau.edu.cn (X.W.); djdshhz@163.com (Y.Z.); zhaojie866@ahau.edu.cn (J.Z.); caoli@ahau.edu.cn (L.C.); zhuleizl@ahau.edu.cn (L.Z.); huangyingying@ahau.edu.cn (Y.H.); chenxiaofang0912@163.com (X.C.); dr_sajid226@yahoo.com (S.U.R.); fengshibin@ahau.edu.cn (S.F.)

* Correspondence: lydhy2014@ahau.edu.cn (Y.L.); wjj@ahau.edu.cn (J.W.)

Received: 12 November 2019; Accepted: 11 December 2019; Published: 16 December 2019

Abstract: The aim of this study was to investigate the effects of deoxynivalenol (DON) exposure on the inflammatory injury nuclear factor kappa-B (NF-κB) pathway in intestinal epithelial cells (IPEC-J2 cells) of pig. The different concentrations of DON (0, 125, 250, 500, 1000, 2000 ng/mL) were added to the culture solution for treatment. The NF-κB pathway inhibitor pyrrolidine dithiocarbamate (PDTC) was used as a reference. The results showed that when the DON concentration increased, the cell density decreased and seemed damaged. With the increase of DON concentration in the culture medium, the action of diamine oxidase (DAO) in the culture supernatant also increased. The activities of IL-6, TNF-α, and NO in the cells were increased with the increasing DON concentration. The relative mRNA expression of *IL-1β* and *IL-6* were increased in the cells. The mRNA relative expression of *NF-κB p65*, *IKKα*, and *IKKβ* were upregulated with the increasing of DON concentration, while the relative expression of *IκB-α* mRNA was downregulated. At the same time, the expression of NF-κB p65 protein increased gradually in the cytoplasm and nucleus with a higher concentration of DON. These results showed that DON could change the morphology of IPEC-J2 cells, destroy its submicroscopic structure, and enhance the permeability of cell membrane, as well as upregulate the transcription of some inflammatory factors and change the expression of NF-κB-related gene or protein in cells.

Keywords: deoxynivalenol; IPEC-J2; cell damage; NF-κB inflammatory signal pathway

Key Contribution: This research discusses the mechanism of DON exposure on NF-κB inflammatory injury pathway in IPEC-J2 cells.

1. Introduction

Mycotoxins are widely found in human and animal foods. Deoxynivalenol (DON) is mainly produced by *Fusarium graminearum* which is prone to crops such as corn and wheat [1–3]. China is widely contaminated by such kind of mycotoxins, especially in the area of Yangtze and Yellow rivers. In different grains the climate is favorable to the growth of mold and their relevant toxin production [4]. Previously, from 2016–2017, 827 complete feed samples of pigs and 724 components of feed were chosen from various provinces of China for analysis, which shows that the last standard rate of DON ingredient was above 74.5%, and 450.0–4381.5 μg/kg was the average concentration [5]. Hence, further studies on DON become even highly significant.

It has been reported that DON can cause animals to food refusal, organ damage, and increase the risk of disease [6–8]. It is generally believed that DON exerts many toxic effects such as neurotoxicity, immuno-toxicity, intestinal toxicity, and cytotoxicity [9,10]. Pigs are the utmost susceptible animal to DON [11]. When DON enters the animal's body, it first acts on the digestive tract. When the animal ingests food or feed containing DON, the intestinal epithelial cells will be exposed to a high concentration of DON, which can potentially affect the intestinal function of the animal [12–14].

Small intestinal epithelial cells can effectively inhibit the propagation of potentially harmful micro-organisms between the inner and lower mucosa [15]. Infection mechanisms of intestinal epithelial cells include bacterial attachment or invasion of cells, proliferation of pathogens, and host cell responses [16]. Experiments have shown that the use of 2000 ng/mL DON to treat pig intestinal epithelial cells (IPEC-J1 and IPEC-J2), decreased ZO-1 protein expression in cells, indicating that DON can make the mechanical integrity of intestinal barrier compromised [17]. At the same time, the transportation and absorption of nutrients in epithelial cells have more impact, through which the substances needed for the growth and development of the body are provided [18]. Maintaining the homeostasis of the intestinal tract, at this point, the structure and function of epithelial cells and lymphocytes in the intestinal mucosa play a key role. For example, lymphocytes in the mucosa can be released by releasing a series of antibody molecules to neutralize toxins [19]. Simultaneously, there is a chemical barrier in the intestines, which is composed of various liquids secreted by epithelial cells mixed with bacteriostatic peptide secreted by the microflora in the intestinal tract to cover the mucus on the surface of epithelial cells [20]. Therefore, the epithelial cells are also considered to be the medium of an early natural immune response [21].

IPEC-J2 is isolated from the jejunal epithelial tissue of newborn piglets and is a small intestine cell model of piglets cultured in vitro [22]. In view of the high homology of the intestinal structure and function of pigs and humans, studies conducted by IPEC-J2 cells can provide a theoretical basis for human medicine [22]. The pig intestinal epithelial cell line IPEC-J2 cells mostly form a cell monolayer when cultured in vitro, and occasionally there are stratified regions. The monolayer consists of cuboid cells interspersed with flat cells. No goblet cells were observed [16]. During cell injury or growth NF-κB plays an important role by acting as a regulatory pathway [23], the classical NF-κB pathway affecting numerous functions of cells that damage cells by enhancing inflammatory related genes expression, growth, and immunity [24].

In this study, we increased the concentration of DON in the culture medium, and selected the NF-κB pathway inhibitor PDTC as a reference, aimed to explore the phenomenon by which DON persuades inflammatory impairment in IPEC-J2 cells via NF-κB signaling pathway by the features of morphological structure, cell viability, cellular inflammatory mediators, and expression of pathway-related gene proteins, etc.

2. Results

2.1. *Effects of DON on Cell Viability Rate*

As can be seen from Figure 1, the viability rate of IPEC-J2 cells decreased with the increasing concentration of DON in the cell culture medium. Compared with the control cells, when the DON concentration reached 250 ng/mL the viability rate of cells in each treatment group was significantly decreased ($p < 0.05$), and further increased to 500 ng/mL and above ($p < 0.01$).

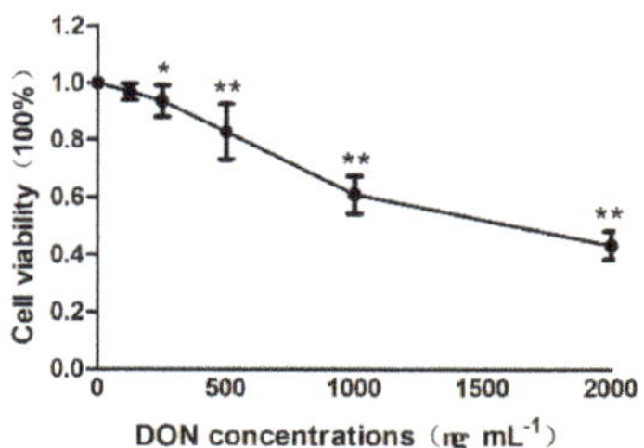

Figure 1. Effects of DON on cell viability rate in cells. The concentrations are 500, 1000, 1500, and 2000 in order. The above data presented as means ± SD of three independent tests (n = 10). Note: * indicates a significant difference compared with the data of the control group ($p < 0.05$), and ** indicates highly significant compared with the control group ($p < 0.01$).

2.2. Effects of DON on Morphological Changes in Intestinal Epithelial Cells

As shown in Figure 2, the cells are in an irregular shape and closely connected when growing normally with a large number of cells (Figure 2A). Subsequently, (125, 250, 500, 1000, and 2000 ng/mL) concentrations of DON were applied to the cells, the cell density steadily reduced, and when the DON concentration enhance in the culture medium, the cells indicate irregular architectures (Figure 2B–F).

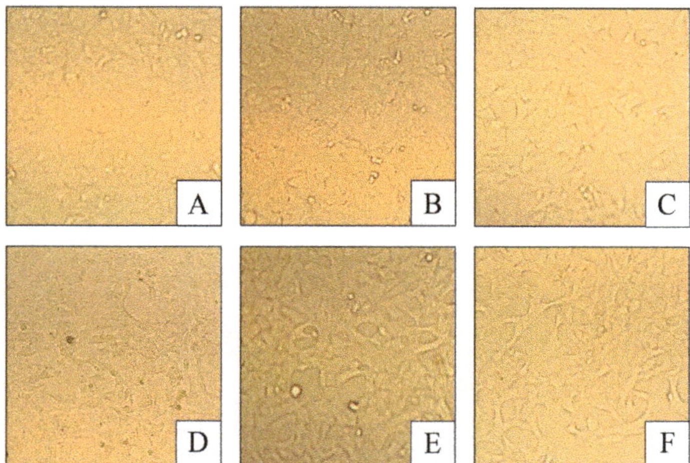

Figure 2. Effect of DON on growth state in cells (400×). (**A**): Control group; (**B**): 125 ng/mL deoxynivalenol (DON) group; (**C**): 250 ng/mL DON group, (**D**): 500 ng/mL DON group, (**E**): 1000 ng/mL DON group, and (**F**): 2000 ng/mL DON group.

2.3. Effect of DON on the Submicroscopic Structure of Cells

Figure 3 shows that, the morphology of the cells in the control group is intact, the organelles in the cytoplasm are normal, and the distribution of chromatin in the nucleus is relatively uniform (Figure 3A1,A2). When the cells were exposed to 125 ng/mL of DON (Figure 3B1,B2), there was no obvious pathological change seen in the nucleus, but rupture of mitochondrial mites seen in the cytoplasm while the number of cristae was also reduced. With the increasing concentration of DON in the culture solution (Figure 3C–E), the ribosome in the cytoplasm gradually decreased, and the organelles such as mitochondria and endoplasmic reticulum gradually vacuolized and, the chromatin marginal aggregation in the nucleus increased. When the concentration reached

2000 ng/mL (Figure 3F1,F2), the chromatin in the nucleus aggregated and disappeared, and the organelle vacuolization becomes severe.

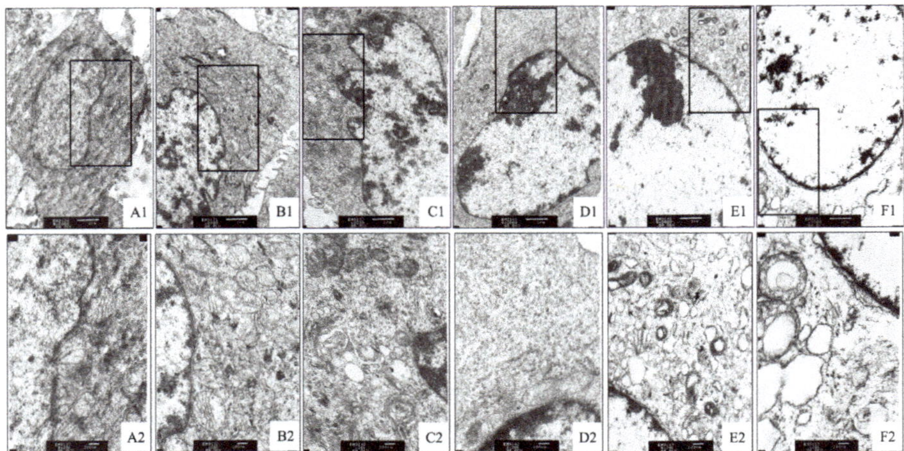

Figure 3. Effect of DON on the ultrastructure of cells. (**A1,A2**: Control group), (**B1,B2**: 125 ng/mL deoxynivalenol (DON) group), (**C1,C2**: 250 ng/mL DON group), (**D1,D2**: 500 ng/mL DON group), (**E1,E2**: 1000 ng/mL DON group), (**F1,F2**: 2000 ng/mL DON group). Image amplification of **A1–F1** is 6000×; image amplification of **A2–F2** is 20,000×. The black frame in A1, B1, C1, D1, E1, and F1, are enlarged to A2, B2, C2, D2, E2, and F2 respectively.

2.4. Effect of DON on DAO Activity in Cell Culture Supernatant

As shown in Figure 4, during comparison with the control group, increasing the concentration of DON in the culture, the amount of DAO was expressively enhanced in the culture supernatant of IPEC-J2 cells ($p < 0.01$). The DAO content in the supernatant of the experimental group was significantly decreased ($p < 0.01$), when the DON was added with PDTC.

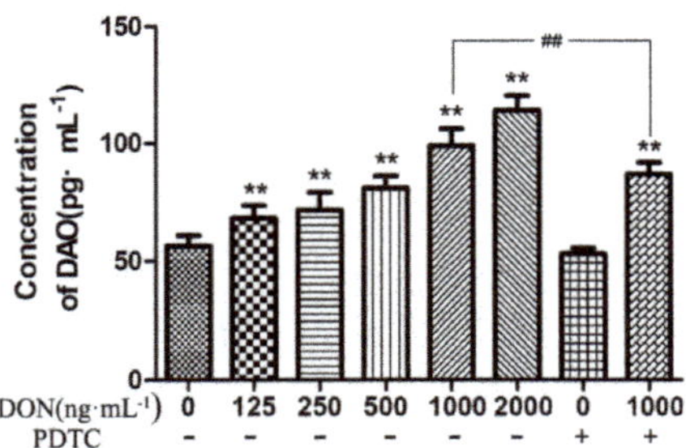

Figure 4. Effects of DON on the activity of DAO in cell culture fluid. The above data shows as means ± SD of three independent tests (n = 10). Note: ** shows extremely significant difference ($p < 0.01$), compared with the control group. ## indicates highly significant ($p < 0.05$) compared with the test group and with the same DON concentration but no addition of PDTC. These rules are also applied for Figures 5–7.

2.5. Effect of DON on the Activity of Inflammatory Factors in Cells

The IL-6, TNF-α, and NO levels in the cells were increased with a higher concentration of DON when matched to the control group, while PDTC reduces the content of these inflammatory mediators presented in Figure 5. Compared to the control group, when the concentration of DON in the culture medium reached 125 ng/mL and above, the concentration of IL-6, TNF-α, and NO in the cells were significantly increased ($p < 0.01$). Compared with the 1000 ng/mL DON group without PDTC, the group after adding PDTC significantly reduced the contents of TNF-α and NO in the cells ($p < 0.01$).

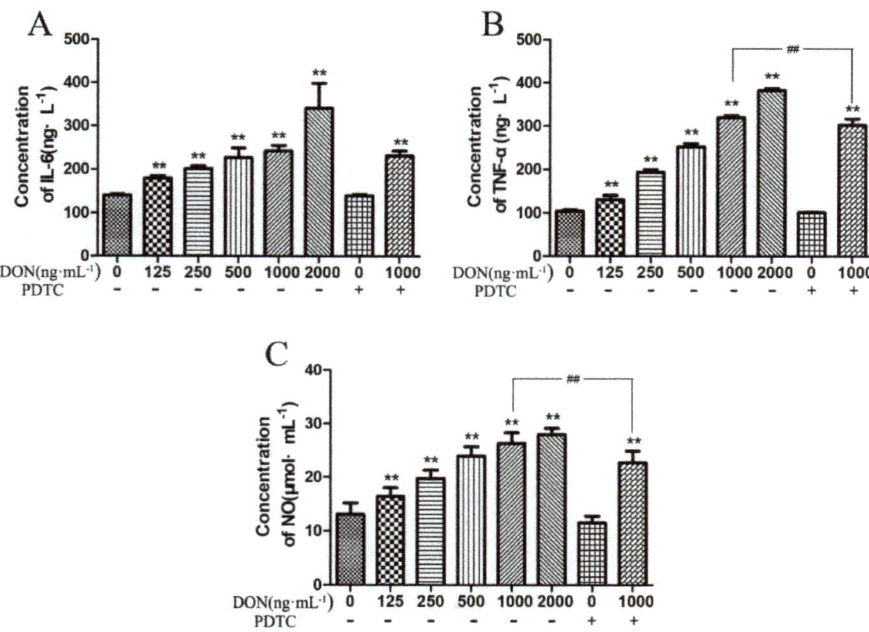

Figure 5. Effect of DON on the concentration of inflammatory factors in cells. (**A–C**) IL-6, TNF-α, and IL-1β expression. The collected data are shown as means ± SD of three independent tests (n = 10). Note: ** shows extremely significant difference ($p < 0.01$), compared with the control group. ## indicates highly significant ($p < 0.05$) compared with the test group and with the same DON concentration but no addition of PDTC.

2.6. DON Effect on the mRNA Expression of Inflammation-Related Genes in Cells

The data in Figure 6 illustrated that DON increased the relative expression of inflammatory mediators in IPEC-J2 cells, and decreased after the addition of pathway inhibitors. Among them, the relative mRNA expression of *IL-1β* in the cells increased with the increasing of DON concentration, and was significantly higher than the control group when the DON concentration was 250 ng/mL and above ($p < 0.01$). The relative expression of *IL-1β* in the test group of 1000 ng/mL DON plus PDTC was significantly lowered than that in the test group of 1000 ng/mL DON without PDTC ($p < 0.05$). In addition, the relative expression of *IL-6* gene in the cells increased significantly with the increasing of DON concentration ($p < 0.01$).

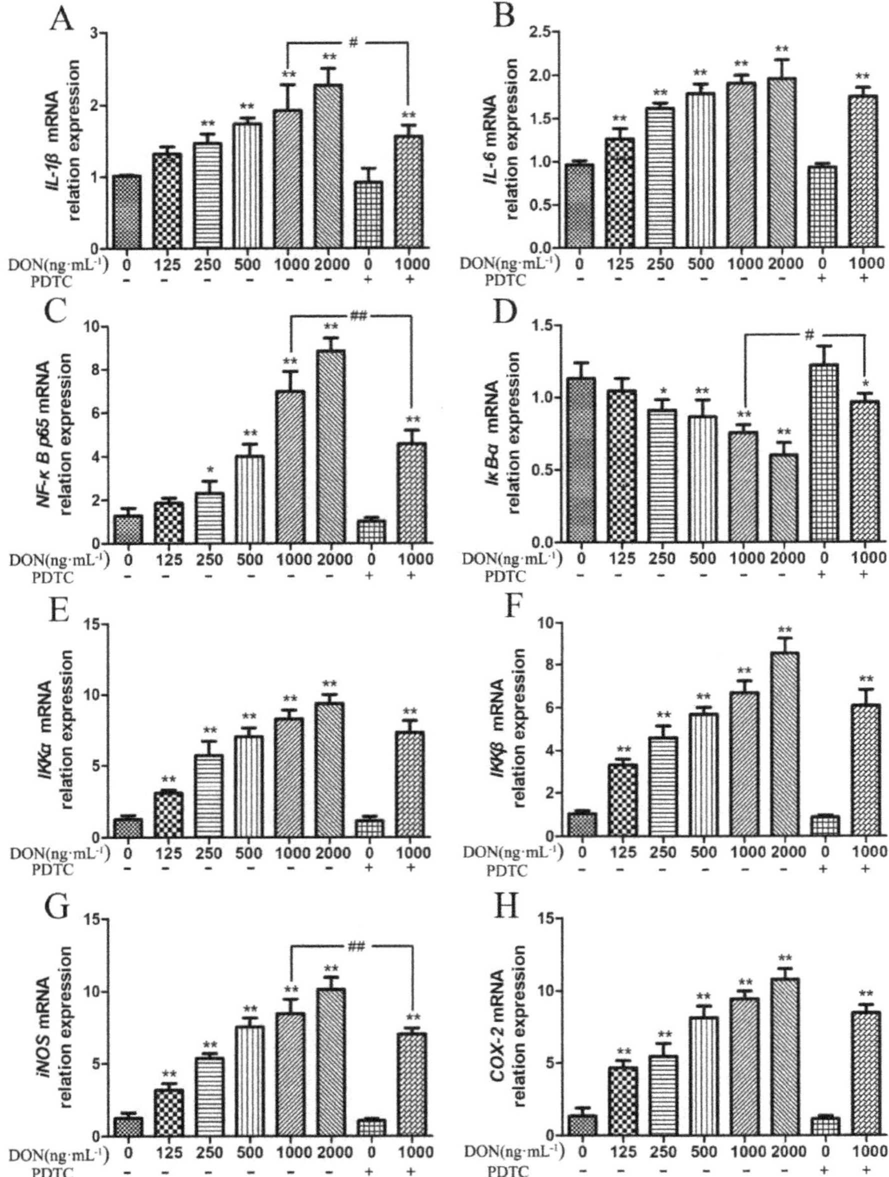

Figure 6. DON influence on the mRNA expression of NF-κB pathway associated genes in cells. (**A–H**) the expression of *IL-1β*, *IL-6*, *NF-κB p65*, *IκB-α*, IKKα, *IKKβ*, *iNOS*, and *COX-2* mRNA. Note: * designates a significant change as compared to the control group ($p < 0.05$), and ** shows extremely significant difference ($p < 0.01$), compared with the control group. # Indicates significant ($p < 0.05$) compared with the test group and with the same DON concentration but no addition of DON. ## Indicates highly significant ($p < 0.01$) compared with the test group and with the same DON concentration but no addition of PDTC.

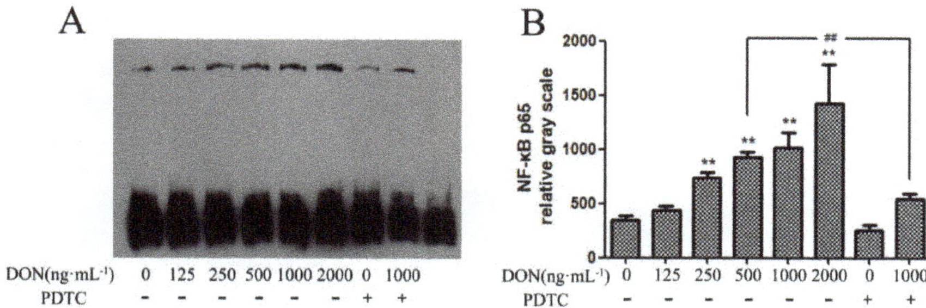

Figure 7. Effect of DON on the transcriptional activity of NF-κB p65 in cells. (**A**) Western blotting showing NF-κB p65. (**B**) Impact of DON on the protein expression of NF-κB p65. Note: ** shows extremely significant difference ($p < 0.01$), compared with the control group. ## indicates highly significant ($p < 0.05$) compared with the test group and with the same DON concentration but no addition of PDTC.

In this experiment, with the increasing of DON contents in the culture medium, the relative expression levels of *NF-κB p65, IKKα, IKKβ, iNOS,* and *COX-2* mRNA in the cells were upregulated to different degrees compared with the control group. Compared with the unadded test group, the relative expression of some genes in the PDTC test group was downregulated by different degrees. When the concentration of DON in the culture medium was 125 ng/mL and above, the mRNA expression levels of *IKKα, IKKβ, iNOS,* and *COX-2* were significantly upregulated compared with the control group ($p < 0.01$). After adding PDTC to the culture medium containing 1000 ng/mL DON, the relative expression of *iNOS* mRNA gene in the cells was significantly decreased ($p < 0.01$). Compared to the control group, when the concentration of DON in culture medium increased to 250 ng/mL, the relative expression of *NF-κB p65* mRNA significantly upregulated ($p < 0.05$), and the relative expression of *IκB-α* mRNA downregulated ($p < 0.05$). When the concentration of DON increased to 500 ng/mL and above, the relative expression of *NF-κB p65* mRNA significantly upregulated compared with the control group ($p < 0.01$) and the relative expression of *IκB-α* mRNA gene was a significant downregulation ($p < 0.01$). After the addition of pathway inhibitors in the experimental group, the relative expression of *NF-κB p65* gene was significantly downregulated ($p < 0.01$), and the relative expression of *IκB-α* mRNA gene was significantly upregulated ($p < 0.05$).

2.7. Effect of DON on the Expression of NF-κB p65 Protein in the Nucleus

In this study, the nuclear expression of NF-κB p65 was detected by EMSA. As shown in Figure 7, with the increasing concentration of DON, the nuclear expression of NF-κB p65 was increased, as well as significantly ($p < 0.01$) increased in DON treatment group, while the expression of NF-κB p65 protein in the 10 μM PDTC-treated cells decreased.

2.8. Effect of DON on the Expression and Distribution of NF-κB p65 Protein in Cells

When the concentration of DON in the culture medium increased, the green fluorescence intensity of NF-κB p65 protein in the cell, especially in the nucleus also increased gradually as shown in Figure 8, indicating that DON can enhance the expression of NF-κB p65 and nuclear expression.

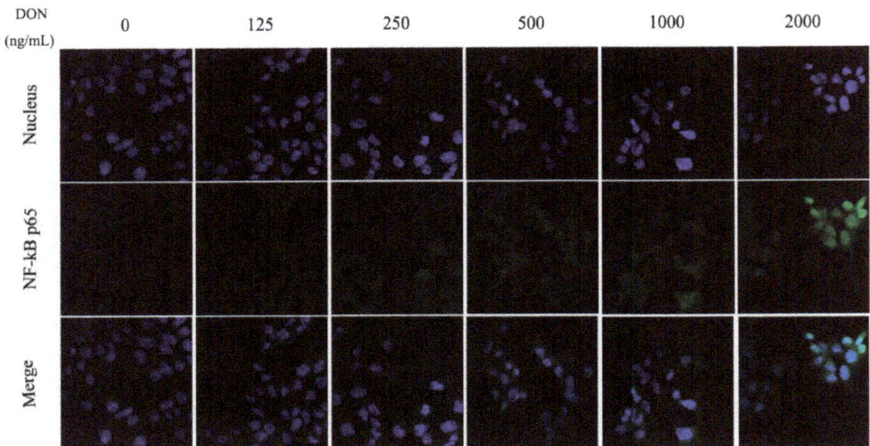

Figure 8. Effect of DON on the nuclear expression of NF-κB p65 protein in cells (800×).

3. Discussion

Many studies have found that zearalenone (ZEA) and DON can reduce the survival rate of IPEC-J2 cells with a certain time and dose dependence manner [13,25]. Research has shown that DON concentrations at 250 and 1000 ng/mL can significantly reduce cell counts in a dose-dependent manner. When the DON concentration is 1000 ng/mL, it caused cell damage, including cell monolayer autolysis, and cell loss [26]. According to the results of this experiment, when the DON concentration in the culture medium was 125 ng/mL, it shows little effect on cell viability. When the concentration increased to 250 ng/mL, the cell survival rate was significantly lower than that of the control group. The cell survival rate decreased significantly, if the culture medium was 250 ng/mL or higher concentration.

Intestinal epithelial cells are not only selectively permeable as a nutrient-absorbing filter, but are also considered to be the first line of defense against foreign antigens, such as pathogens and toxins from the intestinal lumen [27]. DAO is an intracellular enzyme in mammalian intestinal epithelial cells that put forth a protective effect on the intestinal mucosa. When small intestinal mucosal cells are necrotic, DAO in cells will be released [28]. Thus, the extracellular fluid DAO content can reflect the destruction of cells in small intestine. In this experiment, with the higher concentration of DON in the culture solution, the DAO content in the culture supernatant also increased significantly, indicating that DON can damage IPEC-J2 cells and enhance the permeability of the cell membrane [29].

DON influencing intestinal barrier integrity and induced pro-inflammatory cytokines abnormal expression [30]. Previously it was found that DON can cause an increase in the activity of cytokines such as IL-1β, IL-6, and TNF-α in IPEC-J2 cells [31]. In addition, DON also upregulates the activity of TNF-α in IECs cell [30]. According to the test results of the present experiment, the activity of inflammatory mediators in IPEC-J2 cells and the relative expression of *IL-1β* and *IL-6* mRNA in the cells increased with the increasing of DON concentration in the culture medium, while PDTC reduced these trends. This indicated that DON cause an inflammatory response in IPEC-J2 cells and might affect the normal physiological state of the cells via the NF-κB pathway.

Previously, it commonly exists in the shape of a p50/65-IκBα dimer [32]. Different studies showed that when active NF-κB enhanced in a dose-dependent pattern, it initiates to be effective [29,33]. Furthermore, NF-κB pathway regulates COX-2 activation and additional transcription factors [34,35], which demonstrates that this pathway plays a significant role in the inflammatory impairment of the intestine.

In this experiment, the intracellular NO activity and the relative mRNA expression of *iNOS* increased with the dosage of DON, and the relative expression of *COX-2* gene, which is the same as the inducible enzyme, also increased in cells. These results indicate that IPEC-J2 cells are stimulated by DON to produce a corresponding inflammatory response. On the other hand, the NF-κB pathway is primarily thought to be a potential pathogenic factor that exerts deleterious effect on cells when it is excessively or inappropriately activated [24,36]. In our results, the relative mRNA expression of *NF-κB p65*, *IKKα*, and *IKKβ* genes were upregulated, while the mRNA relative expression of *IκB-α* was downregulated to varying degrees. At the same time, the inhibitory changes were inhibited in different degrees.

4. Conclusions

The current study indicated that DON alters IPEC-J2 cells morphology, destroys the submicroscopic structure, boosts cell membrane permeability, and upregulates the transcription of various associated inflammatory factors in cells and alters the expression of NF-κB-related gene or protein in cells. We also concluded that the distribution and content of NF-κB p65 in the intracellular and nucleus further indicated that DON induced inflammatory damage of IPEC-J2 cells through the NF-κB signaling pathway.

5. Materials and Methods

5.1. Chemical and Reagents

DON (CAS No. D0156-5MG) was procured from Sigma (Sigma Chemical Co. St. Louis, MO, USA). Porcine IPEC-J2 cells were obtained from a cell bank in Wuhan academy of agricultural sciences, Wuhan, China. RPMI 1640, SuperScript III kit and Sybr qPCR mix were bought from Thermo Fisher Scientific, Waltham, MA, USA. Fetal bovine serum (FBS) was procured from Clark Bioscience, Richmond, VA, USA. NF-κB inhibitor (Pyrrolidine dithiocarbamate, PDTC) was procured from Beyotime Biotechnology, Shanghai, China. Cell counting kit-8 (CCK-8) kits were obtained from Dojindo Laboratories, Tokyo, Japan. The ELISA kit was bought from Senbeijia Biological Technology, Nanjing, China. BSA was bought from biosharp Company, Beijing, China. The primary NF-κB p65 polyclonal antibody (Product number: 10745-1-AP) was obtained from Proteintech Group, Inc, Rosemont, IL, USA. FITC-goat anti-rabbit IgG antibody (Product number: BA1105) was purchased from Boster Company, Wuhan, China. Trizol reagent was purchased from Invitrogen Biotechnology Co., Ltd., Shanghai, China.

5.2. Cell Culture and Treatments

The intestinal epithelial cells (porcine IPEC-J2) were cultivated in culture bottles (4 × 6 cm) in RPMI 1640 added with 10% (*v*/*v*) FBS, 100 U/mL penicillin, and 100 μg/mL streptomycin, and cultured at 37 °C and a moistened incubator with 5% CO_2. DON solution of 1 mg/mL stock was prepared by liquefying 1 mg DON in 1 mL RPMI 1640 comprising FBS of 10% (*v*/*v*). In this experiment diluted concentrations of 125, 250, 500, 1000, and 2000 ng/mL DON were used [37].

IPEC-J2 cells in logarithmic growing phase (1×10^5 cells/mL) were cultivated in 96-well plates in 100 μL RPMI 1640 for 24 h; and were cured with (0, 125, 250, 500, 1000, and 2000 ng/mL) various DON concentrations for 24 h for the evaluation of cell viability assay. For the estimation of NF-κB pathway in response to DON acquaintance, PDTC, the NF-κB inhibitor was added to the two experimental groups (0, 1000 ng/mL DON) 30 min earlier than the treatment with DON. The supernatants of cell culture were gathered to know DAO releasing. The collected cells were observed for morphological, inflammatory mediator activity, and studies of NF-κB pathway-associated gene or proteins [37].

5.3. IPEC-J2 Cell Morphology by Optical Microscope

The IPEC-J2 cells in the logarithmic growth phase were taken, and the cell density was adjusted to 1×10^5 per mL in a 6-well cell culture plate. After culturing to adherence, the culture solution was changed to a cell culture medium containing different concentrations of DON, and after 24 h, the cell growth of each group was observed under an inverted light microscope (Chongguang Industry Co., Ltd. Chongqing, China).

5.4. IPEC-J2 Cell Morphology by Transmission Electron Microscopy (TEM)

Cells were collected at the bottom of the centrifuge tube and 2.5% glutaraldehyde were used to be fixed in for 4 h, dehydrated, soaked, embedded, ultrathin sections, lead citrate stained, and washed. The cells ultrastructure was detected via a high resolution transmission electron microscope TEOL-2010 (Electronics Corporation, Tokyo, Japan).

5.5. Detection of Cell Viability

IPEC-J2 cells in the logarithmic growth phase were seeded in 96-well plates at about 1000 cells per well, cultured until the cells were attached, and DON was used. After 24 h, 10 μL of CCK-8 reagent was added to each well and cultured for 2 h. The cell viability was calculated by using the absorbance at a wavelength of 450 nm.

5.6. Detection of Inflammatory Mediators and Intestinal Permeability Indicators

The cells were treated with DON and PDTC, and then cultured in an incubator for 24 h. The culture solution was centrifuged at 1000 rpmmin^{-1} for 5 min to obtain a cell culture supernatant. The cells were collected in a cell via a subculture method, and the cell lysate was collected. The NO, IL-6, and TNF-α in the cell lysate were used according to the method of the ELISA kit (Senbeijia Biological Technology, Nanjing, China). The activity and the DAO in the cell culture supernatant was measured, and its activity was calculated according to a self-drawn standard curve.

5.7. Quantitative Real-Time PCR

As per the manufacturer's protocol, total RNA was isolated from cells via a Trizol reagent. Nanodrop lite (Thermo Inc, Waltham, MA, USA) was used to examine the concentrations of RNA. The reverse transcription was accomplished via Super-Script III First-strand cDNA Synthesis Mix (Thermo Inc, USA). Real-time PCR was performed with SybrGreen qPCR Mastermix (Thermo Inc, USA). The overall samples were assayed three times. The reaction mixtures were incubated in a 7900 fast real-time PCR system (Applied Biosystems, Foster City, CA, USA)). The program comprised of 1 cycle at 95 °C for 120 s, 40 cycles at 94 °C for 20 s, 60 °C for 20 s, and 72 °C for 30 s. The gene relative expression levels were calculated according to the $2^{-\Delta\Delta CT}$ method. In real-time PCR analysis, β-actin was used as a housekeeping gene to estimate levels of mRNA for normalization. The primer sequences were synthesized by Sangon Biotech Co., Ltd. (Shanghai, China) and described in Table 1.

Table 1. Parameters of primer for inflammatory cytokines and β-actin genes.

Genes	Accession Number	Primers	Sequences (5′–3′)	Production Size/bp
β-actin	AY_550069.1	Forward	AGATCAAGATCATCGCGCCT	170
		Reverse	ATGCAACTAACAGTCCGCCT	
IL-1β	NW_018085011.1	Forward	TCAGCACCTCTCAAGCAGAA	120
		Reverse	GACCCTCTGGGTATGGCTTT	
IL-6	NC_010451.4	Forward	CTGCAGTCACAGAACGAGTG	131
		Reverse	GACGGCATCAATCTCAGGTG	
NF-κB p65	NM_001114281.1	Forward	GGGGCGATGAGATCTTCCTG	110
		Reverse	CACGTCGGCTTGTGAAAAGG	
IκB-α	NC_010449.5	Forward	GGAGTACGAGCAGATGGTGA	157
		Reverse	TTCCATGGTCAGTGCCTTCT	
iNOS	NC_010454.4	Forward	GGGTCAGAGCTACCATCCTC	114
		Reverse	CGTCCATGCAGAGAACCTTG	

Table 1. *Cont.*

Genes	Accession Number	Primers	Sequences (5′–3′)	Production Size/bp
IKKα	NC_010456.5	Forward Reverse	CACTCTTACAGCGACAGCAC CCACCTTGGGCAGTAGATCA	145
IKKβ	NT_176339.1	Forward Reverse	ACCTGGCTCCCAACGACTT AGATCCCGATGGATGATTCTG	184
COX-2	NC_010451.4	Forward Reverse	TGCGGGAACATAATAGAG GTATCAGCCTGCTCGTCT	90

5.8. Immunofluorescence

Phosphoryl-NF-κB p65 localization was quantified via an immunofluorescence technique. Paraformaldehyde (*v*/*v*, 1/25) was used for IPEC-J2 cells fixation for 30 min at a temperature of 37 °C. After a PBS wash (0.1 mM, pH7.4), permeabilised in 0.5% Triton (Triton×100, Sigma, Harz Lower Saxony, Germany) for 20 min, and for 20 min blocked with 5% BSA, as well as hatched with the anti-phosphoryl-NF-κB p65 antibody (diluted 1:100) for the whole night at a temperature of 4 °C. After washing with PBS (0.1 mM, pH7.4) for the second time, the secondary antibody was used to incubate the cells for 1 h at room temperature. Coverslips were washed two times via PBS (0.1 mM, pH7.4), and hatched with the goat anti-rabbit IgG antibody for 1 h in the absence of light, and hatched in a DAPI staining solution for 10 min. After that it was washed again in PBS. The fluorescence was monitored using an Olympus-fluoview ver.3.1 viewer (Olympus Corporation, Miyazaki Prefecture, Kyushu, Japan) [38].

5.9. Electrophoretic Mobility Shift Assays (EMSAs)

NF-κB DNA-binding activity was examined by EMSA. The cytoplasmic and nuclear protein extraction kit (Jiangsu KeyGEN BioTECH Corp., Ltd., Nanjing, China) was used to prepare nuclear extract. The consensus nucleotide sequence for NF-κB was 5′-AGT TGA GGG GAC TTT CCC AGG C-3′. The EMSA binding reaction was performed by the EMSA kit (Jiangsu KeyGEN BioTECH Corp., Ltd.). A nuclear extract was incubated in a 5× binding reaction buffer containing the biotinylated probe. After incubated at room temperature for 20 min, the reaction mixture was electrophoresed on a non-denaturing 6.5% polyacrylamide gel and then transferred to a nylon membrane. The shifted mixture and the membrane were UV-cross-linked and the ECL kit used to detect was obtained from (Jiangsu KeyGEN BioTECH Corp., Ltd.). For the super shift assay, 1 μg of antibody against NF-κB p65 was added together with the nuclear extract.

5.10. Statistics Analysis

For the calculation of protein expression of average optical density (OD), the Image-Proplus 6.0 Analysis Software (Media Cybernetics, Shanghai, China) were used. The obtained data were presented as means ± SD (n = 10). Statistical analysis was performed by the Statistical Program for Social Sciences (SPSS) software version 19.0 (IBM Corporation, Armonk, NY, USA). Analysis of variance (ANOVA) was performed for the multiple comparison of different groups. The histogram was designed by using the software of Graph Pad Prism version 5.0 (San Diego, California, CA, USA).

Author Contributions: Conceptualization, X.W.; Software, Y.H. and S.U.R.; Formal analysis, L.C.; Investigation, Y.L. and J.W.; Resources, X.W.; Data curation, Y.Z.; Writing—original draft, Y.Z.; Writing—review and editing, J.Z.; Project administration, X.W.; Funding acquisition, L.Z., X.C., and S.F.

Funding: The present study was supported by the National Natural Science Foundation of China (grant No. 31472250) and the Project of Modern Agricultural Industry and Technology System of Anhui Province (grant No. AHCYJSTX-05-07). We wish to thank anonymous reviewers for their kind advice.

Conflicts of Interest: The authors declare that they have no conflict of interest.

References

1. Berthiller, F.; Crews, C.; Dall'Asta, C.; Saeger, S.D.; Haesaert, G.; Karlovsky, P.; Oswald, I.P.; Seefelder, W.; Speijers, G.; Stroka, J. Masked mycotoxins: A review. *Mol. Nutr. Food Res.* **2013**, *57*, 165–186. [CrossRef]
2. Pestka, J. Deoxynivalenol: Toxicity, mechanisms and animal health risks. *Anim. Feed Sci. Technol.* **2007**, *137*, 283–298. [CrossRef]
3. Mesterházy, Á.; Bartók, T.; Mirocha, C.G.; Komoroczy, R. Nature of wheat resistance to fusarium head blight and the role of deoxynivalenol for breeding. *Plant Breed.* **2010**, *118*, 97–110. [CrossRef]
4. Rui, L.I.; Wang, X.U.; Zhou, T.; Yang, D.; Wang, Q.I.; Zhou, Y.U. Occurrence of four mycotoxins in cereal and oil products in Yangtze Delta region of China and their food safety risks. *Food Control* **2014**, *35*, 117–122.
5. Ma, R.; Zhang, L.; Liu, M.; Su, Y.T.; Xie, W.M.; Zhang, N.Y.; Dai, J.F.; Wang, Y.; Rajput, S.A.; Qi, D.S. Individual and combined occurrence of mycotoxins in feed ingredients and complete feeds in China. *Toxins* **2018**, *10*, 113. [CrossRef] [PubMed]
6. Wu, L.; Liao, P.; He, L.; Ren, W.; Yin, J.; Duan, J.; Li, T. Growth performance, serum biochemical profile, jejunal morphology, and the expression of nutrients transporter genes in deoxynivalenol (DON)- challenged growing pigs. *BMC Vet. Res.* **2015**, *11*, 144. [CrossRef] [PubMed]
7. Weaver, A.C.; Todd, S.M.; Hansen, J.A.; Kim, Y.B.; De, S.A.L.P.; Middleton, T.F.; Woo, K.S. The use of feed additives to reduce the effects of aflatoxin and deoxynivalenol on pig growth, organ health and immune status during chronic exposure. *Toxins* **2013**, *5*, 1261–1281. [CrossRef] [PubMed]
8. Mishra, S.; Srivastava, S.; Dewangan, J.; Divakar, A.; Rath, S.K. Global occurrence of deoxynivalenol in food commodities and exposure risk assessment in humans in the last decade: A survey. *Crit. Rev. Food Sci. Nutr.* **2019**, *14*, 1–29. [CrossRef] [PubMed]
9. Wang, X.; Fan, M.; Chu, X.; Zhang, Y.; Rahman, S.U.; Jiang, Y.; Chen, X.; Zhu, D.; Feng, S.; Li, Y. Deoxynivalenol induces toxicity and apoptosis in piglet hippocampal nerve cells via the MAPK signaling pathway. *Toxicon* **2018**, *155*, 1–8. [CrossRef] [PubMed]
10. Manda, G.; Neagu, M.; Constantin, C.; Neagoe, I.; Marin, D.; Taranu, I. Immunotoxicology of mycotoxins produced by fusarium fungi—low concentrations of deoxynivalenol interfere with nucleotide metabolism. *Toxicol. Lett.* **2007**, *172*, 49. [CrossRef]
11. Maresca, M. From the Gut to the Brain: Journey and pathophysiological effects of the Food-Associated trichothecene mycotoxin deoxynivalenol. *Toxins* **2013**, *5*, 784–820. [CrossRef] [PubMed]
12. Rotter, B.A.; Thompson, B.K.; Lessard, M.; Trenholm, H.L.; Tryphonas, H. Influence of low-level exposure to fusarium mycotoxins on selected immunological and hematological parameters in young swine. fundamental & applied toxicology official. *Soc. Toxicol.* **1994**, *23*, 117.
13. Bouhet, S.; Oswald, I.P. The effects of mycotoxins, fungal food contaminants, on the intestinal epithelial cell-derived innate immune response. *Vet. Immunol. Immunopathol.* **2005**, *108*, 199–209. [CrossRef] [PubMed]
14. Zhang, Z.Q.; Wang, S.B.; Wang, R.G.; Zhang, W.; Wang, P.L.; Su, X.O. Phosphoproteome analysis reveals the molecular mechanisms underlying deoxynivalenol-induced intestinal toxicity in IPEC-J2 cells. *Toxins* **2016**, *8*, 270. [CrossRef]
15. Lallès, J.P.; Boudry, G.; Favier, C.; Floc'H, N.L.; Huêrou-Luron, I.L.; Montagne, L.; Oswald, I.P.; Pié, S.; Piel, C.; Sève, B. Gut function and dysfunction in young pigs: Physiology. *Physiology* **2002**, *53*, 301–316. [CrossRef]
16. Schierack, P.; Nordhoff, M.; Pollmann, M.; Weyrauch, K.D.; Amasheh, S.; Lodemann, U.; Jores, J.; Tachu, B.; Kleta, S.; Blikslager, A. Characterization of a porcine intestinal epithelial cell line for in vitro studies of microbial pathogenesis in swine. *Histochem. Cell Biol.* **2006**, *125*, 293–305. [CrossRef] [PubMed]
17. Diesing, A.K.; Nossol, C.; Panther, P.; Walk, N.; Post, A.; Kluess, J.; Kreutzmann, P.; Dänicke, S.; Rothkötter, H.J.; Kahlert, S. Mycotoxin deoxynivalenol (DON) mediates biphasic cellular response in intestinal porcine epithelial cell lines IPEC-1 and IPEC-J2. *Toxicol. Lett.* **2011**, *200*, 8–18. [CrossRef]
18. Okumura, R.; Takeda, K. Roles of intestinal epithelial cells in the maintenance of gut homeostasis. *Exp. Mol. Med.* **2017**, *49*, e338. [CrossRef]
19. Takahashi, I. Mucosal immune system: The second way of the host defense. *Nihon Rinsho* **2007**, *65 Pt 1* (Suppl. 2), 102–108.
20. Welsh, D.A.; Mason, C.M. Host defense in respiratory infections. *Med. Clin. N. Am.* **2001**, *85*, 1329–1347. [CrossRef]

21. Pitman, R.S.; Blumberg, R.S. First line of defense: The role of the intestinal epithelium as an active component of the mucosal immune system. *J. Gastroenterol.* **2000**, *35*, 805–814. [CrossRef] [PubMed]
22. Brosnahan, A.J.; Brown, D.R. Porcine IPEC-J2 intestinal epithelial cells in microbiological investigations. *Vet. Microbiol.* **2012**, *156*, 229–237. [CrossRef] [PubMed]
23. Fan, W.J.; Li, H.P.; Zhu, H.S.; Sui, S.P.; Chen, P.G.; Deng, Y.; Sui, T.M.; Wang, Y.Y. NF-κB is involved in the LPS-mediated proliferation and apoptosis of MAC-T epithelial cells as part of the subacute ruminal acidosis response in cows. *Biotechnol. Lett.* **2016**, *38*, 1839–1849. [CrossRef] [PubMed]
24. Yamamoto, Y.; Gaynor, R.B. Therapeutic potential of inhibition of the NF-κB pathway in the treatment of inflammation and cancer. *J. Clin. Investig.* **2001**, *107*, 135–142. [CrossRef]
25. Li-Ge, Y.; Yang, M.U.; Wei, Z.; Jian-Ping, L.I.; An-Shan, S. Effect of ZEN on mRNA expression and activity of inflammation related factors in IPEC-J2 cells. *Chin. J. Anim. Sci.* **2017**, *53*, 75–81.
26. Awad, W.A.; Aschenbach, J.R.; Zentek, J. Cytotoxicity and metabolic stress induced by deoxynivalenol in the porcine intestinal IPEC-J2 cell line. *J. Anim. Physiol. Anim. Nutr.* **2012**, *96*, 709–716. [CrossRef] [PubMed]
27. Jeong, G.M.; Kwang, S.S.; Sung Moo, P.; In Kyu, L.; Cheol-Heui, Y. Bacillus subtilis protects porcine intestinal barrier from deoxynivalenol via improved zonula occludens-1 expression. *Asian-Australas. Anim. Sci.* **2014**, *27*, 580–586.
28. Namikawa, T.; Fukudome, I.; Kitagawa, H.; Okabayashi, T.; Kobayashi, M.; Hanazaki, K. Plasma diamine oxidase activity is a useful biomarker for evaluating gastrointestinal tract toxicities during chemotherapy with oral fluorouracil anti-cancer drugs in patients with gastric cancer. *Oncology* **2012**, *82*, 147–152. [CrossRef]
29. Wang, X.C.; Zhang, Y.F.; Cao, L.; Zhu, L.; Huang, Y.Y.; Chen, X.F.; Chu, X.Y.; Zhu, D.F.; Sajid, U.R.; Feng, S.B.; et al. Deoxynivalenol induces intestinal damage and inflammatory response through the nuclear factor-κB signaling pathway in piglets. *Toxins* **2019**, *11*, 663. [CrossRef]
30. Vandenbroucke, V.; Croubels, S.; An, M.; Verbrugghe, E.; Goossens, J.; Deun, K.V.; Boyen, F.; Thompson, A.; Shearer, N.; Backer, P.D. The mycotoxin deoxynivalenol potentiates intestinal inflammation by salmonella typhimurium in porcine ileal loops. *PLoS ONE* **2011**, *6*, e23871. [CrossRef]
31. Awad, W.; Ghareeb, K.; Böhm, J.; Zentek, J. The toxicological impacts of the fusarium mycotoxin, deoxynivalenol, in poultry flocks with special reference to immunotoxicity. *Toxins* **2013**, *5*, 912–925. [CrossRef] [PubMed]
32. Giridharan, S.; Srinivasan, M. Mechanisms of NF-κB p65 and strategies for therapeutic manipulation. *Inflamm. Res.* **2018**, *11*, 407–419. [CrossRef] [PubMed]
33. Van De Walle, J.; Romier, B.; Larondelle, Y.; Schneider, Y.J. Influence of deoxynivalenol on NF-κB activation and IL-8 secretion in human intestinal Caco-2 cells. *Toxicol. Lett.* **2008**, *177*, 205–214. [CrossRef] [PubMed]
34. Benitash, S.A.; Valeròn, P.F.; Lacal, J.C. ROCK and nuclear factor-kappa B-dependent activation of cyclooxygenase-2 by Rho GTPases: Effects on tumor growth and therapeutic cinsequences. *Mol. Biol. Cell* **2003**, *14*, 3041–3054. [CrossRef] [PubMed]
35. Yao, C.; Li, G.; Cai, M.; Qian, Y.Y.; Wang, L.Q. Prostate cancer downregulated SIRP-alpha modulates apoptosis and proliferation through p38-MAPK/NF-kappa B/COX-2 signaling. *Oncol. Lett.* **2017**, *13*, 4995–5001. [CrossRef] [PubMed]
36. Adesso, S.; Autore, G.; Quaroni, A.; Popolo, A.; Severino, L.; Marzocco, S. The food contaminants nivalenol and deoxynivalenol induce inflammation in intestinal epithelial cells by regulating reactive oxygen species release. *Nutrients* **2017**, *9*, 1343. [CrossRef] [PubMed]
37. Wang, X.C.; Xu, W.; Fan, M.X.; Meng, T.; Chen, X.; Jiang, Y.; Zhu, D.; Hu, W.; Gong, J.; Feng, S.; et al. Deoxynivalenol induces apoptosis in PC12 cells via the mitochondrial pathway. Environ. *Toxicol. Pharm.* **2016**, *43*, 193–202. [CrossRef] [PubMed]
38. Wan, D.; Wu, Q.H.; Qu, W.; Liu, G.; Wang, X. Pyrrolidine dithiocarbamate (PDTC) inhibits DON-induced mitochondrial dysfunction and apoptosis via the NF-κB/iNOS pathway. *Oxid. Med. Cell. Longev.* **2018**, *2018*, 1324173. [CrossRef]

© 2019 by the authors. Licensee MDPI, Basel, Switzerland. This article is an open access article distributed under the terms and conditions of the Creative Commons Attribution (CC BY) license (http://creativecommons.org/licenses/by/4.0/).

Article

Effects of Deoxynivalenol and Zearalenone on the Histology and Ultrastructure of Pig Liver

Natalia Skiepko [1], Barbara Przybylska-Gornowicz [1], Magdalena Gajęcka [2], Maciej Gajęcki [2] and Bogdan Lewczuk [1,*]

[1] Department of Histology and Embryology, Faculty of Veterinary Medicine, University of Warmia and Mazury in Olsztyn, Oczapowskiego 13, 10-719 Olsztyn, Poland; natalia.skiepko@uwm.edu.pl (N.S.); przybyl@uwm.edu.pl (B.P.-G.)

[2] Department of Veterinary Prevention and Feed Hygiene, Faculty of Veterinary Medicine, University of Warmia and Mazury in Olsztyn, Oczapowskiego 13, 10-719 Olsztyn, Poland; mgaja@uwm.edu.pl (M.G.); gajecki@uwm.edu.pl (M.G.)

* Correspondence: lewczukb@uwm.edu.pl; Tel.: +48-89-523-39-49; Fax: +48-89-523-34-40

Received: 29 June 2020; Accepted: 17 July 2020; Published: 20 July 2020

Abstract: The purpose of this study was to determine the effects of single and combined administrations of deoxynivalenol (DON) and zearalenone (ZEN) on the histology and ultrastructure of pig liver. The study was performed on immature gilts, which were divided into four equal groups. Animals in the experimental groups received DON at a dose of 12 μg/kg body weight (BW) per day, ZEN at 40 μg/kg BW per day, or a mixture of DON (12 μg/kg BW per day) and ZEN (40 μg/kg BW). The control group received vehicle. The animals were killed after 1, 3, and 6 weeks of experiment. Treatment with mycotoxins resulted in several changes in liver histology and ultrastructure, including: (1) an increase in the thickness of the perilobular connective tissue and its penetration to the lobules in gilts receiving DON and DON + ZEN; (2) an increase in the total microscopic liver score (histology activity index (HAI)) in pigs receiving DON and DON + ZEN; (3) dilatation of hepatic sinusoids in pigs receiving ZEN, DON and DON + ZEN; (4) temporary changes in glycogen content in all experimental groups; (5) an increase in iron accumulation in the hepatocytes of gilts treated with ZEN and DON + ZEN; (6) changes in endoplasmic reticulum organization in the hepatocytes of pigs receiving toxins; (7) changes in morphology of Browicz–Kupffer cells after treatment with ZEN, DON, and DON + ZEN. The results show that low doses of mycotoxins used in the present study, even when applied for a short period, affected liver morphology.

Keywords: zearalenone; deoxynivalenol; mycotoxins; histology; ultrastructure; pig; hepatocyte; liver

Key Contribution: Low doses of deoxynivalenol and zearalenone induce significant changes in hepatocytes, Browicz–Kupffer cells, and perilobular connective tissue.

1. Introduction

Mycotoxins, secondary fungal metabolites, are frequent contaminants of cereals and cereal products. Delivered via these plants and products, they pose a serious health threat to humans and animals [1–3]. Whereas mycotoxins are sometimes regarded as stressors [4,5], they have a target-specific mode-of-action and so are true toxins rather than stressors per se [6]. The most common and important mycotoxins in Europe, produced by fungi of the *Fusarium* family, are deoxynivalenol (DON), and zearalenone (ZEN). Pigs are particularly sensitive to DON [7], but they also show high sensitivity to ZEN [8].

In pigs, high doses of DON cause reduced appetite, complete anorexia, vomiting [9], and reproductive disorders [10]. Absorption of DON is rapid, and the toxin reaches the peak plasma concentrations within 30 min of oral administration [11]. The majority of ingested DON is absorbed in the proximal part of the small intestine [12]. In the liver, DON is metabolized into de-epoxy-DON [13]. An evidence showed that chronic ingestion of DON at low doses, which is clinically asymptomatic, alters the mucosal epithelial cells and villi of the small intestine [14–17] and affects the defense mechanisms of the large intestine [18].

ZEN and its metabolites are agonists of estrogen receptors, and they compete with endogenous hormones for the binding sites of estrogen receptors [2,19]. Treatment with ZEN leads to precocious puberty, reproductive disorders, and hyperestrogenism [20]. According to previous studies, domestic pigs are particularly sensitive to the estrogenic effect of ZEN owing to the very rapid and large (approximately 80–85%) absorption of the toxin in their digestive system [21]. In the liver, ZEN is metabolized into α- and β-zearalenol, which are considered more toxic than ZEN. The ratio of these two metabolites to each other is species-specific. Most studies showed that α-zearalenol predominates in pigs [13,21–25]. This form is more active than β-zearalenol, which provides another explanation for the high sensitivity of pigs to ZEN content in feed [21,24]. ZEN at low doses has deleterious effects on the morphology of the small intestine [16,17] and affects the defense mechanisms of the liver [26] and the large intestine [18].

During exposure to environmental mycotoxins, animals often encounter mixtures of toxins, rather than single toxins. Therefore, the toxicity of mycotoxins needs to be addressed in the context of their mixtures to assess their health risk [27–29]. However, studies on the effects of mycotoxin combinations are relatively rare, and their results are very ambiguous [27,28,30,31].

The liver and gastrointestinal parts of the digestive system are major sites of mycotoxin metabolism [32]. However, there is little molecular, metabolic, and histological research on the effects of mycotoxins on the liver. In addition, the outcomes of these studies are not conclusive [15,28]. In pigs, some histological changes have been observed in the liver as a result of the individual or combined effects of DON and ZEN [15,33]. Contrastingly, Renner and coauthors [34] showed that chronic dietary DON exposure (4 weeks, 4.59 μg/kg BW) did not affect the histology of pig liver.

The purpose of this study was to determine the effects of single and combined administrations of DON (12 μg/kg BW) and ZEN (40 μg/kg BW) to pigs for 1, 3, or 6 weeks on the histology and ultrastructure of the liver. The selection of the doses used has been widely discussed in our previously published papers [16–18].

2. Results and Discussion

2.1. Light Microscopy Study

2.1.1. Architecture of the Liver

The livers of control pigs had the typical structural characteristic for this species (Figures 1A and 2A). The hepatic lobules were neatly outlined by an envelope of fibrous connective tissue, which interconnected the portal areas. Within each lobule, hepatocytes were arranged in linear cords radiating from the central vein and separated by sinusoids, which had a uniform diameter over their entire length. The limiting plate of hepatocytes bordered the lobule interior from the connective tissue.

Prominent qualitative differences in the liver architecture were observed between the control group and the DON and DON + ZEN groups. The perilobular connective tissue widened and contained greater amounts of collagen fibers starting from the first week of treatment (Figures 1B and 2B, Table 1). In addition, connective tissue penetrated into the lobule and caused disruption of the limiting plate (Figure 1C). After 6 weeks of treatment, the presence of small scars was pronounced (Figure 1D). The disorganization of hepatic cords was observed in the livers of pigs fed diets contaminated with DON and DON + ZEN for 3 and 6 weeks.

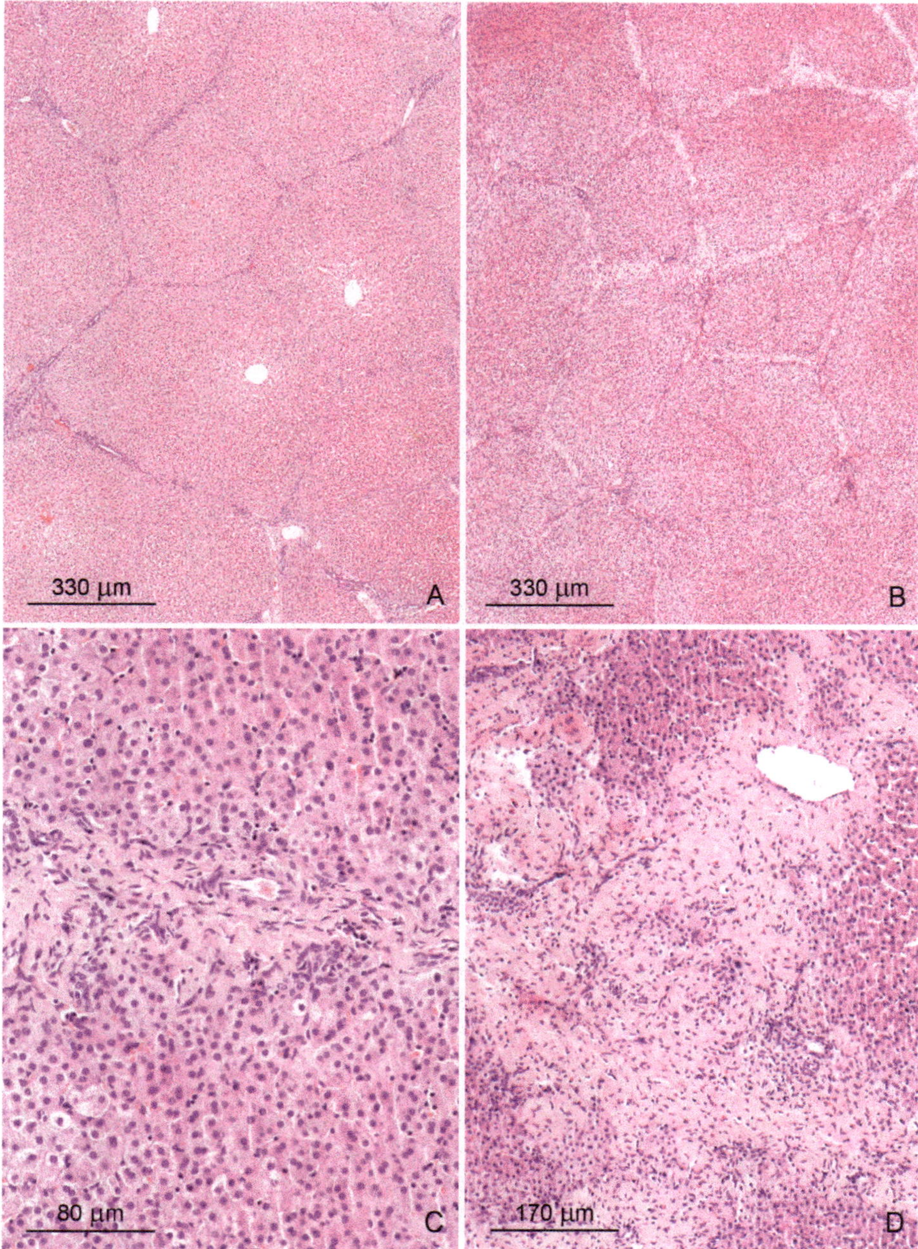

Figure 1. (**A**,**B**) Architecture of the liver in a control pig, 6th week of experiment (**A**) and in a pig receiving deoxynivalenol (DON) for 6 weeks (**B**). Note the thickening of the interlobular septa. (**C**) A strip of connective tissue in the liver lobule. A pig was treated with DON for 3 weeks. (**D**) Focal fibrosis inside the liver lobule of a pig treated with DON for 6 weeks. Note that the central vein is surrounded by connective tissue. Figures (**A**–**D**) show hematoxylin and eosin stained sections.

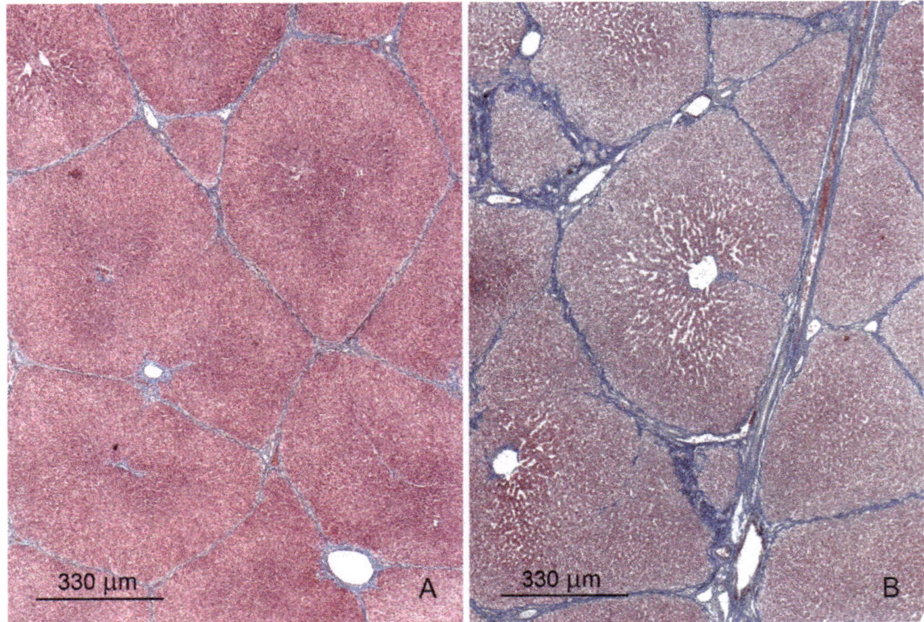

Figure 2. (**A**,**B**) Trichrome staining of the liver in a control pig, 6th week of experiment (**A**) and in a pig receiving DON for 6 weeks (**B**). Pay attention to the thickness of the interlobular septa and the content of collagen fibers in them.

Table 1. Semiquantitative analysis of fibrosis.

Duration of Mycotoxin Treatment	Group				Testing Results $p < 0.05$
	Control (C)	ZEN (Z)	DON (D)	DON + ZEN (M)	
1 week	0	0	1.0	2.67	C, Z < D; D < M
3 weeks	0	0	1.0	2.3	C, Z < D; D < M
6 weeks	0	0	1.33	3.0	C, Z < D; D < M

Description of the score used: 0, no fibrosis, perilobular tissue septa with small number of collagen, characteristic for the pig liver; 1, mild enlargement of portal areas and perilobular tissue septa; 2, large enlargement of portal areas and perilobular tissue septa, limiting plate fibrosis; 3, partial cirrhosis of some lobules. The values presented are means.

The administration of DON and DON + ZEN significantly increased the thickness of perilobular connective tissue in the liver after 3 and 6 weeks of treatment. No significant differences were observed in the thickness of septa after ZEN treatment (Figure 3A). The cross-sectional area of the lobules did not differ between the control group and the groups treated with mycotoxins (Figure 3B).

The data show that DON and DON + ZEN, but not ZEN, affected the qualitative and quantitative characteristics of the liver architecture. The administration of DON and DON + ZEN resulted in an excessive accumulation of extracellular matrix, including collagen, in perilobular connective tissue, starting from the first week of treatment. Moreover, treatment with DON + ZEN for 3 and 6 weeks caused a progressive accumulation of extracellular matrix in the liver parenchyma. We interpreted the intra-lobular presence of mild scars as early fibrosis. The process of liver fibrosis is most often a consequence of chronic diseases, but it can also be caused by many factors that induce acute damage [35,36]. In experimental studies, administration of xenobiotics induced parenchymal fibrosis similar to cirrhosis in rodent liver [36]. The intensity of the fibrosis process caused by external factors shows large differences [37]. Previous studies have shown that mycotoxins affect the histology of pig liver [15,34,38]; however, the authors did not report the process of liver fibrosis.

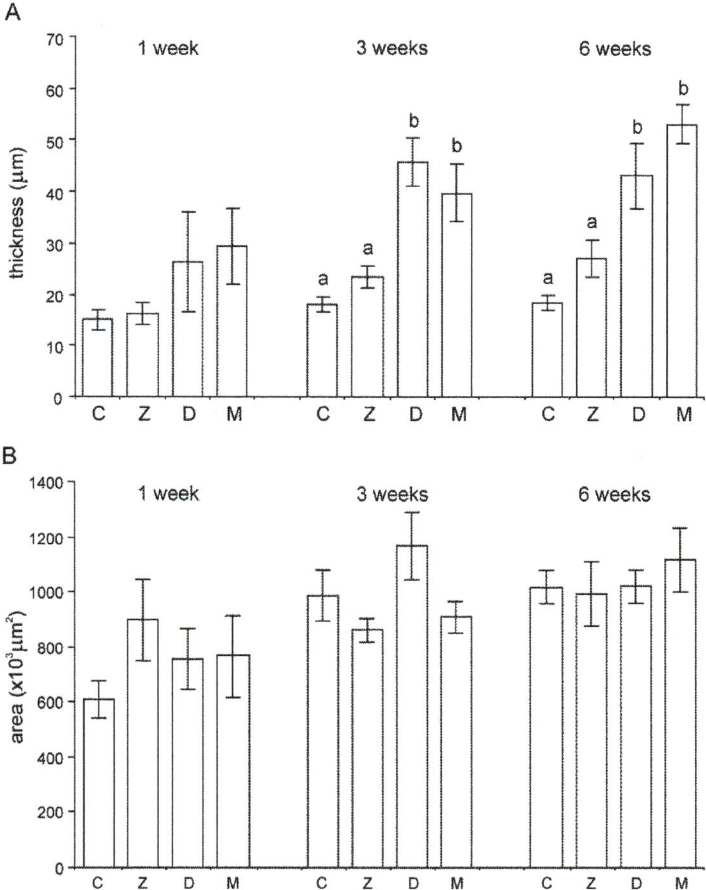

Figure 3. (**A**) Thickness of the connective tissue septa. (**B**) Cross-sectional area of the liver lobules. C—control group, Z—group treated with zearalenone (ZEN), D—group treated with deoxynivalenol (DON), and M—group treated with DON and ZEN. The values are presented as mean ± standard deviation. Bars labeled with different small lower-case letters differ significantly at $p \leq 0.05$.

Disorganization of hepatic cords as one of the main lesions in piglets after chronic exposure to DON was observed by Gerez et al. [15]. In our study, similar changes were found after 6 weeks of DON and DON + ZEN ingestion.

2.1.2. Microscopic Liver Scoring

The analysis using modified microscopic liver scoring (histology activity index (HAI)) comprised six histological criteria. Exemplary microphotographs of changes included in the individual criteria are presented in Figure S1 (Supplementary Material). The cumulative score and the contribution of each criterion are presented in Figure 4. The largest histopathological lesions of the liver were observed in pigs treated with DON for 1 week or DON + ZEN for 1, 3, 6 weeks. The total HAI score for DON resulted mainly from increased portal inflammation and focal lytic necrosis, whereas that for DON +

ZEN resulted from increased portal and periportal inflammation, confluent necrosis, and especially focal lytic necrosis, compared with control pigs. The total HAI scores in pigs receiving DON and DON + ZEN for 1, 3, and 6 weeks were significantly higher than in that in the control group.

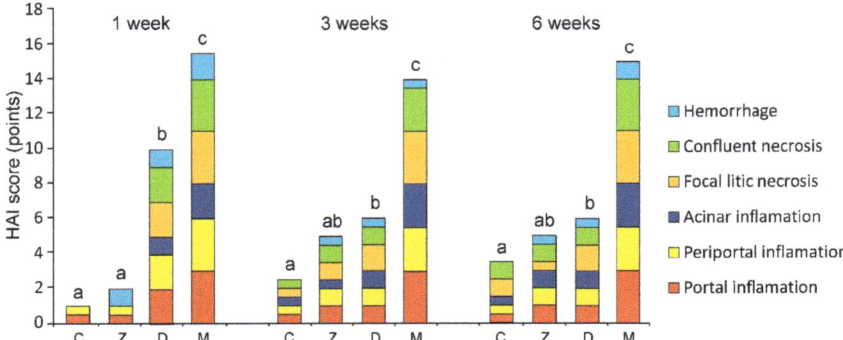

Figure 4. Histopathological score of the examined livers according to the histology activity index (HAI) modified by Stanek et al. [38]. Six histopathological parameters were scored in hematoxylin and eosin (HE)-stained tissues (see Figure S1 in Supplementary material), and the cumulative HAI score was calculated for each experimental group. The mean value of the cumulative HAI scores in each group is represented by the total height of the bar. C—control group, Z—group treated with zearalenone (ZEN), D—group treated with deoxynivalenol (DON), and M—group treated with DON and ZEN. The values of the cumulative HAI score labeled with different lower-case letters above the bars differ significantly at $p \leq 0.05$.

The foci of hepatocyte necrosis with lymphocytic infiltrates, portal, periportal, and acinar inflammation occur in the liver as a result of drug usage, intoxication, or viral and bacterial infections [39]. The occurrence of necrosis is also associated with the activation of immune mechanisms in response to the adverse effects of toxic substances [36,40]. These changes are generally associated with pathological symptoms; however, they also occur in seemingly healthy animals, and, in this case, they are considered as preclinical conditions [41].

The results of our HAI analysis showed that the liver, being the first metabolic station, is affected to varying degrees by the administered toxins. Treatment with ZEN had no significant effect on liver histopathology. However, DON and DON + ZEN significantly affected pig liver regardless of the administration period of toxins. The highest intensity of necrosis foci was observed in the livers of animals receiving both mycotoxins. The results obtained may be interpreted as an effect of the toxic effects of DON and the synergistic toxic effect of the combined toxins. In addition, they may be caused by the stimulation of inflammatory processes in response to the administered mycotoxins. This is due to the pro-inflammatory actions of DON and ZEN [42].

The results of previous studies on the effect of mycotoxins on the histology of the liver parenchyma are diverse. Histological changes, including the disorganization of hepatic cords, the cytoplasmic vacuolization of hepatocytes, megalocytosis, and focal necrosis, were reported in pigs subjected to 28 days of diet contaminated with DON (3 mg/kg) or DON (3 mg/kg) + NIV (1,5 mg/kg) + ZEN (1,5 mg/kg) [15]. Acute exposure to DON at a dose of 1 mg/kg BW for 6 and 24 h led to apoptosis of hepatocytes [33]. On the other hand, exposure to DON at a concentration of 4.59 mg/kg feed for 28 days [34] or 3.1 mg/kg feed for 37 days did not affect liver HAI [38].

Megalocytosis and cytoplasmic vacuolization were the main histological lesions reported in piglets subjected to chronic exposure to DON [15]. Magalocytosis potentially indicates irreversible hepatocyte injury [43]. However, in our study, this phenomenon concerned only a few hepatocytes, and there were no differences in their presence between control and experimental pigs.

2.1.3. Hepatic Sinusoids

The livers of pigs from all experimental groups were characterized by significant dilatation of the hepatic sinusoids in Zone III (drainage) of acinus compared with those of control animals (Table 2; Supplementary material, Figure S2). The largest sinusoidal dilatation was observed in pigs receiving DON + ZEN. In these animals, significant sinusoidal dilatation occurred after 1, 3, and 6 weeks of mycotoxin administration. In the groups treated with ZEN alone and DON alone, sinusoidal dilatation decreased after 3 and 6 weeks (Table 2).

Table 2. Semiquantitative analysis of sinusoidal dilatation.

Duration of Mycotoxin Treatment	Group				Testing Results $p < 0.05$
	Control (C)	ZEN (Z)	DON (D)	DON + ZEN (M)	
1 week	0	3.00	3.67	3.33	C < Z, D, M
3 weeks	0.66	2.33	3.33	3.67	C < Z, D, M; Z < M
6 weeks	0.33	2.0	2.33	3.33	C < Z, D, M; Z, D < M

Description of the score used in estimating changes: 0, absence; 1, sporadic presence; 2, few presence; 3, middle presence; and 4, numerous presence.

Under physiological conditions, the liver sinusoids are characterized by a constant, uniform diameter. Sinusoidal dilatation in Zone I has been observed in pregnancy, long-term steroid administration, and hepatomegaly [44]. This phenomenon also occurs after exposure to vinyl chloride and arsenic [40], and may also be the result of the administration of certain chemotherapy drugs [45]. Sinusoidal dilatation in Zone III was observed in the case of local disorders of blood flow in the portal veins or hepatic veins [40]. In our study, pronounced sinusoidal dilatation in Zone III was observed as a result of DON and ZEN administration, which suggested a disruptive effect of mycotoxins on blood flow in the liver. However, the histological findings did not provide any suggestions about the possible cause of this disturbance.

2.1.4. Glycogen Storage

We used periodic acid-Schiff (PAS) staining to evaluate changes in glycogen content. An increase in glycogen deposits was observed in the groups treated with ZEN, DON and DON + ZEN compared with those in the control group after the first week of experiment (Figure 5, Table 3). The staining was particularly intense in Zone I. After 3 and 6 weeks of treatment, there were no differences in glycogen content between the control and experimental groups.

Table 3. Semiquantitative analysis of glycogen content in the central and peripheral parts of liver lobules.

Time of Mycotoxin Treatment	Groups				Testing Results
	Control (C)	ZEN (Z)	DON (D)	DON + ZEN (M)	
		Central part			
1 week	0	3.00	2.33	2.33	C < Z, D, M
3 weeks	2.66	3.33	2.33	3.00	–
6 weeks	3.00	2.66	3.00	3.33	–
		Peripheral part			
1 week	0	4.00	4.00	3.33	C < Z, D, M
3 weeks	3.66	4.00	4.00	4.00	-
6 weeks	4.00	3.66	4.00	444	-

For a description of the score, see Table 2.

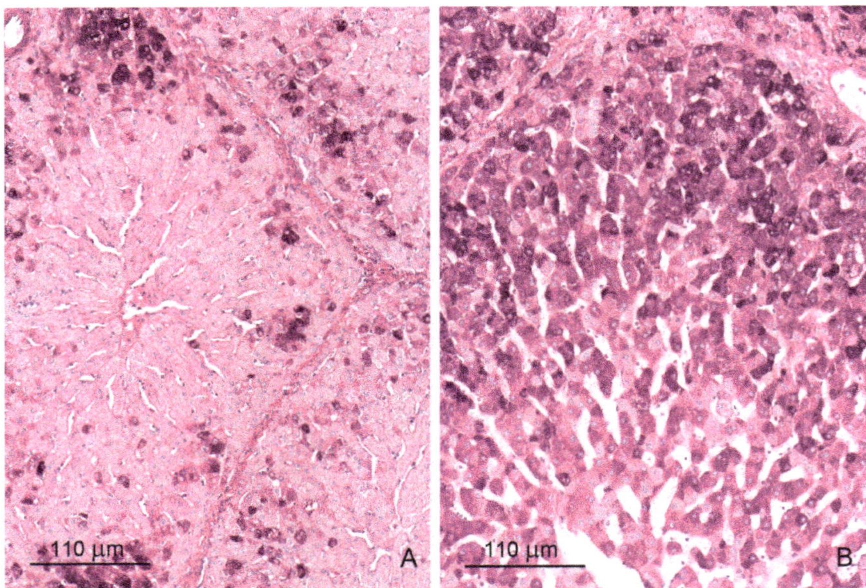

Figure 5. (**A**,**B**) Distribution of glycogen in hepatocytes in a control pig (**A**) and a pig treated with ZEN + DON (**B**). The first week of experiment. Periodic acid-Schiff reaction. Note the increase in the amount of glycogen in a pig treated with mycotoxins.

The liver is a major site of glycogen accumulation, and the main role of glycogen in the liver is to store glucose for release during fasting. The amount of stored glycogen depends on nutritional and living factors and shows large individual fluctuations [36,40]. In previous studies, no effects of mycotoxins on glycogen storage were observed in pig liver [7]. The results of our current studies showed that the effect of DON and ZEN on liver glycogen was temporary.

2.1.5. Iron Deposits

Prussian blue (PB) staining was used for the visualization of iron deposits. In control pigs, small amounts of iron deposits were observed in hepatocytes and Browicz–Kupffer cells in the form of very fine (pollen) granules. Iron-loaded cells are either isolated or grouped together without any lobular systematization. In connective tissues, positive staining for iron corresponds to deposition within fibrocytes or macrophages. In experimental animals, increased amounts of iron deposits were observed both in the liver lobules and perilobular connective tissues, especially after 6 weeks of ZEN and DON + ZEN administration (Figure 6, Table 4).

Table 4. Semiquantitative analysis of iron deposits.

Time of Mycotoxin Treatment	Groups				Testing Results
	Control (C)	ZEN (Z)	DON (D)	DON + ZEN (M)	
One week	1.00	1.00	1.00	1.00	-
Three weeks	1.00	1.66	1.00	2.00	C, D < Z, M
Six weeks	1.33	3.33	1.33	3.33	C, D < Z, M

For a description of the score, see Table 2.

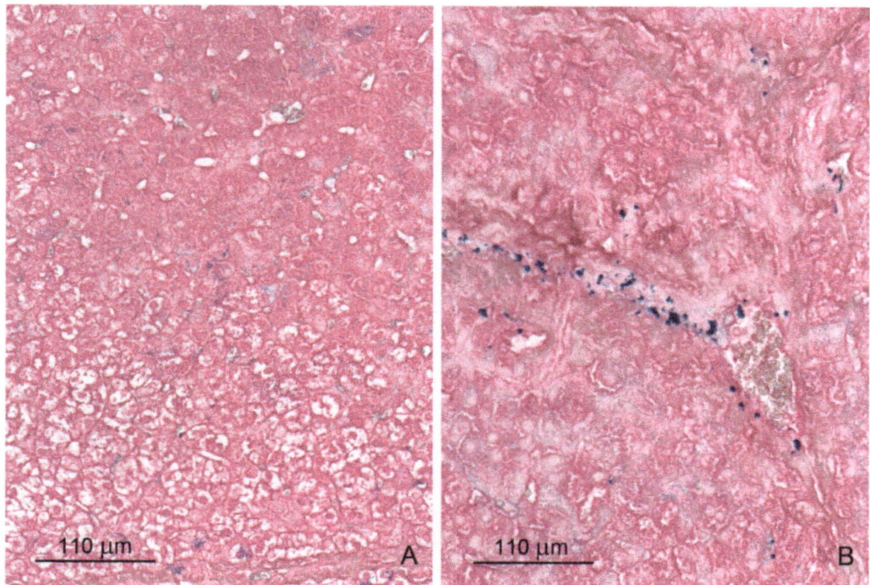

Figure 6. (**A**,**B**) Iron deposits in the liver of pigs receiving ZEN + DON for six weeks. (**A**) Iron deposits in hepatocytes in the form of fine granules. (**B**) Iron deposits with variable sizes in perilobular connective tissue. Prussian blue staining.

Physiologically, iron deposits are present within hepatocytes as fine granules at the biliary pole of cells and are distributed throughout the lobule according to a decreasing gradient from periportal to centrilobular areas. Iron excess is associated with chronic liver diseases of various causes [46–48]. It is known that in states of persistent iron overload, the liver could be seriously affected [49], mainly because of the extremely reactive hydroxyl radical formed during Fe^{2+} metabolism (by Fenton reaction) in lysosomes [50].

Previous studies on the effect of mycotoxins on pig liver did not consider the effect of ZEN on iron storage [7]. Based on the results obtained in our study, it can be stated that ZEN induced duration-dependent iron accumulation in hepatocytes, Browicz–Kupffer cells, and connective tissue cells. This indicated that disruption of iron metabolism in the liver worsened along with the duration of ZEN administration. One possible mechanism is the effect of ZEN on hepcidin, a liver-derived peptide hormone, which is a master regulator of iron metabolism [50]. Our results show a similar effect of food contaminated with DON + ZEN on iron accumulation in the liver, as in the case of ZEN, which suggested that DON had no effect on iron accumulation.

2.2. Ultrastructural Study

2.2.1. Hepatocytes

The hepatocytes of control pigs (Figure 7A) with polygonal or rectangular cross sections were arranged in regular rows. Their nuclei were distinctly round and centrally located, with a predominance of euchromatin. Hepatocytes had abundant amounts of both rough and smooth endoplasmic reticulum (ER). Medium-length cisterns of rough ER were located close to the nucleus and around the mitochondria (Figure 7B). The smooth ER was in the form of branched tubules and vesicles (Figure 7B). Mitochondria with numerous cristae and an electron-dense matrix were distributed in the cytoplasm in clusters or individually (Figure 7A). Lysosomes were localized mainly in the bile pole of hepatocytes. The content of glycogen deposits varied significantly between individual hepatocytes. After the first week of experiment, glycogen deposits were rather sparse; however, they were more numerous after three and six weeks. Lipid droplets were few and randomly distributed in the cytoplasm of hepatocytes. The vascular domain surface of hepatocytes was covered by numerous microvilli.

The differences in hepatocyte ultrastructure in experimental pigs compared with control animals were evident after 1, 3, and 6 weeks of treatment. The hepatocytes of ZEN-treated animals were characterized by a specific ER (Figure 7C,D). The smooth ER consisted of a network of closely located very narrow tubules and small vesicles, and occupied most of the cross sections of the hepatocytes. The poorly developed rough ER created several clusters of cisterns arranged in parallel. The mitochondria were located near these cisterns. Changes in the ER were observed from the first week of experiment; however, the intensity increased markedly after 3 and 6 weeks of treatment with ZEN. The ultrastructure of the mitochondria and lysosomes in ZEN-treated pigs did not differ significantly from that in the control animals. However, occasionally, some damaged mitochondria were observed. The amount of glycogen deposits in hepatocytes was usually low, except in the samples taken after the first week of the experiment. Some necrotic hepatocytes were noted, especially after 6 weeks of treatment.

In hepatocytes of DON-treated pigs, both types of ER were formed by abundant short, dilated cisterns and vesicles (Figure 7E,F), starting from the first week of the experiment. Both types of ER were quite evenly distributed in the cytoplasm of hepatocytes. In some cells, very dilated vacuoles of rough ER filled with protein micelles were observed. Large autophagosomes were present in the cytoplasm of numerous hepatocytes. Necrotic cells were occasionally observed.

The ultrastructure of hepatocytes in DON + ZEN-treated pigs did not differ from that in DON-treated pigs, except that necrotic hepatocytes were more frequently observed.

Changes in hepatocyte ultrastructure accompany numerous subclinical and clinical intoxications, both acute and chronic. The nature of these changes depends on the chemical properties of the toxin, dose, and duration of exposure. The most common changes are smooth and rough ER swelling, damage of mitochondria, and modification in the amount and distribution of lipid drops and glycogen deposits [51]. To date, data on the effect of mycotoxins on the ultrastructure of hepatocytes in pigs are scarce [7].

Our results show that the mycotoxins used in the experiment affected the ultrastructure of hepatocytes, especially the ER. ZEN administration resulted in conspicuous smooth ER proliferation and rough ER marginalization. Smooth ER is involved in the metabolism of various chemicals, and cells exposed to these chemicals showed hypertrophy of the smooth ER as an adaptive response. For this reason, smooth ER hypertrophy is considered a sensitive toxicological parameter [52].

The administration of DON and DON + ZEN resulted in the dilatation of ER cisterns and no clear difference between the two types of reticulum. The dilatation of ER cisternae, which indicates a loss of ER homeostasis, is known as ER stress [53]. ER stress is associated with liver injury and fibrosis. The hepatic factors that regulate ER stress remain unknown [54]. The fact that DON and DON + ZEN induced ER dilatation suggested that chronic ingestion of low doses of these mycotoxins caused injury in pig hepatocytes.

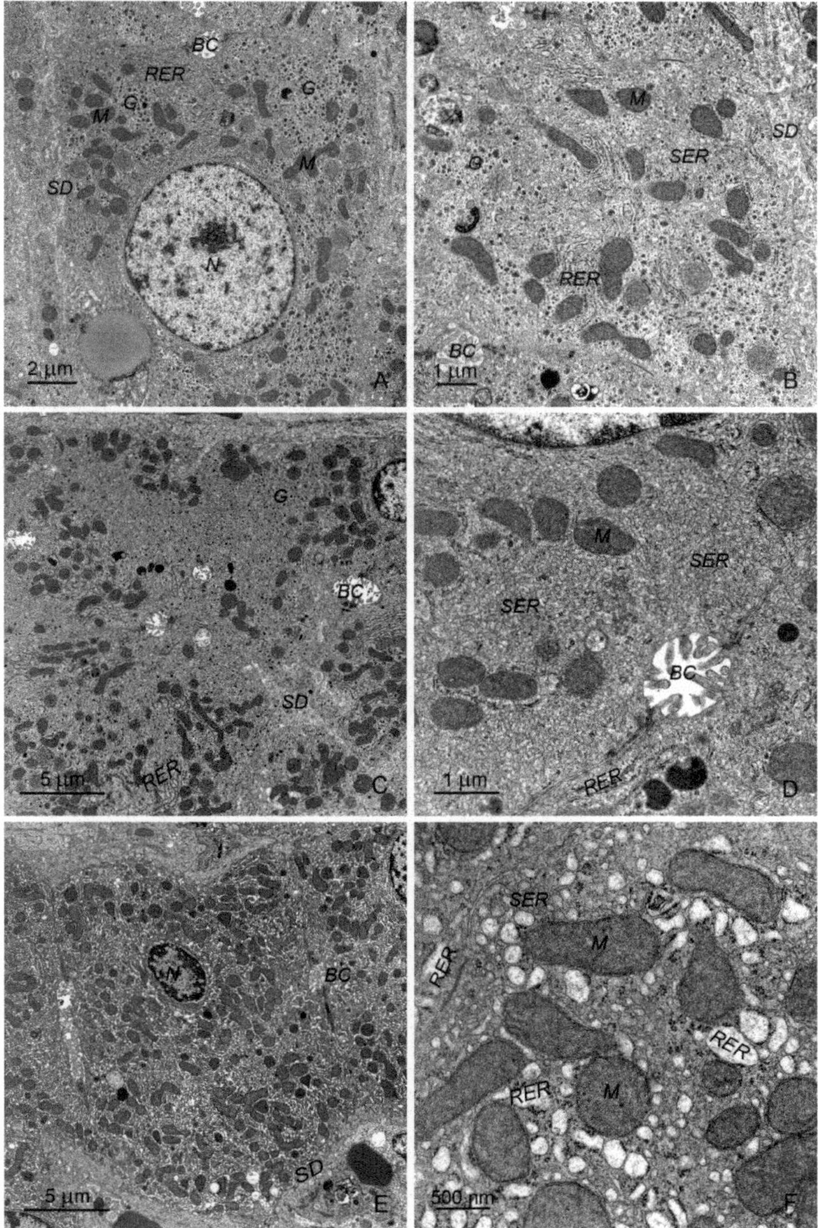

Figure 7. (**A–F**) Ultrastructure of hepatocytes in control pigs, 6th week of the experiment (**A**,**B**), pigs treated with ZEN for 6 weeks (**C**,**D**) and pigs treated with DON for 6 weeks (**E**,**F**). Note the specific organization of endoplasmic reticulum with the predominance of smooth reticulum over rough reticulum in pigs treated with ZEN. In DON-treated animals, both types of endoplasmic reticulum were formed by abundant short, dilated cisterns and vesicles. SD, space of Disse; BC, bile canaliculus; N, nucleus; M, mitochondria; RER, rough endoplasmic reticulum; SER, smooth endoplasmic reticulum; G, glycogen particles.

2.2.2. Sinusoids and Perivascular Species

Hepatic sinusoids were formed by flat endothelial cells with prominent fenestrations, usually lacking the basal lamina. Browicz–Kupffer cells with electron-lucent cytoplasm and numerous granules with variable appearance were located inside the sinusoids. Ito cells situated outside the vessels were numerous and usually contained one lipid droplet of moderate size. Pit cells situated inside the vessels were sporadically observed.

There were no differences in the ultrastructure of endothelial cells between control pigs and animals receiving mycotoxins. Browicz–Kupffer cells appeared to be more numerous in pigs receiving ZEN for 1 week as well as in pigs treated with DON and DON + ZEN for 1, 3, or 6 weeks than those in the control animals (Figure 8A–C). In these animals, Browicz–Kupffer cells showed prominent cell processes and occupied a large part of the vessel lumen. Their cytoplasm contained numerous phagosomes and rest bodies. Pit cells were more frequently observed in pigs treated with DON and DON + ZEN for 3 and 6 weeks than in the control animals (Figure 8C). In pigs treated with DON and DON + ZEN for 1, 3, or 6 weeks Ito cells were characterized by the presence of prominent cisterns of the rough ER (Figure 8D).

The focal accumulation of collagen fibers was observed in the perivascular spaces of pigs receiving DON and DON + ZEN for 3 and 6 weeks (Figure 9A). Their number increased with the duration of treatment. The penetration of collagen fibers between hepatocytes and foci of fibrosis was noted in pigs receiving DON + ZEN for 3 and 6 weeks (Figure 9B).

Our data shows, for the first time, that the administration of mycotoxins, especially DON, induced changes in the population of Browicz–Kupffer cells, which suggested their activation. These cells play an important role in the clearance of toxins from the portal blood [55]. Several cytokines, chemokines, and reactive nitrogen and oxygen species are released by activated Browicz–Kupffer cells, allowing these cells to modulate microvascular responses and the functions of hepatocytes and Ito cells [55,56]. Browicz–Kupffer cells show large plasticity, adopting changes in local metabolic and immune environment. They can play a protective role through their tolerogenic phenotype in toxin-induced liver injury, but can also shift to a pathologically activated state and contribute to liver inflammation. It could be considered that the changes in sinus diameters observed in our histological studies were due to the activation of Browicz–Kupffer cells. Because of their location and shape, these cells can interact with blood flow. Activation of Browicz–Kupffer cells is probably responsible for the more frequent occurrence of pit cells in liver sinuses [55,57].

Ito cells are "quiescent" in the normal liver, having received no stimuli to transform into a myofibroblastic state [56]. In our study, treatment with DON and DON + ZEN for 1, 3, or 6 weeks resulted in the presence of Ito cells with prominent cisterns of rough ER. The changes in Ito cells were probably related to their activation and transformation. They correlated with the occurrence of collagen fibers in the spaces of Disse.

The obtained data show that the administration of low doses of DON and ZEN resulted in prominent changes in the ultrastructure and histology of pig liver. In view of our results, even low levels of these mycotoxins, being below or close to No Observed Adverse Effect Level (NOAEL) values, should be considered as affecting the pig liver. The published data on the effects of such doses of DON and ZEN on the liver biochemistry are limited; however, they showed rather weak effects of these toxins on the serum activity of hepatic enzymes [58–60]. These findings agree with our morphological data showing that the responses of hepatocytes to intoxication with ZEN and DON seem to be adaptive. However, the changes in ER caused by ZEN and DON can affect the response of hepatocytes to other stressors or toxins [7,51]. In our opinion, special attention should be paid to the influence of ZEN and DON of the perilobular connective tissue, Browicz–Kupffer cells and Ito cells, which may lead to chronic damage of the liver. These effects cannot be detected with routine laboratory diagnostic methods.

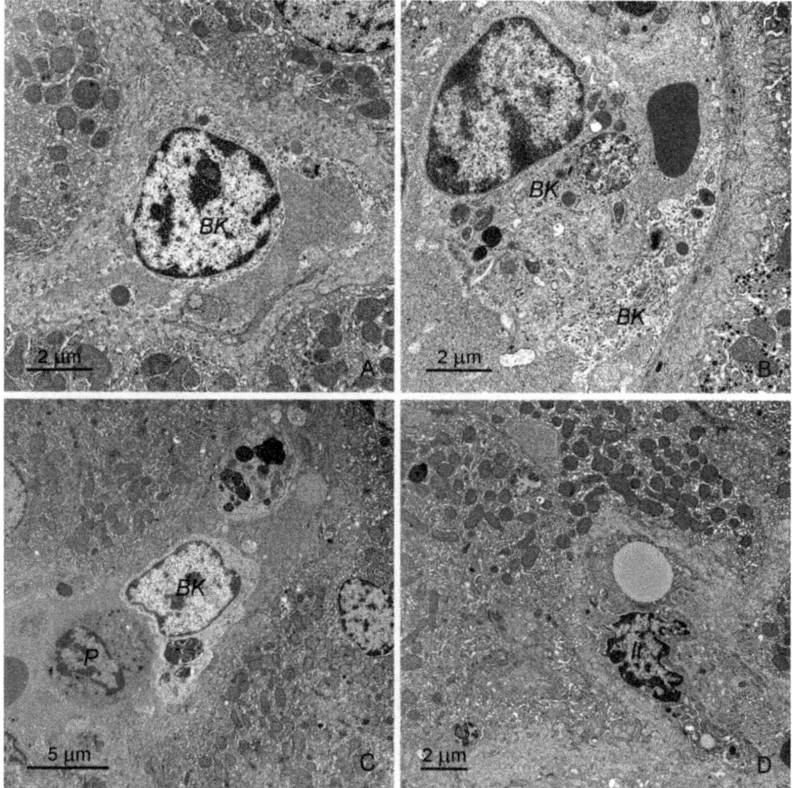

Figure 8. (**A**,**B**) Browicz–Kupffer cells (BK) in the sinusoid of pigs treated with DON for 3 weeks (**A**) and 6 weeks (**B**). Note the presence of numerous processes in Figure (**A**) and lysosomes and rest bodies in Figure (**B**,**C**) BK and pit (P) cells in the sinusoid of a pig treated with DON + ZEN for 6 weeks. (**D**) Ito cell (It) in the perivascular space of pigs treated with DON + ZEN for 6 weeks. Note the cisterns of the rough endoplasmic reticulum.

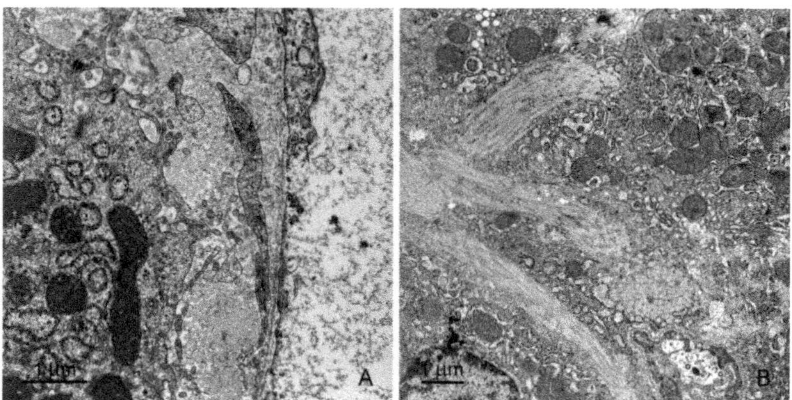

Figure 9. (**A**) Accumulation of collagen fiber in the space of Disse in a pig receiving DON for 6 weeks. (**B**) Penetration of collagen fibers between hepatocytes in a pig treated with DON + ZEN for 6 weeks.

3. Conclusions

Our current data provided strong evidence that the administration of low doses of DON and ZEN affected the ultrastructure and histology of pig liver. The changes caused by mycotoxins varied depending on the toxin and duration of intoxication. Both mycotoxins induced prominent modifications in the hepatocyte ER. The hepatocytes of ZEN-treated animals were characterized by extremely well-developed smooth ER, which comprised a dense network of narrow tubules. In the hepatocytes of DON-treated pigs, both rough and smooth ER were formed by an abundance of very short, dilated cisterns and vesicles. The effect of DON was greater than that of ZEN, as the ultrastructure of hepatocytes in pigs treated with DON + ZEN did not differ from that in DON-treated pigs. The total HAI scores in pigs receiving DON and DON + ZEN were significantly higher than that in the control animals, mainly because of the greater degree of hepatocyte necrosis and focal infiltration of inflammatory cells. Treatment with DON significantly increased the thickness of the perilobular connective tissue. The focal accumulation of collagen fibers was also observed in the perivascular spaces around the sinusoids, as shown by electron microscopy. These changes pointed to fibrosis as a potential effect of DON intoxication. Browicz–Kupffer cells were more frequently found and formed more prominent processes in pigs receiving mycotoxins, especially DON, and this phenomenon could be responsible for the dilation of sinusoids. Pit cells were more frequently observed in pigs treated with DON and DON + ZEN than in control animals. Examinations of glycogen and deposits of iron show the minor effect of the examined mycotoxins on these features. Our data demonstrate that DON and ZEN act both on hepatocytes and on liver immune/connective tissue cells. The response of hepatocytes to low doses of these toxins seems to be adaptive, and it is not related to serious cell damage. The influence of toxins on immune processes in the liver requires special attention because it can be destructive to the liver. Further molecular and biochemical studies are necessary to clarify the mechanisms of ZEN and DON toxicity in the liver.

4. Material and Methods

4.1. Animals, Toxins, and Experimental Design

The study was performed on 36 clinically healthy gilts of mixed breed (White Polish Big x Polish White Earhanging), with body weights of 25 ± 2 kg at the beginning of the experiment. The animals were purchased from a farm where they received feed without detectable amounts of ZEN, DON, α-zearalenol, aflatoxin, and ochratoxin. The pigs were fed twice daily and had free access to water.

The animals were divided into three experimental groups (D, Z, and M; $n = 9$ in each group) and a control group (C; $n = 9$). Group D received DON at a dose of 12 µg/kg BW per day, group Z received ZEN at a dose of 40 µg/kg BW per day, and group M received a mixture of DON + ZEN (ZEN 40 µg/kg BW + DON 12 µg/kg BW per day). The mycotoxins were synthesized and standardized at the Department of Chemistry, Faculty of Wood Technology, Poznań University of Life Sciences, Poland. The mycotoxins were administered orally during morning feeding in water-soluble capsules containing oat bran as a vehicle. The gilts were weighed every week to establish the amount of DON and ZEN administered to each animal. The animals in group C received capsules without mycotoxins.

The animals from the control and experimental groups were killed by intravenous administration of sodium pentobarbital (Vetbutal, Biowet, Poland) at a dose of 140–150 mg/kg and by exsanguination after 1, 3, and 6 weeks of experiment. Tissue samples were taken no longer than 3 min after cardiac arrest.

All procedures were carried out in compliance with Polish legal regulations for the determination of the terms and methods for performing experiments on animals and with the European Community Directive for the ethical use of experimental animals. The protocol was approved by the Local Ethical Council in Olsztyn (opinion No. 88/N of 16 December, 2009).

4.2. Histological Examination

Tissue samples (approximately 1 × 0.5 cm) were cut from the middle part of the liver. They were fixed in 4% paraformaldehyde in 0.1 M phosphate buffer (pH 7.4) for 48 h, dehydrated in ethanol (TP 1020; Leica, Wetzlar, Germany), and embedded in paraffin (EG1150; Wetzlar, Leica). Next, 4-µm-thick sections were prepared using an HM 340E microtome (Microm, Lugo, Spain) and stained with hematoxylin and eosin (HE), Mallory's trichrome, PAS, and PB to detect iron using an automated multistainer ST 5020 (Leica, Wetzlar, Germany). For histological evaluation, the sections were scanned using a Mirax Desk scanner (Carl Zeiss, Oberkochen, Germany). The slides were signed in a way that prevented the people involved in the microscopic analysis from knowing the kind and duration of the animal treatment.

We evaluated the architectural changes in the liver, noting the organization of hepatic cords, the appearance of perilobular connective tissue, and the penetration of collagen into the parenchyma on Mallory- and HE-stained specimens.

Histological changes in the liver parenchyma were analyzed on HE-stained slides. The following parameters were evaluated:

- Ishak modified histology activity index (HAI) based on Ishak et al. [61] and modified by Stanek et al. [38]; parameters taken in consideration include portal, periportal, and acinar inflammation, focal or confluent necrosis, and hemorrhages;
- The presence of lymphoid follicles, steatosis, hepatocellular dysplasia, karyomegaly;
- The dilatation of hepatic sinusoids using the semi-counted method.

For morphometrical evaluations, the following parameters were determined:

- The cross-sectional area of lobules;
- The thickness of the interlobular connective tissue septa.

Measurements were performed on HE-stained sections using the Pannoramic Viewer 1.15 software (3D-Histech, Budapest, Hungary).

Glycogen content and iron accumulation in hepatocytes was evaluated in PAS- and PB-stained sections, respectively.

4.3. Ultrastructural Examination

Samples of liver tissue were collected from sites adjacent to the sampling sites for histological examination, and then immersion-fixed in a mixture of 1% paraformaldehyde and 2.5% glutaraldehyde in 0.2 M phosphate buffer (pH 7.4) for 2 h at 4 °C. Next, they were washed and post-fixed in 2% OsO_4 in 0.2 M phosphate buffer (pH 7.4) for 2 h. After dehydration, the samples were embedded in Epon 812. Semi-thin sections were cut from each block of tissue, stained with 1% toluidine blue, and examined under a light microscope to choose the sites for preparing ultrathin sections. Ultrathin sections were cut using a Leica Ultracut III ultramicrotome (Leica, Wetzlar, Germany) and contrasted with uranyl acetate and lead citrate. They were examined with a Tecnai 12 Spirit G2 BioTwin transmission electron microscope (FEI, Hillsboro, OR, USA) equipped with two digital cameras: Veleta (Olympus, Tokyo, Japan) and Eage 4k (FEI, Hillsboro, OR, USA).

4.4. Statistical Analysis

The data from morphometric investigations were analyzed using one-way ANOVA with a Duncan test as a post-hoc procedure, and the data from semiquantitative analyses were analyzed using Kruskal–Wallis non-parametric ANOVA. Statistical analyses were performed using Statistica 10.0 software (StatSoft Polska, Cracow, Poland).

Supplementary Materials: The following are available online at http://www.mdpi.com/2072-6651/12/7/463/s1. Figure S1. Microphotographs of the histopathological parameters taken into account in HAI A. Portal inflammation. B. Periportal inflammation C. Acinar inflammation. D. Confluent necrosis. E. Focal lytic necrosis F. Hemorrhage, Figure S2. Dilatation of hepatic sinusoids in zone III (periportal) of acinus in a pig receiving DON + ZEN for 3 weeks. HE staining.

Author Contributions: Conceptualization, M.G. (Maciej Gajęcki) and B.L.; methodology, M.G. (Maciej Gajęcki) and B.P.-G.; investigation, M.G. (Magdalena Gajęcka), B.P.-G., N.S., and B.L.; writing—original draft preparation, N.S., B.P-G.; writing—review and editing, B.L.; visualization, N.S.; supervision, B.L.; project administration, M.G. (Maciej Gajęcki) and B.L.; funding acquisition, M.G. (Maciej Gajęcki) All authors have read and agreed to the published version of the manuscript.

Funding: This study was supported by research grant No NR12-0080-10 from the Polish National Centre for Research and Development. Publication costs were covered by the Ministry of Science and Higher Education under the program entitled "Regional Initiative of Excellence for the years 2019-2022", Project No. 010/RID/2018/19; the amount of funding was 12.000.000 PLN.

Acknowledgments: The authors would like to thank Jacek Sztorc and Krystyna Targońska for their skillful technical assistance.

Conflicts of Interest: The authors declare no conflict of interest.

References

1. Cavret, C.; Lecoeur, S. Fusariotoxin transfer in animals. *Food Chem. Toxicol.* **2006**, *44*, 444–453. [CrossRef]
2. Buszewska-Forajta, M. Mycotoxins, invisible danger of feedstuff with toxic effect on animals. *Toxicon* **2020**, *182*, 34–53. [CrossRef]
3. Nayakwadi, S.; Ramu, R.; Kumar Sharma, A.; Kumar Gupta, V.; Rajukumar, K.; Kumar, V.; Shirahatti, P.S.; Rashmi, L.; Basalingappa, K.M. Toxicopathological studies on the effects of T-2 mycotoxin and their interaction in juvenile goats. *PLoS ONE* **2020**, *15*, e229463. [CrossRef]
4. Koselski, M.; Dziubińska, H.; Trębacz, K.; Sieprawska, A.; Filek, M. The Role of SV Ion Channels under the Stress of Mycotoxins Induced in Wheat Cells—Protective Action of Selenium Ions. *J. Plant Growth Regul.* **2019**, *38*, 1255–1259. [CrossRef]
5. Sun, Y.; Wen, J.; Chen, R.; Deng, Y. Variable protein homeostasis in housekeeping and non-housekeeping pathways under mycotoxins stress. *Sci. Rep.* **2019**, *9*, 7819. [CrossRef]
6. Hallsworth, J.E. Stress-free microbes lack vitality. *Fungal Biol.* **2018**, *122*, 379–385. [CrossRef] [PubMed]
7. Tiemann, U.; Brüssow, K.P.; Dannenberger, D.; Jonas, L.; Pöhland, R.; Jäger, K.; Dänicke, S.; Hagemann, E. The effect of feeding a diet naturally contaminated with deoxynivalenol (DON) and zearalenone (ZON) on the spleen and liver of sow and fetus from day 35 to 70 of gestation. *Toxicol. Lett.* **2008**, *179*, 113–117. [CrossRef] [PubMed]
8. Jiang, S.Z.; Yang, Z.B.; Gao, J.; Liu, F.X.; Broomhead, J.; Chi, F. Effects of purified zearalenone on growth performance, organ size, serum metabolities, and oxidative stress in postweaning gilts. *J. Anim. Sci.* **2011**, *89*, 3008–3015. [CrossRef] [PubMed]
9. Conková, E.; Laciaková, A.; Kovác, A.; Seidel, H. Fusorial toxins and their role in animal disease. *Vet. J.* **2003**, *165*, 214–220. [CrossRef]
10. Alm, H.; Brussow, K.P.; Vanselow, J.; Tomek, W.; Danicke, S.; Tiemann, U. Influence of Fusarium-toxins contaminated feed on initial quality and meiotic competence of gilt oocytes. *Reprod. Toxicol.* **2006**, *22*, 44–50. [CrossRef]
11. Prelusky, D.B.; Hartin, K.E.; Trenholm, H.L.; Miller, J.D. Pharmacokinetic fate of 14C-labeled deoxynivalenol in swine. *Fundam. Appl. Toxicol.* **1988**, *10*, 276–286. [CrossRef]
12. Dänicke, S.; Valenta, S.; Döll, S. On the toxicokinetics and the metabolism of deoxynivalenol (DON) in the pig. *Arch. Anim. Nutr.* **2004**, *58*, 169–180. [CrossRef] [PubMed]
13. Dänicke, S.; Swiech, E.; Buraczewska, L.; Ueberschar, K.H. Kinetics and metabolism of zearalenone in young female pigs. *J. Anim. Physiol. Anim. Nutr.* **2005**, *89*, 268–276. [CrossRef]
14. Pinton, P.; Oswald, I.P. Effects of deoxynivalenol and other Type B trichothecenes on the intestine: A review. *Toxins* **2014**, *6*, 1615–1643. [CrossRef]
15. Gerez, J.R.; Pinton, P.; Callu, P.; Grosjean, F.; Oswald, I.P.; Bracarense, A.P. Deoxynivalenol alone or in combination with nivalenol and zearalenone induce systemic histological changes in pigs. *Exp. Tox. Pathol.* **2015**, *67*, 89–98. [CrossRef]

16. Lewczuk, B.; Przybylska-Gornowicz, B.; Gajęcka, M.; Targońska, K.; Ziółkowska, N.; Prusik, M.; Gajęcki, M. Histological structure of duodenum in gilts receiving low doses of zearalenone and deoxynivalenol in feed. *Exp. Toxicol. Pathol.* **2016**, *68*, 157–166. [CrossRef]
17. Przybylska-Gornowicz, B.; Tarasiuk, M.; Lewczuk, B.; Prusik, M.; Ziółkowska, N.; Zielonka, Ł.; Gajęcki, M.; Gajęcka, M. The effects of low doses of two Fusarium toxins, zearalenone and deoxynivalenol, on the pig jejunum. A light and electron microscopic study. *Toxins* **2015**, *7*, 4684–4705. [CrossRef]
18. Przybylska-Gornowicz, B.; Lewczuk, B.; Prusik, M.; Hanuszewska, M.; Petrusewicz-Kosińska, M.; Gajęcka, M.; Zielonka, Ł.; Gajęcki, M. The effects of Deoxynivalenol and Zearalenone on the pig large intestine. A light and electron microscopy study. *Toxins* **2018**, *10*, 148. [CrossRef] [PubMed]
19. Doll, S.; Danicke, S. The fusarium toxins deoxynivalenol (DON) and zearalenone (ZON) in animal feeding. *Prev. Vet. Med.* **2011**, *102*, 132–145. [CrossRef] [PubMed]
20. Gajęcki, M. Zearalenone—Ndesirable substances in feed. *Pol. J. Vet. Sci.* **2002**, *5*, 117–122.
21. Malekinejad, H.; Maas-Bakker, R.F.; Fink-Gremmels, J. Bioactivation of zearalenone by porcine hepatic biotransformation. *Vet. Res.* **2005**, *36*, 799–810. [CrossRef] [PubMed]
22. Olsen, M.; Kiessling, K.H. Species differences in zearalenone-reducing activity in subcellular fractions of liver from female domestic animals. *Acta Pharmacol. Toxicol.* **1983**, *52*, 287–291. [CrossRef] [PubMed]
23. Miles, C.O.; Alistair, L.W.; Wilkins, A.L.; Neal, R.T.; Barry, L.S.; Ian, G.; Bryan, G.S.; Richard, H.P. Ovine metabolism of zearalenone to alpha-zearalenol (Zeranol). *J. Agric. Food Chem.* **1996**, *44*, 3244–3250. [CrossRef]
24. Malekinejad, H.; Maas-Bakker, R.F.; Fink-Gremmels, J. Species differences in the hepatic biotransformation of zearalenone. *Vet. J.* **2006**, *172*, 96–102. [CrossRef]
25. Dong, M.; Tulayakul, P.; LI, J.Y.; Manabe, N.; Kumagai, S. Metabolic conversion of zearalenone to alpha-zearalenol by goat tissues. *J. Vet. Med. Sci.* **2010**, *72*, 307–312. [CrossRef]
26. Yang, C.; Song, G.; Lim, W. Effects of endocrine disrupting chemicals in pigs. *Environ. Pollut.* **2020**, 114505. [CrossRef]
27. Wan, L.Y.M.; Turner, P.C.; El-Nezami, H. Individual and combined cytotoxic effects of Fusarium toxins (deoxynivalenol, nivalenol, zearalenone and fumonisins B1) on swine jejunal epithelium cells. *Food Chem. Tox.* **2013**, *57*, 276–283. [CrossRef]
28. Ji, J.; Cui, F.; Pi, F.; Zhang, Y.; Li, Y.; Wang, J.; Sun, X. The antagonistic effect of mycotoxins deoxynivalenol and zearalenone on metabolic profiling in serum and liver mice. *Toxins* **2017**, *9*, 28. [CrossRef]
29. Van Le Thanh, B.; Lemay, M.; Bastien, A.; Lapointe, J.; Lessard, M.; Chorfi, Y.; Guay, F. The potential effects of antioxidant feed additives in mitigating the adverse effects of corn naturally contaminated with fusarium mycotoxins on antioxidant systems in the intestinal mucosa, plasma, and liver in weaned pigs. *Mycotoxin Res.* **2016**, *32*, 99–116. [CrossRef] [PubMed]
30. Alassane-Kpembi, I.; Kolf-Clauw, M.; Gauthier, T.; Abrami, R.; Abiola, F.A.; Oswald, P.; Puel, O. New insights into mycotoxin mixtures: The toxicity of low doses of Type B trichothecenes on intestinal epithelial cells is synergistic. *Toxicol. Appl. Pharmacol.* **2013**, *272*, 191–198. [CrossRef]
31. Reddy, K.E.; Jeong, J.; Lee, Y.; Lee, H.-J.; Kim, M.S.; Kim, D.W.; Jung, H.-J.; Choe, C.; Oh, Y.K.; Lee, S.D. Deoxynivalenol- and zearalnone-contaminated feeds alter gene expression profiles in the livers of piglets. *Asian-Australas. J. Anim. Sci.* **2018**, *31*, 595–606. [CrossRef] [PubMed]
32. Zollner, P.; Jodlauber, J.; Kleinova, M.; Kahlbacher, H.; Kuhn, T.; Hochsteiner, W.; Lindner, W. Concentrations levels of zearalenone and its metabolities in urine, muscle tissue, and liver samples of pigs fed with mycotoxins-contaminated oats. *J. Agric. Food Chem.* **2002**, *50*, 2494–2501. [CrossRef]
33. Mikami, O.; Yamaguchi, H.; Murata, H.; Nakajima, Y.; Miyazaki, S. Induction of apoptotic lesions in liver and lymphoid tissues and modulation of cytokine mRNA expression by acute exposure to deoxynivalenol in piglets. *J. Vet. Sci.* **2010**, *12*, 107–113. [CrossRef]
34. Renner, L.; Kahlert, S.; Tesch, T.; Bannert, E.; Frahm, J.; Barta-Böszörményi, A.; Kluess, J.; Kersten, S.; Schönfeld, P.; Rothkötter, H.J.; et al. Chronic DON exposure and acute LPS challenge: Effects on porcine liver morphology and function. *Mycotoxin Res.* **2017**, *33*, 207–218. [CrossRef] [PubMed]
35. Povero, D.; Busletta, C.; Novo, E.; di Bonzo, L.V.; Cannito, S.; Paternostro, C.; Parola, M. Liver fibrosis: A dynamic and potentially reversible process. *Histol. Histopathol.* **2010**, *25*, 1075–1091. [CrossRef] [PubMed]
36. Greaves, P. *Histopathology of Preclinical Toxicity Studies: Interpretation and Relevance in Drug Safety Evaluation*, 4th ed.; Elsevier Inc.: Amsterdam, The Netherlands, 2012; pp. 433–479.

37. Weber, S.N.; Wasmuth, H.E. Liver fibrosis: From animal models to mapping of human risk variants. *Best Pract. Res. Clin. Gastrooenterol.* **2010**, *24*, 635–646. [CrossRef]
38. Stanek, C.; Reinhardt, N.; Diesing, A.-K.; Nossol, C.; Kahlert, S.; Panther, P.; Kluess, J.; Rothkotter, H.-J.; Kuester, D.; Brosig, B.; et al. A chronic oral exposure of pigs with deoxynivalenol partially prevents the acute effects of lipopolisaccharides on hepatic histopathology and blood clinical chemistry. *Toxicol. Res.* **2012**, *215*, 193–200. [CrossRef]
39. Friedman, S.L.; McQuaid, K.R.; Grendell, J.H. *Current Diagnosis & Treatment in Gastroenterology*, 2nd ed.; Lang Medical Books/McGraw-Hill: New York, NY, USA, 2003; pp. 664–679.
40. Mills, S.E. *Histology for Pathologists*, 3rd ed.; W.B. Saunders Company: Sydney, Australia, 2007; pp. 685–703.
41. Foster, J.R. Spontaneous and drug-induced hepatic pathology of the laboratory beagle dog, the cynomolgus macaque and the marmoset. *Toxicol. Pathol.* **2005**, *33*, 63–74. [CrossRef]
42. Van Amersfoort, E.S.; Van Berkel, T.J.C.; Kuiper, J. Receptors, Mediators, and Mechanisms Involved in Bacterial Sepsis and Septic Shock. *Clin. Microbiol. Rev.* **2003**, *16*, 379–414. [CrossRef]
43. Coulombe, R.A., Jr. Pyrrolizidine alkaloids in foods. *Adv. Food Nutr. Res.* **2003**, *45*, 61–99. [CrossRef]
44. Winkler, K.; Poulsen, H. Liver disease with periportal sinusoidal dilatation. A possible complication to contraceptive steroids. *Scand. J. Gastroenterol.* **1975**, *10*, 699–704.
45. Ribero, D.; Wang, H.; Donadon, M.; Zorzi, D.; Thomas, M.B.; Eng, C.; Chang, D.Z.; Curley, S.A.; Abdalla, E.K.; Ellis, L.M.; et al. Bevacizumab improves pathologic response and protects against hepatic injury in patients treated with oxaliplatin-based chemotherapy for colorectal liver metastases. *Cancer* **2007**, *110*, 2761–2767. [CrossRef]
46. Halliday, J.W.; Searle, J. Hepatic iron deposition in human disease and animal models. *Biometals* **1996**, *9*, 205–209. [CrossRef]
47. Graham, R.M.; Chua, A.C.; Herbison, C.E.; Olynyk, J.K.; Trinder, D. Liver iron transport. *World J. Gastroenterol.* **2007**, *13*, 4725–4736. [CrossRef]
48. Deugnier, Y.; Turlin, B. Pathology of hepatic iron overload. *World J. Gastroenterol.* **2007**, *13*, 475–4760. [CrossRef]
49. Pietrangelo, A. Mechanisms of iron hepatotoxicity. *J. Hepatol.* **2016**, *65*, 226–227. [CrossRef] [PubMed]
50. Yiannikourides, A.; Latunde-Dada, G.O. A Short Review of Iron Metabolism and Pathophysiology of Iron Disorders. *Medicines* **2019**, *6*, 85. [CrossRef] [PubMed]
51. Cheville, N.F. Ultrastructural pathology. In *The Comparative Cellular Basis of Diseases*, 2nd ed.; Willey-Blackwell: Hoboken, NJ, USA, 2009; pp. 1–1000.
52. Hutterer, F.; Schaffner, F.; Klion, F.M.; Popper, H. Hypertrophic, Hypoactive Smooth Endoplasmic Reticulum: A Sensitive Indicator of Hepatotoxicity Exemplified by Dieldrin. *Science* **1968**, *161*, 1017–1019. [CrossRef] [PubMed]
53. Burton, G.J.; Yung, H.W. Endoplasmic reticulum stress in the pathogenesis of early-onset pre-eclampsia. *Pregnancy Hypertens.* **2011**, *1*, 72–78. [CrossRef] [PubMed]
54. Han, C.Y.; Rho, H.S.; Kim, A.; Kim, T.H.; Jang, K.; Jun, D.W.; Kim, J.W.; Kim, B.; Kim, S.G. FXR inhibits endoplasmic reticulum stress-induced NLRP3 inflammasome in hepatocytes and ameliorates liver injury. *Cell Rep.* **2018**, *24*, 2985–2999. [CrossRef]
55. Crawford, J.M.; Bioulac-Sage, P.; Hytiroglou, P. Structure, Function, and Responses to Injury. In *Macsween's Pathology of the Liver*, 7th ed.; Burt, A.D., Ferrell, L.D., Hübscher, S.G., Eds.; Elsevier: Amsterdam, The Netherlands, 2018; pp. 1–87. [CrossRef]
56. Hasegawa, D.; Wallace, M.C.; Friedman, S.C. Chapter 4—Stellate Cells and Hepatic Fibrosis. In *Stellate Cells in Health and Disease*; Gandhi, C.R., Pinzani, M., Eds.; Academic Press: Cambridge, MA, USA, 2015; pp. 41–62. [CrossRef]
57. Peng, H.; Wisse, E.; Tian, Z. Liver natural killer cells: Subsets and roles in liver immunity. *Cell Mol. Immunol.* **2016**, *13*, 328–336. [CrossRef]
58. Gajęcka, M.; Tarasiuk, M.; Zielonka, Ł.; Dąbrowski, M.; Gajęcki, M. Risk assessment for changes in the metabolic profile and body weights of pre-pubertal gilts during long-term monotonic exposure to low doses of zearalenone (ZEN). *Res. Vet. Sci.* **2016**, *109*, 169–180. [CrossRef]
59. Nicpoń, J.; Sławuta, P.; Nicpoń, J. Effect of zearalenone toxicosis on the complete blood cell count and serum biochemical analysis in wild boars. *Vet. Med.* **2016**, *72*, 250–254.

60. Holanda, D.M.; Kim, S.W. Efficacy of Mycotoxin Detoxifiers on Health and Growth of Newly-Weaned Pigs under Chronic Dietary Challenge of Deoxynivalenol. *Toxins* **2020**, *12*, 311. [CrossRef] [PubMed]
61. Ishak, K.; Baptista, A.; Bianchi, L.; Callea, F.; De Groote, J.; Gudat, F.; Denk, H.; Desmet, V.; Korb, G.; MacSween, R.N.M.; et al. Histological grading and staging of chronic hepatitis. *J. Hepatol.* **1995**, *22*, 696–699. [CrossRef]

© 2020 by the authors. Licensee MDPI, Basel, Switzerland. This article is an open access article distributed under the terms and conditions of the Creative Commons Attribution (CC BY) license (http://creativecommons.org/licenses/by/4.0/).

Article

Imbalance in the Blood Concentrations of Selected Steroids in Pre-pubertal Gilts Depending on the Time of Exposure to Low Doses of Zearalenone

Anna Rykaczewska [1], Magdalena Gajęcka [1,*], Ewa Onyszek [2], Katarzyna Cieplińska [3], Michał Dąbrowski [1], Sylwia Lisieska-Żołnierczyk [4], Maria Bulińska [5], Andrzej Babuchowski [2], Maciej T. Gajęcki [1] and Łukasz Zielonka [1]

1 Department of Veterinary Prevention and Feed Hygiene, Faculty of Veterinary Medicine, University of Warmia and Mazury in Olsztyn, Oczapowskiego 13/29, 10-718 Olsztyn, Poland; anna.rykaczewska@uwm.edu.pl (A.R.); michal.dabrowski@uwm.edu.pl (M.D.); gajecki@uwm.edu.pl (M.T.G.); lukaszz@uwm.edu.pl (Ł.Z.)

2 Dairy Industry Innovation Institute Ltd., 11-700 Mrągowo, Poland; ewa.onyszek@iipm.pl (E.O.); andrzej.babuchowski@iipm.pl (A.B.)

3 Microbiology Laboratory, Non-Public Health Care Centre, ul. Limanowskiego 31A, 10-342 Olsztyn, Poland; kasiacieplinska@gmail.com

4 Independent Public Health Care Centre of the Ministry of the Interior and Administration, and the Warmia and Mazury Oncology Centre in Olsztyn, Wojska Polskiego 37, 10-228 Olsztyn, Poland; lisieska@wp.pl

5 Department of Discrete Mathematics and Theoretical Computer Science, Faculty of Mathematics and Computer Science, University of Warmia and Mazury in Olsztyn, Słoneczna 34, 10-710 Olsztyn, Poland; bulma@uwm.edu.pl

* Correspondence: mgaja@uwm.edu.pl

Received: 12 August 2019; Accepted: 22 September 2019; Published: 25 September 2019

Abstract: Zearalenone (ZEN) is a mycotoxin that not only binds to estrogen receptors, but also interacts with steroidogenic enzymes and acts as an endocrine disruptor. The aim of this study was to verify the hypothesis that low doses, minimal anticipated biological effect level (MABEL), no-observed-adverse-effect level (NOAEL) and lowest-adverse-effect level (LOAEL), of ZEN administered orally for 42 days can induce changes in the peripheral blood concentrations of selected steroid hormones (estradiol, progesterone and testosterone) in pre-pubertal gilts. The experiment was performed on 60 clinically healthy gilts with average BW of 14.5 ± 2 kg, divided into three experimental groups and a control group. Group ZEN5 animals were orally administered ZEN at 5 μg ZEN/kg BW, group ZEN10 — at 10 μg ZEN/kg BW, group ZEN15 — at 15 μg ZEN/kg BW, whereas group C received a placebo. Five gilts from every group were euthanized on analytical dates 1, 2 and 3 (days 7, 14 and 42 of the experiment). Qualitative and quantitative changes in the biotransformation of low ZEN doses were observed. These processes were least pronounced in group ZEN5 (MABEL dose) where ZEN metabolites were not detected on the first analytical date, and where β-ZEL was the predominant metabolite on successive dates. The above was accompanied by an increase in the concentration of estradiol (E_2) which, together with "free ZEN", probably suppressed progesterone (P_4) and testosterone (T) levels.

Keywords: zearalenone; low doses; steroid hormones; biotransformation; pre-pubertal gilts

Key Contribution: A comparison of the present findings with the results of a previous study performed on the same gilts indicates that the MABEL dose contributes to accelerating somatic development and delays sexual maturity.

1. Introduction

Zearalenone (ZEN) is a macrocyclic lactone of β-resorcylic acid with clear estrogen activity. This non-steroidal estrogenic mycotoxin is produced by several species of the genus *Fusarium*. Zearalenone is metabolized to numerous derivatives by microorganisms, plants, animals and humans. Previous research into the metabolism of ZEN revealed the presence of reducing metabolites, in particular α-zearalenol (α-ZEL) and its stereoisomer, β-zearalenol (β-ZEL) [1]. When these catechol metabolites are synthesized, the activity of ZEN resembles that of endogenous estrogens, such as estradiol (E_2). As a result, ZEN can affect reproduction in pre-pubertal gilts, the expression of hydroxysteroid dehydrogenases (HSDs) and the synthesis and secretion of sex hormones, including E_2, progesterone (P_4) and testosterone (T) [2,3]. These unverified facts have become the reason for the studies. This was accompanied by other doubts of tape – that this mycotoxin should not be tested using the minimal anticipated biological effect level (MABEL) dose, the no-observed-adverse-effect level (NOAEL) dose [4] and/or the lowest-adverse-effect level (LOAEL) dose [5]. Mammals have adapted to prolonged exposure to low monotonic doses of ZEN [6], or have even learned to exploit this mycotoxin in their physiological processes [7–9].

At the beginning of this decade, the classical dose-response paradigm was undermined by the "low dose hypothesis", in particular, with regard to hormonally active substances [6] which act as endocrine disruptors (EDs) and/or disrupt paracrine and endocrine signaling [1]. These processes can be observed during exposure to low doses of undesirable substances which are present in food and feed [10,11] and which induce differential responses in macroorganisms (hormesis [12]). These dose-response interactions remain relatively unexplored, and the risks (clinical symptoms or laboratory results) associated with high doses cannot be clearly extrapolated to low doses [13,14] that deliver counterintuitive effects.

From the point of view of biomedical practice [15], the MABEL concept is garnering increasing interest because the clinical picture is influenced by numerous endogenous factors. The clinical picture reflects not only disruptions in the steroid hormone balance or the quantity and quality of exogenous hormone-like substances [16], but also other somatic responses, including reproductive behavior [1], immune responses [17] and changes in the metabolic profile of peripheral blood [9]. A thorough understanding of the relevant mechanisms of action and the final effects supports sound decision-making [14]. Substances that disrupt hormonal homeostasis have undermined long-standing paradigms in toxicology, in particular the "dose makes the poison" concept [18]. Low doses of ZEN and its metabolites (ZELs) induce specific changes [7–9] that are not encountered during exposure to high doses. Research into natural hormones and EDs has demonstrated that low doses produce ambiguous responses (pro-inflammatory, anti-inflammatory, increase or decrease in proliferative activity [19]). The above also applies to ZEN [1,20,21].

In view of the above, the aim of this study was to validate the hypothesis that MABEL, NOAEL and LOAEL doses of ZEN administered orally for 42 days can induce changes in peripheral blood concentrations of selected steroid hormones (estradiol, progesterone and testosterone) in pre-pubertal gilts.

2. Results

2.1. Experimental Feed

The analyzed feed did not contain mycotoxins, or its mycotoxin content was below the sensitivity of the method (VBS). The concentrations of modified and masked mycotoxins were not analyzed.

2.2. Clinical Observations

Clinical signs of ZEN mycotoxicosis were not observed throughout the experiment. However, changes in specific tissues or cells were frequently observed in analyses of the serum biochemical profile, caecal water genotoxicity and intestinal microbiome parameters in samples collected from the same animals and in those animals' growth performance. The results of these analyses were published in a different paper [7–9].

2.3. The Effect of Various Doses of Zearalenone on Hormone Secretion

2.3.1. The Effect of Estradiol

The concentration of E_2 in the blood of pre-pubertal gilts (see Figure 1) ranged from 2.9 to 19.9 pg/mL. Significant differences (see Figure 1) were observed between analytical dates in groups and between groups on different dates. In the control group (C), E_2 levels increased by 5.7 pg/mL during the experiment. Estrogen concentration also increased in all experimental groups during the entire period of exposure (by 6.2 pg/mL in group ZEN5; by 1.9 pg/mL in group ZEN10 – a value lower than in group C; by 14.3 pg/mL in group ZEN15), but the reference level on D1 was higher in the experimental groups than in group C. On D2, the concentration of E_2 decreased in groups ZEN5 and ZEN10, relative to D1.

On different analytical dates (see Figure 1), the concentration of E_2 increased in the experimental groups relative to group C (from 1.8 pg/mL on D1, through 3.8 pg/mL on D2 to 10.2 pg/mL on D3), proportionally to the applied mycotoxin dose. Estrogen levels decreased in groups ZEN5 (difference of 2.2 pg/mL) and ZEN10 (difference of 2.3 pg/mL) on date D2, and in group ZEN10 (difference of 2.0 pg/mL) on date D3, relative to group C.

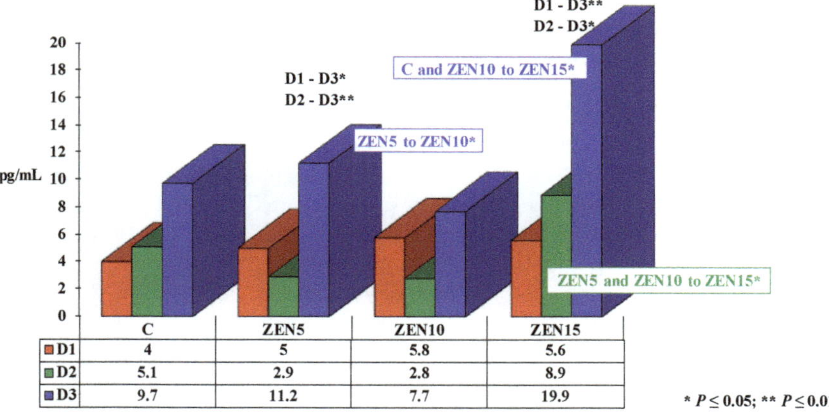

Figure 1. The effect of zearalenone (ZEN) on estradiol (E_2) concentration in the blood of pre-pubertal gilts: arithmetic means ($\overline{x}$) of five samples collected on each analytical date (D1, D2 and D3) in every group (control (C), ZEN5, ZEN10 and ZEN15). Statistically significant differences were determined at $^{*}P \leq 0.05$ and $^{**}P \leq 0.01$.

A comparison of the mean ($\overline{x}$) concentrations of E_2 in groups during the entire experiment revealed the lowest values in group ZEN10 (5.4 pg/mL). In the remaining experimental groups, E_2 levels were higher than in group C, proportionally to the administered dose of ZEN (from 6.2 pg/mL in group C and 6.4 pg/mL in group ZEN5 to 11.5 pg/mL in group ZEN15).

2.3.2. The Effect of Progesterone

Progesterone concentrations ranged from 0.1 to 0.9 ng/mL during the experiment (Figure 2).

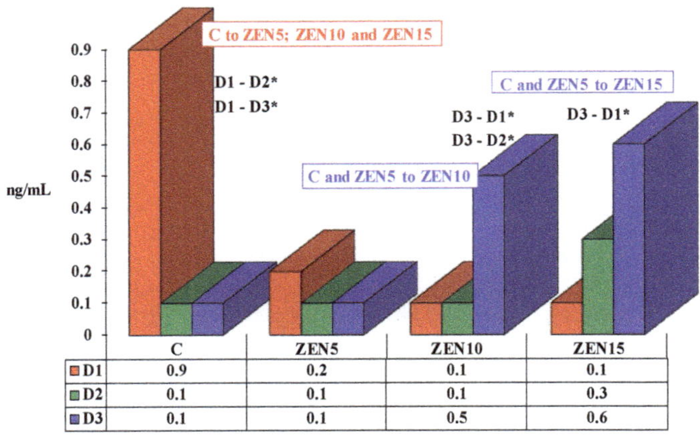

Figure 2. The effect of ZEN on progesterone (P_4) concentration in the blood of pre-pubertal gilts: arithmetic means ($\overline{x}$) of five samples collected on each analytical date (D1, D2 and D3) in every group (C, ZEN5, ZEN10 and ZEN15). Statistically significant differences were determined at $^*P \leq 0.05$.

Significant differences in progesterone concentrations are presented in Figure 2. In group C, P_4 levels decreased by 0.7 ng/mL and 0.8 ng/mL between D1 versus D2 and D3, respectively. In group ZEN5, P_4 concentration remained stable at approximately 0.1 ng/mL. In the remaining experimental groups, a steady increase in P_4 levels was observed over time (to 0.5 ng/mL in group ZEN10; to 0.6 ng/mL in group ZEN15).

On D1, progesterone concentrations were much lower in the experimental groups than in group C (by 0.7; 0.8 and 0.8 ng/mL, respectively). On D2, P_4 values were similar in group C and groups ZEN5 and ZEN10 (0.1 ng/mL lower in both groups than in group C). In group ZEN15, the concentration of P_4 increased to 0.3 ng/mL, and it was 0.2 ng/mL higher than in group C. On D3, P_4 levels were low and similar in groups C and ZEN5 at around 0.1 ng/mL. In groups ZEN10 and ZEN15, the noted values were much higher than in group C (by 0.5 and 0.6 ng/mL, respectively).

The mean concentration of P_4 in each group was very low on each analytical date. The mean values of P_4 in all experimental groups were lower than in group C (0.3 ng/mL in group C; 0.1 ng/mL in group ZEN5; 0.2 ng/mL in group ZEN10; 0.3 ng/mL in group ZEN15), proportionally to the administered dose of ZEN.

2.3.3. The Effect of Testosterone

Significant differences in testosterone concentrations are presented in Figure 3. Testosterone levels ranged from 0.03 to 0.14 ng/mL during the experiment.

In all groups, testosterone levels were lowest at the beginning of the experiment, and they increased over time of exposure to ZEN. Similar changes were observed in groups C, ZEN5 and ZEN10, but a rapid increase in T levels was noted on D2. Testosterone concentrations increased gradually over time from 0.08 ng/mL on D1, to 0.09 ng/mL on D2 and 0.14 ng/mL on D3 only in group ZEN15.

On D1 and D3, T levels were highly similar at 0.03–0.04 ng/mL, excluding in group ZEN15 where they were very high at 0.08 ng/mL on D1 and 0.14 ng/mL on D3. On D2, T concentrations ranged from 0.08 to 0.09 ng/mL in all groups, and the highest value was noted in group C.

During the experiment, the mean concentration of T was highest in group ZEN15 (0.102 ng/mL). Testosterone levels were lower in the remaining groups (from 0.05 ng/mL in group C and 0.05 ng/mL in group ZEN5 to 0.05 ng/mL in group ZEN10).

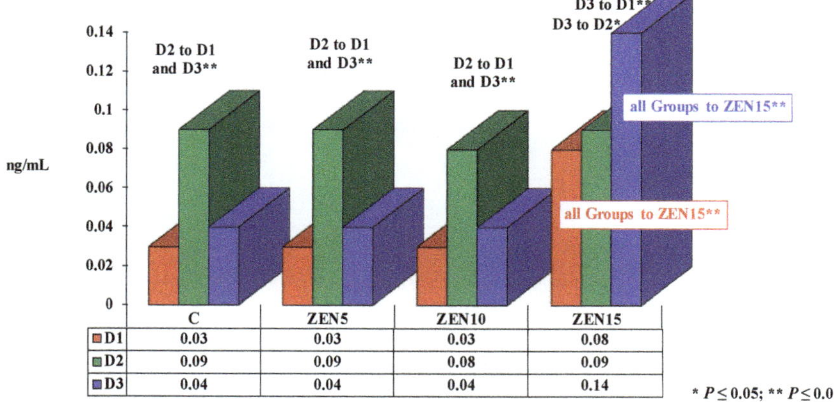

Figure 3. The effect of ZEN on testosterone (T) concentration in the blood of pre-pubertal gilts: arithmetic means ($\bar{x}$) of five samples collected on each analytical date (D1, D2 and D3) in every group (C, ZEN5, ZEN10 and ZEN15). Statistically significant differences were determined at $^{*}P \leq 0.05$ and $^{**}P \leq 0.01$.

2.3.4. Pearson's Correlation Coefficient (r)

The values of r were calculated based on the concentrations of steroid hormones (estradiol – E_2, progesterone – P_4 and testosterone – T) in the blood of pre-pubertal gilts on different analytical dates and in different groups. The coefficients of correlations between the concentrations of E_2 and P_4, E_2 and T, and P_4 and T are presented in Tables 1–3, respectively [22]. A positive correlation ($r > 0$) was determined when an increase in the concentration of one hormone led to a rise in the concentration of another hormone. The strength of positive correlations was determined on the following scale: $r < 0.2$ — no correlation, $r = 0.2$ to 0.4 — weak correlation, $r = 0.4$ to 0.7 — moderate correlation, $r = 0.7$ to 0.9 — relatively strong correlation, and $r > 0.9$ – very strong correlation. The correlation coefficient was determined at 0.0 only when the concentration of P_4 on a given analytical date was below the sensitivity of the method. A negative correlation ($r < 0$) was determined when an increase in the concentration of one hormone led to a decrease in the concentration of another hormone. The strength of negative correlations was determined on the following scale: $r = 0.0$ to -0.2 — no correlation, $r = -0.2$ to -0.4 — weak correlation, $r = -0.4$ to -0.7 — moderate correlation, $r = -0.7$ to -0.9 — relatively strong correlation, and $r = -0.9$ to -1.0 — very strong correlation.

Table 1. Coefficients of correlations (r) between the concentrations of estradiol (E_2) and progesterone (P_4).

Analytical Date	Group C	Group ZEN5	Group ZEN10	Group ZEN15
D1	0.105	0.445	−0.483	0.0
D2	0.767	0.0	0.0	0.594
D3	0.244	−0.231	−0.147	0.511

Key: Strength of linear correlations between the concentrations of E2 and P4 in the blood of pre-pubertal gilts on different analytical dates (D1 — exposure day 7; D2 — exposure day 21; D3 — exposure day 42) and in different groups where ZEN was administered once daily before the morning feeding (group C — placebo; group ZEN5 — 5 µg ZEN/kg BW; group ZEN10 — 10 µg ZEN/kg BW; group ZEN15 — 15 µg ZEN/kg BW).

In group C, the correlations between the concentrations of E_2 and P_4 (Table 1) were absent, relatively strong and weak on successive analytical dates (these values were used as the reference in further analyses). In group ZEN5, the above correlations were evaluated as moderate on D1 and as absent on D2 and D3. In group ZEN10, a moderate negative correlation was noted on D1, and the absence of correlations was determined on D2 as well as D3. In group ZEN15, E_2 and P_4 concentrations were not correlated on D1, whereas moderate correlations were noted on D2 and D3.

Table 2. Coefficients of correlations (r) between the concentrations of E_2 and testosterone (T).

Analytical Date	Group C	Group ZEN5	Group ZEN10	Group ZEN15
D1	0.326	0.408	−0.596	−0.015
D2	0.076	0.452	0.069	0.117
D3	−0.203	−0.492	−0.128	−0.472

Key: Strength of linear correlations between the concentrations of E2 and T in the blood of pre-pubertal gilts on different analytical dates (D1 — exposure day 7; D2 — exposure day 21; D3 — exposure day 42) and in different groups where ZEN was administered once daily before the morning feeding (group C — placebo; group ZEN5 — 5 µg ZEN/kg BW; group ZEN10 — 10 µg ZEN/kg BW; group ZEN15 — 15 µg ZEN/kg BW).

In group C, the correlations between the concentrations of E_2 and T (Table 2) were determined as weak, absent, and weak negative on successive analytical dates (these values were used as the reference in further analyses). In group ZEN5, the analyzed correlations were moderate on successive dates, and a moderate negative correlation was noted on D3. In group ZEN10, a moderate negative correlation was found on D1, the absence of a negative correlation was observed on D3, and the absence of a positive correlation was noted on D2 (negative value). In group ZEN15, the absence of a negative correlation was found on D1, a moderate negative correlation was observed on D3, and the absence of a positive correlation was noted on D2.

Table 3. Coefficients of correlations (r) between the concentrations of P_4 and T.

Analytical Date	Group C	Group ZEN5	Group ZEN10	Group ZEN15
D1	0.220	0.998	0.730	0.0
D2	0.576	0.0	0.0	0.694
D3	0.522	0.0	0.148	−0.372

Key: Strength of linear correlations between the concentrations of P4 and T in the blood of pre-pubertal gilts on different analytical dates (D1 — exposure day 7; D2 — exposure day 21; D3 — exposure day 42) and in different groups where ZEN was administered once daily before the morning feeding (group C — placebo; group ZEN5 — 5 µg ZEN/kg BW; group ZEN10 — 10 µg ZEN/kg BW; group ZEN15 — 15 µg ZEN/kg BW).

In group C, the concentrations of P_4 and T (Table 3) were not correlated on D1, whereas moderate correlations were noted on D2 and D3 (these values were used as the reference in further analyses). In group ZEN5, a very strong correlation was observed on D1, whereas the values noted on D2 and D3 were not correlated. In group ZEN10, a relatively strong correlation was determined on D1, and the absence of correlations was noted on D2 (r = 0.00) and D3. In group ZEN15, no correlations were determined on D1 (r = 0.0), a moderate correlation was observed on D2, and a weak negative correlation was found on D3.

2.4. Concentrations of Zearalenone and its Metabolites in Peripheral Blood

Zearalenone concentrations in the peripheral blood of pre-pubertal gilts did not differ significantly between analytical dates or groups (see Figure 4). However, considerable differences were noted in mean values. On D1, mean ZEN levels differed by 0.6 ng/mL between groups ZEN5 and ZEN10 and by 0.5 ng/mL between groups ZEN5 and ZEN15. On D2, the corresponding differences were determined at 1.3 and 2.6 ng/mL. The smallest differences between the above groups were observed on D3 at 0.0 and 1.1 ng/mL, respectively.

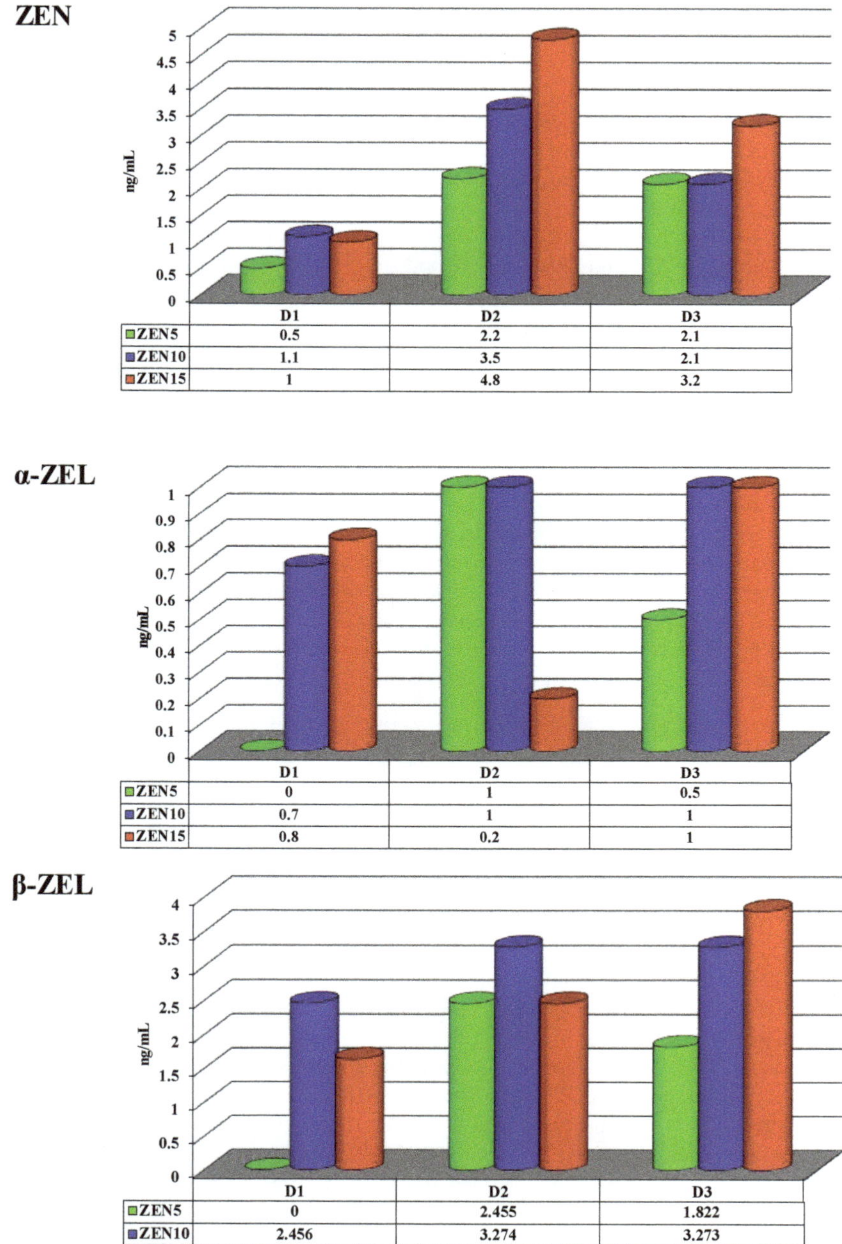

Figure 4. Mean ($\overline{x}$) concentrations of ZEN and its metabolites (α-ZEL and β-ZEL) (ng/mL) in the peripheral blood of pre-pubertal gilts on different analytical dates (D1 — exposure day 7; D2 — exposure day 21; D3 — exposure day 42) and in the experimental groups (group ZEN5 — 5 μg ZEN/kg BW; group ZEN10 — 10 μg ZEN/kg BW; group ZEN15 – 15 μg ZEN/kg BW). Limits of detection (LOD) > values below the limit of detection were regarded as equal to 0.

Similar observations were made in an analysis of ZEN metabolites, i.e., α-ZEL and β-ZEL. However, the mean levels of both metabolites differed considerably between analytical dates (see Figure 4).

On D1, the concentrations of α-ZEL differed between group ZEN5 (values below limits of detection (LOD), regarded as equal to 0) and groups ZEN10 and ZEN15 by 0.7 and 0.8 ng/mL, respectively (see Figure 4). On D2, α-ZEL levels were lowest in group ZEN15 relative to groups ZEN5 and ZEN10 (0.8 ng/mL). On D3, the differences in α-ZEL concentrations between group ZEN5 and groups ZEN10 and ZEN15 reached 0.5 ng/mL.

On D1, the concentrations of β-ZEL differed between group ZEN5 (values below LOD, regarded as equal to 0) and groups ZEN10 and ZEN15 by 2.5 and 1.6 ng/mL, respectively (see Figure 1). On D2, the corresponding differences were determined at 0.8 and 0.0 ng/mL, respectively. On D3, the respective differences were clearly pronounced at 1.5 and 2.0 ng/mL, respectively.

3. Discussion

The results of the present experiment validated the hypothesis that low doses of ZEN affect the concentrations of E_2, P_4 and T in pre-pubertal gilts on different days of exposure.

The results obtained on the first analytical date (D1) reflect the stimulatory effects of ZEN (undesirable substance) administered over a period of seven days. In the studied gilts, the effect of adaptive mechanisms, accompanied by considerable loss of energy and protein [9], was manifested on D2 [23]. The above could also be accompanied by an increase in Ca^{2+} deposition, in particular in the mitochondria [24,25], or changes in the activity of selected enzymes, such as hydroxysteroid dehydrogenases [26], which can disrupt steroidogenesis [14]. Hyperestrogenism induced by excess ZEN (that was not biotransformed or was recovered from enterohepatic circulation) probably took place on D3. "Free ZEN" can probably be utilized in specific life processes [14].

3.1. Estradiol

A significant or highly significant increase in E_2 concentrations (see Figure 1) responsible for hyperestrogenism or supraphysiological hormonal levels [27] was observed in the experimental groups [2,14] relative to group C, excluding groups ZEN5 and ZEN10 on D2. The observed increase occurred as a counter-reaction to the administered doses of ZEN. On D1, E_2 concentrations in the experimental groups relative to group C (physiological levels) were indicative of supraphysiological hormonal levels rather than hypoestrogenism (in relation to the physiological deficiency of endogenous E_2 in group C). "Free ZEN" was captured by estrogen receptors (ERs) in the gastrointestinal tract (own study, unpublished data), and it stimulated qualitative changes (activation?) in ERs. These processes were manifested by changes in the expression of ERs, in particular ER*ß* in the descending colon, with a simultaneous quantitative increase in microbiota and an increase in genotoxicity under exposure to higher ZEN doses [7,8]. Estradiol concentrations in the peripheral blood and the metabolic profile of the studied gilts [9] indicate that "free ZEN" could: (i) induce changes during steroidogenesis (see Figure 1, group ZEN15), thus confirming that ZEN modifies the expression of enzymes such as HSD at the pre-receptor level, inversely to the applied ZEN dose [26,28], and is capable of converting T to E_2. (ii) Increases feed intake and the accumulation of body energy reserves in gilts [9] for reproductive processes in the future [29]. The strength of linear correlations between E_2 levels and the concentrations of P_4 and T in the blood of pre-pubertal gilts (see Tables 1 and 2 – moderate correlations or absence of correlations) testifies to a minor and inversely proportional (negative) effect of "free ZEN" in groups ZEN5 (MABEL dose) and ZEN10 (NOAEL dose). As a result, ZEN suppressed the concentrations of the two remaining hormones [30]. This is a significant consideration in the treatment of autoimmune disorders [23] and, more importantly, in the production of pork (sexual maturity is delayed in gilts [31,32]). In contrast, E_2 levels in group ZEN15 (LOAEL dose) were not correlated with the concentration of P_4 (see Table 1) on D1, whereas moderate correlations indicative of ZEN's stimulatory effects were noted on the remaining analytical dates. In group ZEN15, the concentrations of E_2 and T (see Table 2) were not correlated or were bound by a moderate correlation (positive correlation on D2, negative correlation on D1 and D3), which suggests that ZEN exerted weak and suppressive effects.

3.2. Progesterone

An analysis of P_4 levels in the blood of pre-pubertal gilts exposed to different doses of ZEN (see Figure 2) revealed low and similar values (0.1 ng/mL) in groups C and ZEN5. Progesterone levels were determined at 0.9 ng/mL in group C only on D1. These variations are difficult to interpret [27,33]. In groups ZEN10 and ZEN15, P_4 concentrations continued to increase throughout the experiment, which indicates that a premature increase in P_4 levels can enhance endometrial receptivity and induce morphological changes in the reproductive system [31]. The above was accompanied by a rise in the levels of endogenous E_2, but the correlation between E_2 and P_4 concentrations was negative (inversely proportional) in group ZEN10 (see Table 1). In group ZEN15, the correlation was positive. These findings suggest that the presence of MABEL and NOAEL doses of ZEN in the diet contributes to supraphysiological hormonal levels [27] and impairs P_4 synthesis in pre-pubertal gilts long before the first estrus in pre-pubertal gilts (on D3, BW values ranged from 34 kg in group C to 41 kg in group ZEN15 [9]). In groups ZEN5 and ZEN10, P_4 suppressed the production of antibodies, in particular asymmetrically glycosylated antibodies that are incapable of triggering immune effector mechanisms [23]. The above could be attributed to the fact that endogenous hormones and ZEN (an endocrine-disrupting chemical [6]) demonstrate hormonal activity by binding to nuclear hormone receptors (a rivalry mechanism). The consequences of the unexpected interactions [24] that take place outside the endocrine system are difficult to predict.

3.3. Testosterone

Testosterone was the third analyzed hormone in the present experiment (see Figure 3). Endogenous T plays a key role in female health because it exerts direct androgenic effects or is converted to E_2 [3]. Testosterone concentrations are characterized by significant diurnal variations [34]. They are also affected by the phase of the reproductive cycle as well as the strength and duration of exposure to stressors [35]. Testosterone regulates (i) sexual differentiation, (ii) muscle and bone mass, and (iii) erythropoietic and metabolic processes. It can exert direct and indirect biological effects through conversion to E_2 [36]. In pre-pubertal gilts, an increase in the blood concentrations of T was directly correlated with an increase in the animals' body weights [9]. In groups ZEN10 and ZEN15, E_2 and T levels were bound by negative (inversely proportional) correlations (moderate correlation in group ZEN10, absence of correlations in group ZEN15) on D1. On D3, negative correlations were noted in all groups, and the highest negative values (indicative of moderate correlations) were observed in groups ZEN5 and ZEN15, whereas weak correlations were noted in the groups ZEN10 and C. These findings contradict the results reported by Kanakis et al. [34] and White et al. [35] who argued that the concentration of E_2 in pre-pubertal gilts is influenced by the rate at which T is converted to E_2. In pre-pubertal gilts, the levels of endogenous T do not compensate for physiological demand, and the resulting deficit is covered by "free ZEN" which changes the expression of enzymes such as HSDs [26]. However, the above does not explain the high concentration of T in all groups on D2 (see Figure 3). According to van Anders et al. [37], dominance hierarchy and animal behavior, such as maintenance of leadership and status in the herd, are affected by T, regardless of sex. This is highly probable, but in this study, ZEN was tested only in gilts of similar age and body weight that had been housed without boars. The observed differences ($P \leq 0.01$) indicate that ZEN was a stimulating factor only in group ZEN15 (see Figure 3). Considerable variations were noted in the linear correlations between P_4 and T levels (see Table 2). No correlations or moderate correlations between P_4 and T levels were noted in group C. The experimental groups were characterized by extreme variations in r values, denoting an absence of correlations to the presence of very strong correlations, regardless of group or analytical date. These data are very difficult to interpret. It could be postulated that a LOAEL dose of ZEN increases the concentration of T on all analytical dates (see Figure 3), without inducing any correlations or promoting only weak correlations between E_2 and T levels. However, the interactions between these hormones increase muscle mass and decrease adipose tissue mass [38]. Similar results were reported by Rykaczewska et al. [9].

3.4. Zearalenone (ZEN) and its Metabolites

In the current study, the absorption and biotransformation of ZEN and its metabolites in pre-pubertal gilts varied on an individual basis. Significant differences were not observed due to considerable variations in SD values (see Figure 4).

Our study demonstrated that even trace amounts of ZEN and its metabolites in peripheral blood can affect the levels of selected steroid hormones. This observation is supported by the concentrations of ZEN and its metabolites on D1 (see Figure 4) in all experimental groups. The noted levels of ZEN metabolites have not been documented in the literature. According to most studies, the biotransformation of ZEN in pigs (where 3α-HSD is more active than in other animal species) produces more α-ZEL than β-ZEL [1]. However, the noted concentrations of ZEN and its metabolites in peripheral blood could be indicative of biotransformation processes that do not induce hyperestrogenism, but alleviate the deficiency of endogenous estrogen in prepubertal gilts [27]. In group ZEN5 (MABEL dose), ZEN metabolites were not detected on D1, which can probably be attributed to the low supply of endogenous steroid hormones and the presence of exogenous ZEN. In the first case, the present results (see Figure 4) contradict other authors' findings suggesting that the predominant ZEN metabolite in the peripheral blood of pigs is α-ZEL, rather than β-ZEL [1,14]. The above could be attributed to the higher demand for compounds with estrogenic activity, such as α-ZEL and ZEN, but not β-ZEL [39]. The noted levels of ZEN metabolites could also represent physiological values that are essential for vital life processes [2]. Our findings point to a predominance of detoxification processes in pigs that were exposed to very low doses of ZEN. It should also be noted that: (i) the seventh day of exposure to an undesirable compound such as ZEN is the final date of adaptive processes, in particular adaptive immunity [23], (ii) ZEN was utilized as a substrate (inversely proportional) regulating the expression of HSDs genes which act as molecular switches for the modulation of steroid hormones at the pre-receptor level [26,28,40], (iii) undesirable substances undergo enterohepatic circulation before they are excreted [14,41], and/or (iv) exposure to ZEN induces a specific response from intestinal microbiota [7]. The above processes, alone or in combination, can influence the peripheral blood levels of ZEN and its metabolites. Zearalenone is a promiscuous compound that can inhibit the synthesis and secretion of the follicle-stimulating hormone (FSH, [1]) via negative feedback, which decreases the production of steroid hormones [16,42].

Similar correlations between the mean concentrations of ZEN and its metabolites in peripheral blood were observed on D2 and D3 relative to D1 (see Figure 4). Zearalenone and α-ZEL levels were higher, but still low or very low. The mean concentrations of β-ZEL were similar to the values noted for ZEN on the analyzed dates. These results should be interpreted similarly to the observations made on D1. However, two differences were noted (see Figure 4). Firstly, both ZEN metabolites were present in all experimental groups. Secondly, the values of all analyzed indicators were higher, which can probably be attributed to the accumulation of ZEN and its metabolites (due to the saturation of active estrogen receptors and other factors that affect the concentrations of steroid hormones [1,43]) throughout the experiment. In the literature, the biotransformation of ZEN in peripheral blood was studied in animals exposed to higher doses of ZEN [14,44,45]. According to the hormesis principle, exposure to very low doses of ZEN [12,14,46] influences the synthesis and secretion of sex steroids [1,14,47]. The above implies that very low doses of ZEN were biotransformed in an identical manner, but the parent compound (ZEN) and its metabolites were utilized completely or to a much greater extent (which was the case in group ZEN5). The interactions between endogenous and environmental (exogenous) steroids could also be affected by other endogenous factors [15].

3.5. Summary

This study produced interesting observations regarding the biotransformation of ZEN in pre-pubertal gilts that were orally administered low doses of zearalenone (MABEL, NOAEL and LOAEL) over a period of 42 days. On D1, ZEN metabolites were not detected in the peripheral blood of pigs exposed to the MABEL dose. On D2 and D3 (NOAEL and LOAEL doses), the average concentration of β-ZEL in all experimental groups was three to four times higher than the concentration of α-ZEL. As a result:

- a minor increase was noted in peripheral blood levels of E2 and "free ZEN", proportionally to the ZEN dose and analytical date;
- the concentration of E2 in peripheral blood decreased on D1 in all experimental groups and it increased on D2 and D3 in selected experimental groups (in group ZEN15 on D2, and in groups ZEN10 and ZEN15 on D3);
- testosterone levels increased significantly on all analytical dates in response to the LOAEL dose, and the concentrations of E2 and T were not correlated or were bound by weak linear correlations.

The observed endocrine effects differed in all groups due to qualitative and quantitative changes in the biotransformation of low doses of ZEN. These processes were least pronounced in the group exposed to the MABEL dose: ZEN metabolites were not detected on D1, whereas β-ZEL was the predominant metabolite on D2 and D3. The above was accompanied by an increase in the concentration of E_2 which, together with "free ZEN", probably suppressed P_4 and T levels.

4. Materials and Methods

4.1. General Information

All experimental procedures involving animals were carried out in compliance with Polish regulations setting forth the terms and conditions of animal experimentation (Opinions No. 12/2016 and 45/2016/DLZ of the Local Ethics Committee for Animal Experimentation the University of Warmia and Mazury in Olsztyn, Poland in of 27 April 2016 and 30 November 2016).

4.2. Experimental Animals and Feed

The in vivo experiment was performed at the Department of Veterinary Prevention and Feed Hygiene of the Faculty of Veterinary Medicine at the University of Warmia and Mazury in Olsztyn on 60 clinically healthy pre-pubertal gilts with initial BW of 14.5 ± 2 kg [9]. The animals were housed in pens with free access to water. All groups of gilts received the same feed throughout the experiment. They were randomly assigned to three experimental groups (group ZEN5, group ZEN10 and group ZEN15; $n = 15$) and a control group (group C; $n = 15$ — control group) [48,49]. Group ZEN5 gilts were orally administered ZEN (Sigma-Aldrich Z2125-26MG, St. Louis, MO, USA) at 5 μg ZEN/kg BW, group ZEN10 pigs — at 10 μg ZEN/kg BW, and group ZEN15 pigs — at 15 μg ZEN/kg BW. Analytical samples of ZEN were dissolved in 96 μL of 96% ethanol (SWW 2442-90, Polskie Odczynniki SA, Poland) in weight-appropriate doses. Feed containing different doses of ZEN in an alcohol solution was placed in gel capsules. The capsules were stored at room temperature before administration to evaporate the alcohol. In the experimental groups, ZEN was administered daily in gel capsules before morning feeding. The animals were weighed at weekly intervals, and the results were used to adjust individual mycotoxin doses. Feed was the carrier, and group C pre-pubertal gilts were administered the same gel capsules, but without mycotoxins [7–9].

The feed administered to all experimental animals was supplied by the same producer. Friable feed was provided ad libitum twice daily, at 8:00 a.m. and 5:00 p.m., throughout the experiment. The composition of the complete diet, as declared by the manufacturer, is presented in Table 4.

Table 4. Declared composition of the complete diet.

Parameters	Composition Declared by The Manufacturer (%)
Soybean meal	16
Wheat	55
Barley	22
Wheat bran	4.0
Chalk	0.3
Zitrosan	0.2
Vitamin-mineral premix[1]	2.5

[1]Composition of the vitamin-mineral premix per kg: vitamin A — 500,000 IU; iron — 5000 mg; vitamin D3 — 100,000 IU; zinc — 5000 mg; vitamin E (alpha-tocopherol) — 2000 mg; manganese — 3000 mg; vitamin K — 150 mg; copper (CuSO4·5H2O) — 500 mg; vitamin B1 — 100 mg; cobalt — 20 mg; vitamin B2 — 300 mg; iodine — 40 mg; vitamin B6 — 150 mg; selenium — 15 mg; vitamin B12 — 1500 µg; L-lysine — 9.4 g; niacin — 1200 mg; DL-methionine + cystine — 3.7 g; pantothenic acid — 600 mg; L-threonine — 2.3 g; folic acid — 50 mg; tryptophan — 1.1 g; biotin — 7500 µg; phytase + choline — 10 g; ToyoCerin probiotic + calcium — 250 g; antioxidant + mineral phosphorus and released phosphorus — 60 g; magnesium — 5 g; sodium; calcium — 51 g.

The proximate chemical composition of diets fed to pigs in groups C, ZEN5, ZEN10, and ZEN15 was determined using the NIRS™ DS2500 F feed analyzer (FOSS, Hillerød, Denmark), a monochromator-based NIR reflectance and transflectance analyzer with a scanning range of 850–2500 nm.

4.3. Determination of Mycotoxins in Feed

Feed was analyzed for the presence of mycotoxins and their metabolites: ZEN, α-ZEL and DON. Mycotoxin concentrations in feed were determined by separation in immunoaffinity columns (Zearala-Test™ Zearalenone Testing System, G1012, VICAM, Watertown, MA, USA; DON-Test™ DON Testing System, VICAM, Watertown, MA, USA) and high-performance liquid chromatography (HPLC system, Hewlett Packard type 1050 and 1100) — mass spectrometry (MS) and chromatographic column (Atlantis T3 3 µm 3.0 × 150 mm Column No. 186003723, Waters, AN Etten-Leur, Ireland). The mobile phase was a water and acetonitrile mixture with an 80:10 solvent ratio and 2 ml of CH3 COOH. The flow rate was 0.4 mL/min. The obtained values did not exceed the limits of quantitation (LoQ) of 2 ng/g for ZEN and 5 ng/g for DON. The analyzed compounds were quantified at the Department [50].

4.4. Blood Sampling

Blood was sampled from 5 gilts from every group on three analytical dates: exposure day 7 (D1), exposure day 21 (D2) and exposure day 42 (D3). Directly before slaughter, blood samples of 20 mL each were collected from all gilts (blood was sampled within 20 s after immobilization [51]) by jugular venipuncture into syringe containing 0.5 mL of heparin solution. Blood was centrifuged at 3000 rpm for 20 min at 4 °C. The obtained plasma samples were stored at – 18 °C until the analyses of ZEN, α-ZEL, β-ZEL, estradiol (E_2), progesterone (P_4) and testosterone (T) concentrations.

4.5. Determination of Hormone Concentrations

4.5.1. Estradiol

Estradiol concentration was determined at the Institute of Animal Reproduction and Food Research of the Polish Academy of Sciences in Olsztyn, Poland. Blood plasma concentrations of E_2 were analyzed by the radioimmunoassay (RIA) method with a commercially available kit (ESTR-US-CT, CIS BIO ASSAYS), as described previously [52,53]. All measurements were performed in duplicate for every cultured probe. The extraction yield for E_2 was 90.67% ± 0.73%. The radioactivity of samples with J^{125} was measured with the Wallac 1470 WIZARDÆ automatic gamma scintillation counter (Perkin Elmer, Waltham, MA, USA). Radioactivity was determined within 1 min with a Geiger counter (counting efficiency, 75%). The sensitivity of the E_2 assay was 1.36 pg/mL. The standard curve range was from 2.72 to 550 pg/mL. The intra- and inter-assay coefficients of variation were 5% and 5%, respectively.

4.5.2. Progesterone and Testosterone

Progesterone and testosterone were quantified at the Analytical Laboratory of the Municipal Hospital with Polyclinic in Olsztyn, Poland, by the ECLIA electrochemiluminescence assay with the use of Elecsys Progesterone II and Elecsys Testosterone II assays and the Cobas c6000 analyzer (Hitachi, Tokyo, Japan). In the first stage, the samples were incubated with biotinylated monoclonal antibodies specific for P_4 and T and for P_4 and T derivatives labeled with a ruthenium complex. The extent to which the hormones were bound to antibodies was determined by their concentrations. Streptavidin-coated microspheres were added in the second stage, and the complex was bound to the solid phase during the interactions between biotin and streptavidin. The quantity of labeled P_4 bound to the solid phase was inversely proportional to the concentration of P_4 in the sample. The reaction mix was sucked into the measuring cell where microspheres were magnetically captured on the surface of the electrode. Unbound compounds were removed with the ProCell. Voltage was applied to the electrode, and the resulting chemiluminescence was measured with a photomultiplier. The results were read from a two-point calibration curve and a standard curve developed with a barcode verifier. The analytical range of the method was determined by the lower limit of detection and the highest point on the calibration curve at 0.03–60 ng/mL for P_4 and 0.025–15 ng/mL for T. All determinations were performed in accordance with the manufacturer's instructions.

4.5.3. Statistical Analysis

Hormone concentrations were measured in three experimental groups and the control group on three analytical dates. The results were expressed as mean values ($\bar{x}$) and standard deviation (SD) for each sample. The following assays were performed for every hormone: (i) the differences between means were analyzed for the experimental groups and the control groups on fixed dates and (ii) the differences between means were analyzed in a fixed group on each analytical date. In both cases, the determinations were made by one-way ANOVA. If the differences between group means were statistically significant, the differences between pairs of means were determined by Tukey's multiple comparison test. The equality of variances in the compared groups was evaluated with Levene's test and the Brown–Forsythe test. If the equal variance hypothesis was rejected in both tests, the significance of differences was evaluated with the Kruskal–Wallis non-parametric test. In each analysis, the tested values were regarded as highly significant at $P < 0.01$ (**) and as significant at $0.01 < P < 0.05$ (*). Linear correlations between the concentrations of steroid hormones in fixed groups were determined based on the values of the Pearson's correlation coefficient [22]. Data were processed in Statistica v. 13 (TIBCO Software Inc., Silicon Valley, CA, USA, 2017).

4.6. Extraction Procedure

The presence of zearalenone, α-ZEL and β-ZEL in the blood plasma were determined with the use of immuno-affinity columns (Zearala-TestTM Zearalenone Testing System, G1012, VICAM, Watertown, MA, USA) and different protocols for each compound. All extraction procedures were performed according to the recommendations of column manufacturers. After extraction, the eluates were placed in a water bath at 50 °C, and the solvent was evaporated in a stream of nitrogen. Next, 0.5 mL of 99.8% methanol was added to dry residues to dissolve the mycotoxin.

4.6.1. Quantification of ZEN and Its Metabolites

The presence of ZEN, α-ZEL and β-ZEL in the blood plasma was determined by various separation methods with the use of immuno-affinity columns (Zearala-TestTM Zearalenone Testing System, G1012, VICAM, Watertown, MA, USA) and the Agilent 1100 series liquid chromatography (LC)/mass spectrometry (MS) system. The prepared sample was estimated with the use of a chromatographic column (Atlantis T3 3 μm 3.0 × 150 mm Column No. 186003723, Waters, AN Etten-Leur, Ireland). The mobile phase consisted of 70% acetonitrile (LiChrosolvTM, No. 984 730 109, Merck-Hitachi, Mannheim, Germany), 20% methanol (LiChrosolvTM, No. 1.06 007, Merck-Hitachi, Mannheim, Germany) and 10% deionized water (Milipore Water Purification System, Millipore S.A. Molsheim-France, 2 mL of CH3 COOH). The immunoaffinity bed in the column was washed with demineralized water (Millipore Water Purification System, Millipore S.A., Molsheim, France). The flow rate was 0.4 mL/min., and the temperature of the oven column was 40 °C. The chromatographic analysis was completed in 4 min. The column was eluted with 99.8% methanol (LIChrosolvTM, No. 1.06 007, Merck-Hitachi, Mannheim, Germany) to remove the bound mycotoxin. The eluates were placed in a water bath at 50 °C, and the solvent was evaporated in a stream of nitrogen. In the next step, 0.5 mL of 99.8% methanol was added to dry residues to dissolve the mycotoxin. Mycotoxin concentrations were determined according to the external standard and were expressed in ppb (ng/mL).

Matrix-matched calibration standards were used for quantification to avoid matrix effects which can reduce sensitivity. The calibration standards were dissolved in the sample matrix prepared according to the same procedure as the remaining samples. The material used for the preparation of calibration standards was mycotoxin-free. The limits of detection (LODs) for individual mycotoxins were determined as the concentrations at which the signal-to-noise ratio decreased to 3. Alpha-ZEL and beta-ZEL were also determined. Derivative concentrations were below the LODs, and they were separated from their respective parent compounds during purification.

4.6.2. Statistical Analysis

The concentrations of ZEN and its metabolites in the blood plasma of prepubertal gilts were analyzed in the control group and in three experimental groups on three analytical dates. The results were expressed as mean values ($\overline{x}$) and standard deviation (SD) for each sample. The following assays were performed for every hormone: (i) the differences between means were analyzed for the three ZEN doses (experimental groups) and the control groups on fixed dates, and (ii) the differences between means were analyzed for a fixed ZEN dose (group) on each analytical date. In both cases, the differences between means were determined by one-way ANOVA. If the differences between group means were statistically significant, the differences between pairs of means were determined with Tukey's multiple comparison test. If all values were below the limit of detection (mean and variance equal to zero) in any group, one-way ANOVA was performed for the remaining groups (if the number of the remaining groups was higher than two), and the means of these groups were compared against zero with the use of Student's *t*-test. The differences between groups were determined with Student's *t*-test. In each analysis, the tested values were regarded as highly significant at $P < 0.01$ (**) and as significant at $0.01 < P < 0.05$ (*). Data were processed in Statistica v. 13 (TIBCO Software Inc., Silicon Valley, CA, USA, 2017).

Author Contributions: The experiments were conceived and designed by M.G. and M.T.G. The experiments were performed by K.C., A.R., E.O. and M.D. Data were analyzed and interpreted by K.C., M.D., M.B., S.L.-Z. and M.G. The manuscript was drafted by K.C. and M.G. and critically edited by A.B., Ł.Z. and M.T.G.

Funding: The study was supported by the "Healthy Animal - Safe Food" Scientific Consortium of the Leading National Research Centre (KNOW) pursuant to a decision of the Ministry of Science and Higher Education No. 05-1/KNOW2/2015.

Conflicts of Interest: The authors declare no conflict of interest.

References

1. Zheng, W.; Feng, N.; Wang, Y.; Noll, L.; Xu, S.; Liu, X.; Lu, N.; Zou, H.; Gu, J.; Yuan, Y.; et al. Effects of zearalenone and its derivatives on the synthesis and secretion of mammalian sex steroid hormones: A review. *Food Chem. Toxicol.* **2019**, *126*, 262–276. [CrossRef] [PubMed]
2. Kowalska, K.; Habrowska-Górczyńska, D.E.; Piastowska-Ciesielska, A. Zearalenone as an endocrine disruptor in humans. *Environ. Toxicol. Pharmacol.* **2016**, *48*, 141–149. [CrossRef] [PubMed]

3. Islam, R.M.; Bell, R.J.; Green, S.; Davis, S.R. Effects of testosterone therapy for women: A systematic review and meta-analysis protocol. *Syst. Rev.* **2019**, *8*, 19. [CrossRef] [PubMed]
4. Knutsen, H.-K.; Alexander, J.; Barregård, L.; Bignami, M.; Brüschweiler, B.; Ceccatelli, S.; Cottrill, B.; Dinovi, M.; Edler, L.; Grasl-Kraupp, B.; et al. Risks for animal health related to the presence of zearalenone and its modified forms in feed. *EFSA J.* **2017**, *15*, 4851. [CrossRef]
5. Pastoor, T.P.; Bachman, A.N.; Bell, D.R.; Cohen, S.M.; Dellarco, M.; Dewhurst, I.C.; Doe, J.E.; Doerrer, N.G.; Embry, M.R.; Hines, R.N.; et al. A 21st century roadmap for human health risk assessment. *Crit. Rev. Toxicol.* **2014**, *44*, 1–5. [CrossRef]
6. Vandenberg, L.N.; Colborn, T.; Hayes, T.B.; Heindel, J.J.; Jacobs, D.R.; Lee, D.-H.; Shioda, T.; Soto, A.M.; vom Saal, F.S.; Welshons, W.V.; et al. Hormones and endocrine-disrupting chemicals: Low-dose effects and nonmonotonic dose responses. *Endocr. Rev.* **2012**, *33*, 378–455. [CrossRef]
7. Cieplińska, K.; Gajęcka, M.; Dąbrowski, M.; Rykaczewska, A.; Zielonka, Ł.; Lisieska-Żołnierczyk, S.; Bulińska, M.; Gajęcki, M.T. Time-dependent changes in the intestinal microbiome of gilts exposed to low zearalenone doses. *Toxins* **2019**, *11*, 296. [CrossRef]
8. Cieplińska, K.; Gajęcka, M.; Nowak, A.; Dąbrowski, M.; Zielonka, Ł.; Gajęcki, M.T. The gentoxicity of caecal water in gilts exposed to low doses of zearalenone. *Toxins* **2018**, *10*, 350. [CrossRef]
9. Rykaczewska, A.; Gajęcka, M.; Dąbrowski, M.; Wiśniewska, A.; Szcześniewska, J.; Gajęcki, M.T.; Zielonka, Ł. Growth performance, selected blood biochemical parameters and body weight of pre-pubertal gilts fed diets supplemented with different doses of zearalenone (ZEN). *Toxicon* **2018**, *152*, 84–94. [CrossRef]
10. Gruber-Dorninger, C.; Jenkins, T.; Schatzmayr, G. Global Mycotoxin Occurrence in Feed: A Ten-Year Survey. *Toxins* **2019**, *11*, 375. [CrossRef]
11. Pereira, C.S.; Cunha, S.C.; Fernandes, J.O. Prevalent Mycotoxins in Animal Feed: Occurrence and Analytical Methods. *Toxins* **2019**, *11*, 290. [CrossRef] [PubMed]
12. Calabrese, E.J. Paradigm lost, paradigm found: The re-emergence of hormesis as a fundamental dose response model in the toxicological sciences. *Environ. Pollut.* **2005**, *138*, 378–411. [CrossRef] [PubMed]
13. Gajęcka, M.; Zielonka, Ł.; Gajęcki, M. The effect of low monotonic doses of zearalenone on selected reproductive tissues in pre-pubertal female dogs—A review. *Molecules* **2015**, *20*, 20669–20687. [CrossRef] [PubMed]
14. Gajęcka, M.; Zielonka, Ł.; Gajęcki, M. Activity of zearalenone in the porcine intestinal tract. *Molecules* **2017**, *22*, 18. [CrossRef] [PubMed]
15. Luisetto, M.; Almukhtar, N.; Ahmadabadi, B.N.; Hamid, G.A.; Mashori, G.R.; Khan, K.R.; Khan, F.A.; Cabianca, L. Endogenus Toxicology: Modern Physio-Pathological Aspects and Relationship with New Therapeutic Strategies. An Integrative Discipline Incorporating Concepts from Different Research Discipline Like Biochemistry, Pharmacology and Toxicology. *Clin. Pathol.* **2019**, *3*, 000113. [CrossRef]
16. He, J.; Wei, C.; Li, Y.; Liu, Y.; Wang, Y.; Pan, J.; Liu, J.; Wu, Y.; Cui, S. Zearalenone and alpha-zearalenol inhibit the synthesis and secretion of pig follicle stimulating hormone via the non-classical estrogen membrane receptor GPR30. *Mol. Cell. Endocrinol.* **2018**, *461*, 43–54. [CrossRef] [PubMed]
17. Dąbrowski, M.; Obremski, K.; Gajęcka, M.; Gajęcki, M.T.; Zielonka, Ł. Changes in the Subpopulations of Porcine Peripheral Blood Lymphocytes Induced by Exposure to Low Doses of Zearalenone (ZEN) and Deoxynivalenol (DON). *Molecules* **2016**, *21*, 557. [CrossRef]
18. Chen, L.; Giesy, J.P.; Xie, P. The dose makes the poison. *Sci. Total Environ.* **2018**, *621*, 649–653. [CrossRef] [PubMed]
19. Cole, T.J.; Short, K.L.; Hooper, S.B. The science of steroids. *Semin. Fetal Neonat. Med.* **2019**, *24*, 170–175. [CrossRef]
20. Przybylska-Gornowicz, B.; Lewczuk, B.; Prusik, M.; Hanuszewska, M.; Petrusewicz-Kosińska, M.; Gajęcka, M.; Zielonka, Ł.; Gajęcki, M. The Effects of Deoxynivalenol and Zearalenone on the Pig Large Intestine. A Light and Electron Microscopy Study. *Toxins* **2018**, *10*, 148. [CrossRef]
21. Przybylska-Gornowicz, B.; Tarasiuk, M.; Lewczuk, B.; Prusik, M.; Ziółkowska, N.; Zielonka, Ł.; Gajęcki, M.; Gajęcka, M. The Effects of Low Doses of Two Fusarium Toxins, Zearalenone and Deoxynivalenol, on the Pig Jejunum. A Light and Electron Microscopic Study. *Toxins* **2015**, *7*, 4684–4705. [CrossRef] [PubMed]
22. Williams, M.S.; Ebel, E.D. Estimating correlation of prevalence at two locations in the farm-to-table continuum using qualitative test data. *Int. J. Food Microbiol.* **2017**, *245*, 29–37. [CrossRef] [PubMed]

23. Benagiano, M.; Bianchi, P.; D'Elios, M.M.; Brosens, I.; Benagiano, G. Autoimmune diseases: Role of steroid hormones. *Best Pract. Res. Clin. Obstet. Gynaecol.* **2019**, in press. [CrossRef] [PubMed]
24. Gajęcka, M.; Przybylska-Gornowicz, B. The low doses effect of experimental zearalenone (ZEN) intoxication on the presence of Ca2+ in selected ovarian cells from pre-pubertal bitches. *Pol. J. Vet. Sci.* **2012**, *15*, 711–720. [CrossRef] [PubMed]
25. Wilkenfeld, S.R.; Linc, C.; Frigo, D.E. Communication between genomic and non-genomic signaling events coordinate steroid hormone actions. *Steroids* **2018**, *133*, 2–7. [CrossRef] [PubMed]
26. Gajęcka, M.; Otrocka-Domagała, I. Immunocytochemical expression of 3β- and 17β-hydroxysteroid dehydrogenase in bitch ovaries exposed to low doses of zearalenone. *Pol. J. Vet. Sci.* **2013**, *16*, 55–62. [CrossRef] [PubMed]
27. Lawrenz, B.; Melado, L.; Fatemi, H. Premature progesterone rise in ART-cycles. *Reprod. Biol.* **2018**, *18*, 1–4. [CrossRef] [PubMed]
28. Schoevers, E.J.; Santos, R.R.; Colenbrander, B.; Fink-Gremmels, J.; Roelen, B.A.J. Transgenerational toxicity of Zearalenone in pigs. *Reprod. Toxicol.* **2012**, *34*, 110–119. [CrossRef] [PubMed]
29. Rivera, H.M.; Stincic, T.L. Estradiol and the control of feeding behavior. *Steroids* **2018**, *133*, 44–52. [CrossRef]
30. Jiang, X.; Lu, N.; Xue, Y.; Liu, S.; Lei, H.; Tu, W.; Lu, Y.; Xia, D. Crude fiber modulates the fecal microbiome and steroid hormones in pregnant Meishan sows. *Gen. Comp. Endocrinol.* **2019**, *277*, 141–147. [CrossRef]
31. Jakimiuk, E.; Kuciel-Lisieska, G.; Zwierzchowski, W.; Gajęcka, M.; Obremski, K.; Zielonka, Ł.; Skorska-Wyszyńska, E.; Gajęcki, M. Morphometric changes of the reproductive system in gilts during zearalenone mycotoxicosis. *Med. Weter.* **2006**, *62*, 99–102.
32. Goldstein, I.; Kim, N.N.; Clayton, A.H.; DeRogatis, L.R.; Giraldi, A.; Parish, S.J.; Pfaus, J.; Simon, J.A.; Kingsberg, S.A.; Meston, C.; et al. Hypoactive Sexual Desire Disorder: International Society for the Study of Women's Sexual Health (ISSWSH) Expert Consensus Panel Review. *Mayo Clin. Proc.* **2017**, *92*, 114–128. [CrossRef] [PubMed]
33. Blitek, A.; Szymanska, M.; Pieczywek, M.; Morawska-Pucinska, E. Luteal P4 synthesis in early pregnant gilts after induction of estrus with PMSG/hCG. *Anim. Reprod. Sci.* **2016**, *166*, 28–35. [CrossRef] [PubMed]
34. Kanakis, G.A.; Tsametis, C.P.; Goulis, DG. Measuring testosterone in women and men. *Maturitas* **2019**, in press. [CrossRef] [PubMed]
35. White, S.F.; Lee, Y.; Phan, J.M.; Moody, S.N.; Shirtcliff, E.A. Putting the flight in "fight-or-flight": Testosterone reactivity to skydiving is modulated by autonomic activation. *Biol. Psychol.* **2019**, *143*, 93–102. [CrossRef] [PubMed]
36. Bhasin, S.; Jasuja, R. Reproductive and Nonreproductive Actions of Testosterone. *Enc. Endocr. Dis.* **2019**, *2*, 721–734.
37. Van Anders, S.M.; Steiger, J.; Goldey, K.L. Effects of gendered behavior on testosterone in women and men. *Proc. Natl. Acad. Sci. USA* **2015**, *112*, 13805–13810. [CrossRef]
38. Tyagi, V.; Scordo, M.; Yoon, R.S.; Liporace, F.A.; Greene, L.W. Revisiting the role of testosterone: Are we missing something? *Rev. Urol.* **2017**, *19*, 16–24. [CrossRef]
39. Yang, D.; Jiang, T.; Lin, P.; Chen, H.; Wang, L.; Wang, N.; Zhao, F.; Tang, K.; Zhou, D.; Wang, A.; et al. Apoptosis inducing factor gene depletion inhibits zearalenone-induced cell death in a goat Leydig cell line. *Reprod. Toxicol.* **2017**, *67*, 129–139. [CrossRef]
40. Gajęcka, M.; Rybarczyk, L.; Zwierzchowski, W.; Jakimiuk, E.; Zielonka, Ł.; Obremski, K.; Gajęcki, M. The effect of experimental, long-term exposure to low-dose zearalenone mycotoxicosis on the histological condition of ovaries in sexually immature gilts. *Theriogenology* **2011**, *75*, 1085–1094. [CrossRef]
41. Hennig-Pauka, I.; Koch, F.J.; Schaumberger, S.; Woechtl, B.; Novak, J.; Sulyok, M.; Nagl, V. Current challenges in the diagnosis of zearalenone toxicosis as illustrated by a field case of hyperestrogenism in suckling piglets. *Porc. Health Manag.* **2018**, *4*, 1–9. [CrossRef] [PubMed]
42. Rogowska, A.; Pomastowski, P.; Sagandykova, G.; Buszewski, B. Zearalenone and its metabolites: Effect on human health, metabolism and neutralisation methods. *Toxicon* **2019**, *162*, 46–56. [CrossRef] [PubMed]
43. Zielonka, Ł.; Waśkiewicz, A.; Beszterda, M.; Kostecki, M.; Dąbrowski, M.; Obremski, K.; Goliński, P.; Gajęcki, M. Zearalenone in the Intestinal Tissues of Immature Gilts Exposed per os to Mycotoxins. *Toxins* **2015**, *7*, 3210–3223. [CrossRef] [PubMed]

44. Demaegdt, H.; Daminet, B.; Evrard, A.; Scippo, M.L.; Muller, M.; Pussemier, L.; Callebaut, A.; Vandermeiren, K. Endocrine activity of mycotoxins and mycotoxin mixtures. *Food Chem. Toxicol.* **2016**, *96*, 107–116. [CrossRef] [PubMed]
45. Gajęcka, M.; Sławuta, P.; Nicpoń, J.; Kołacz, R.; Kiełbowicz, Z.; Zielonka, Ł.; Dąbrowski, M.; Szweda, W.; Gajęcki, M.; Nicpoń, J. Zearalenone and its metabolites in the tissues of female wild boars exposed per os to mycotoxins. *Toxicon* **2016**, *114*, 1–12. [CrossRef] [PubMed]
46. Parveen, M.; Zhu, Y.; Kiyama, R. Expression profiling of the genes responding to zearalenone and its analogues using estrogen-responsive genes. *FEBS Lett.* **2009**, *583*, 2377–2384. [CrossRef] [PubMed]
47. Gajęcka, M.; Zielonka, Ł.; Dąbrowski, M.; Mróz, M.; Gajęcki, M. The effect of low doses of zearalenone and its metabolites on progesterone and 17β-estradiol concentrations in peripheral blood and body weights of pre-pubertal female Beagle dogs. *Toxicon* **2013**, *76*, 260–269. [CrossRef]
48. Heberer, T.; Lahrssen-Wiederholt, M.; Schafft, H.; Abraham, K.; Pzyrembel, H.; Henning, K.J.; Schauzu, M.; Braeunig, J.; Goetz, M.; Niemann, L.; et al. Zero tolerances in food and animal feed—Are there any scientific alternatives? A European point of view on an international controversy. *Toxicol. Lett.* **2007**, *175*, 118–135. [CrossRef]
49. Smith, D.; Combes, R.; Depelchin, O.; Jacobsen, S.D.; Hack, R.; Luft, J.; Lammens, L.; von Landenberg, F.; Phillips, B.; Pfister, R.; et al. Optimising the design of preliminary toxicity studies for pharmaceutical safety testing in the dog. *Regul. Toxicol. Pharmacol.* **2005**, *41*, 95–101. [CrossRef]
50. Gajęcki, M. *The Effect of Experimentally Induced Fusarium Mycotoxicosis on Selected Diagnostic and Morphological Parameters of the Porcine Digestive Tract*; Development Project NR12-0080-10 entitled; The National Centre for Research and Development: Warsaw, Poland, 30 November 2013; pp. 1–180.
51. Kowalski, A.; Kaleczyc, J.; Gajęcki, M.; Zieliński, H. Adrenaline, noradrenaline and cortisol levels in pigs during blood collection (In Polish). *Med. Weter.* **1996**, *52*, 716–718.
52. Gajęcka, M.; Woźny, M.; Brzuzan, P.; Zielonka, Ł.; Gajęcki, M. Expression of CYPscc and 3β-HSD mRNA in bitches ovary after long-term exposure to zearalenone. *Bull. Vet. Inst. Pulawy* **2011**, *55*, 777–780.
53. Stanczyk, F.Z.; Xu, X.; Sluss, P.M.; Brinton, L.A.; McGlynn, K.A. Do metabolites account for higher serum steroid hormone levels measured by RIA compared to mass spectrometry? *Clin. Chim. Acta* **2018**, *484*, 223–225. [CrossRef] [PubMed]

© 2019 by the authors. Licensee MDPI, Basel, Switzerland. This article is an open access article distributed under the terms and conditions of the Creative Commons Attribution (CC BY) license (http://creativecommons.org/licenses/by/4.0/).

Article

The Effect of Zearalenone on the Cytokine Environment, Oxidoreductive Balance and Metabolism in Porcine Ileal Peyer's Patches

Kazimierz Obremski [1,*], Wojciech Trybowski [2], Paweł Wojtacha [3,*], Magdalena Gajęcka [1], Józef Tyburski [4] and Łukasz Zielonka [1]

1 Department of Veterinary Prevention and Feed Hygiene, Faculty of Veterinary Medicine, University of Warmia and Mazury in Olsztyn, Oczapowskiego 13/29, 10-718 Olsztyn, Poland; mgaja@uwm.edu.pl (M.G.); lukaszz@uwm.edu.pl (Ł.Z.)
2 Regional Veterinary Inspectorate in Gdańsk, Na Stoku 50, 80-958 Gdańsk, Poland; wtrybowski@gmail.com
3 Department of Industrial and Food Microbiology, Faculty of Food Science, University of Warmia and Mazury in Olsztyn, Pl. Cieszyński 1, 10-726 Olsztyn, Poland
4 Department of Agroecosystems, Faculty of Environmental Management and Agriculture, University of Warmia and Mazury in Olsztyn, Pl. Łódzki 3, 10-719 Olsztyn, Poland; jozef.tyburski@uwm.edu.pl
* Correspondence: kazimierz.obremski@uwm.edu.pl (K.O.); pawel.wojtacha@uwm.edu.pl (P.W.)

Received: 17 April 2020; Accepted: 25 May 2020; Published: 27 May 2020

Abstract: The aim of the present study was to determine the effect of zearalenone (ZEN), administered *per os* to gilts at doses equivalent to 50%, 100%, and 150% of no-observed-adverse-effect level (NOAEL) values for 14, 28, and 42 days during weaning, on changes in the parameters of the oxidoreductive balance, cytokine secretion, and basal metabolism in ileal Payer's patches. Immunoenzymatic ELISA tests and biochemical methods were used to measure the concentrations of interleukin 1α, interleukin 1β, interleukin 12/23p40, interleukin 2, interferon γ, interleukin 4, interleukin 6, interleukin 8, tumor necrosis factor α, interleukin 10, transforming growth factor β, malondialdehyde, sulfhydryl groups, fructose, glucose, and proline, as well as the activity of peroxidase, superoxide dismutase and catalase. The study demonstrated that ZEN doses corresponding to 50%, 100%, and 150% of NOAEL values, i.e., 5 µg, 10 µg, and 15 µg ZEN/kg BW, respectively, have proinflammatory properties, exacerbate oxidative stress responses, and disrupt basal metabolism in ileal Payer's patches in gilts.

Keywords: zearalenone; pre-pubertal gilts; GALT; oxidative stress; cytokine; metabolism

Key Contribution: This is the first ever study to investigate the influence of ZEN administered *per os* at NOAEL concentrations on the maintenance of homeostasis in the intestinal immune system of gilts during weaning. The results indicate that ZEN exerts proinflammatory effects, initiates the mechanism of oxidative stress in GALT, and modulates intestinal epithelial function by inducing changes in the metabolism of sugars and amino acids.

---

1. Introduction

Zearalenone (ZEN) is regarded as one of the most frequently occurring mycotoxins in the world [1–3]. The structure of ZEN molecules enables the mycotoxin to bind to estrogen receptors and induce estrogenic effects. Cereals are the most frequent vectors of ZEN transmission [4], and according to the literature, 32% of 5010 mixed cereal samples analyzed in Europe were found to be ZEN positive [5]. Zearalenone is detected not only in fresh plants, but also in processed products, including pelleted feed, which can be attributed to the mycotoxin's stability under exposure to increased temperature and pressure [6,7].

Pigs are regarded as a livestock that is very sensitive to ZEN. The above applies particularly to pre-pubertal gilts where ZEN induces symptoms of ovarian cysts, uterine edema, and early maturation of ovarian follicles [8,9]. In pre-pubertal gilts, exposure to ZEN causes edema and thickening of the vaginal wall and the lining of the vulva, increases uterus weight, and leads to ovarian atrophy and intensified proliferation of the epithelium of the uterine mucosa within 3 to 7 days [10]. In pregnant and lactating pigs, exposure to ZEN reduces the number of follicles in F1-generation piglets, which can lead to the premature recruitment of the oocyte pool in subsequent stages of development [11].

The gastrointestinal tract, a very important system in the body that digests and absorbs nutrients, is the first physiological barrier to ZEN [12,13]. Considerable differences in ZEN absorption have been observed in different segments of the small intestine, where 70%–80% of the mycotoxin was absorbed in the jejunum, and only 15%–30% in the ileum [14]. After crossing the intestinal barrier, ZEN enters the bloodstream and reaches the liver, where it is mainly biotransformed in hepatocytes in the presence of reducing factors such as NADPH, which leads to the transformation of ZEN to α-zearalenol (α-ZEL), β-zearalenol (β-ZEL), α-zearalanol (α-ZAL), and β-zearalanol (β-ZAL). Despite the fact that ZEN metabolites have different affinity for estrogen receptors ERα and ERβ [15], they disrupt endocrine functions in various animal species, both male and female [16]. Zearalenone's estrogenic activity is 80 to 160 times weaker than that of 17-β-estradiol, but ZEN effectively competes with 17-β-estradiol, and the formed complex activates the transcription of estrogen-sensitive genes [16,17].

During continuous exposure to ZEN, the tissues of the porcine gastrointestinal tract remain under the constant influence of ZEN and its metabolites due to enterohepatic circulation of metabolites in pigs [18,19]. Zearalenone's tropism for estrogen receptors plays a key role in the mycotoxin's influence on the functions of gut-associated lymphoid tissue (GALT). Zearalenone exerts its effects mainly through ERα on T lymphocytes, natural killer (NK) cells, and macrophages, as well as ERβ that occur mainly on B lymphocytes and monocytes, and these processes are closely related to the mycotoxin's effect on metabolic transformations in immunocompetent cells. Zearalenone also disrupts the oxidoreductive balance, which can lead to changes in cytokine expression in GALT.

Previous studies have demonstrated that progressing oxidative stress induced by ZEN in the gastrointestinal system increases the peroxidation of macromolecules, leading to changes to lipid membranes and proteins, and, consequently, a decrease in the barrier properties of the intestinal epithelium [20]. Reactive oxygen species (ROS) produced during lipid peroxidation promote the formation of 4-hydroxy-2-nonenal (4-HNE), malondialdehyde (MDA), propanal, and hexanal [21]. Malondialdehyde is the most mutagenic product, but 4-HNE is most toxic because it reacts with protein thiol and amino groups. Moreover, 4-HNE is metabolized to acetaldehyde, and together with MDA, it is bound to macromolecules via the amino groups of lysine, histidine, and arginine residues. The Schiff base is regrouped, which leads to the synthesis of lipid peroxidation end products. The same reactions take place in DNA, which induce the formation of protein/MDA/DNA intermolecular cross-links. These molecules induce inflammation, initiate complement activation, and play a role in the atherosclerotic process. By binding to mitochondrial proteins, MDA also modifies the activity of enzymes, membrane transport proteins, and cytoskeletal proteins.

GALT comprises different structures in the small intestine. It is composed of organized lymphoid follicles in the jejunum, whereas the ileum contains a continuous layer of aggregated lymphoid nodules (Peyer's patches) that stretches from the ileum to the colon [22,23]. The functional significance of the organizational structure of Peyer's patches remains unknown. It could be speculated that the massive accumulation of lymphoid tissue protects the boundary between the small intestine (jejunum and ileum) with moderate numbers of bacteria and the large intestine, which has an abundant microflora and contains potentially pathogenic microorganisms. The aim of the present experiment was to determine the effect of ZEN doses equivalent to 50% no-observed-adverse-effect level (NOAEL)—the no observable effect level (5 μg/kg BW), 100% NOAEL (10 μg/kg BW), and 150% NOAEL (15 μg/kg BW) values on changes in the oxidoreductive balance, cytokine secretion, and metabolic markers in ileal Peyer's patches in pre-pubertal gilts during weaning.

2. Results

2.1. The Effect of Zearalenone on Cytokine Secretion

2.1.1. Proinflammatory Cytokines

The concentration of IFN-γ in porcine ileal Peyer's patches ranged from 9.52 to 52.45 pg/mg (Table 1). Significant differences were observed between groups and analytical dates. In groups ZEN II and ZEN III, the secretion of IFN-γ tended to increase with prolonged exposure to ZEN, and IFN-γ concentration increased by 29.73 and 11.03 ng/mg, respectively, relative to day 14 of the experiment. The noted increase was determined not only by the duration of exposure, but also by the administered ZEN dose. The concentration of IFN-γ peaked at 52.45 pg/mg on day 42 in response to a ZEN dose of 10 μg ZEN/kg BW (Table 1). On day 42, the concentration of IFN-γ decreased by 18.47 and 9.58 pg/mg in the control group and group ZEN I, respectively, relative to day 14.

Table 1. Changes in IFN-γ content in the ileum. Experimental group and zearalenone (ZEN) concentration: Control Group, ZEN I—5 μg/kg BW, ZEN II—10 μg/kg BW, ZEN III—15 μg/kg BW.

Group	Experimental Day 14		28		42	
	mean	SEM	mean	SEM	mean	SEM
Control	35.98	9.66	17.33	2.84	17.51^{A}	2.49
ZEN I	19.10	7.88	10.60	1.69	9.52^{B}	0.70
ZEN II	22.72	3.31	16.55	2.59	52.45^{C}	10.37
ZEN III	23.66	5.39	28.61	7.28	34.69	3.17

The results are expressed as means ± SEM (standard error of the mean). Statistically significant differences: 42 day, ^{A} Control Group vs. ZEN II, $p < 0.01$; ^{B} ZEN I vs. ZEN II, $p < 0.001$; ^{C} ZEN I vs. ZEN III, $p < 0.05$.

In the remaining groups of proinflammatory cytokines, similar trends were noted in the secretion of interleukin (IL)-1α, IL-1β, and IL-2 (Tables 2–4). Significant differences were observed between experimental and control gilts and analytical dates. The greatest increase in IL-1α secretion was noted in response to ZEN doses of 10 μg/kg BW (ZEN II) and 15 μg/kg BW (ZEN III), in particular on days 14 and 42, and IL-1α levels increased by 322.70 and 351.90 pg/mg (ZEN II) and by 155.80 and 287.30 pg/mg (ZEN III), respectively, relative to the control group (Table 2).

Table 2. Changes in IL-1α content in the ileum. Experimental group and ZEN concentration: Control Group, ZEN I—5 μg/kg BW, ZEN II—10 μg/kg BW, ZEN III—15 μg/kg BW.

Group	Experimental day 14		28		42	
	mean	SEM	mean	SEM	mean	SEM
Control	307.30^{A}	12.12	352.70	30.74	$281.00^{C,D}$	16.87
ZEN I	283.10^{B}	32.76	267.50	20.80	$161.20^{E,F}$	8.16
ZEN II	630.00	72.20	390.70	33.47	632.90	107.90
ZEN III	463.10	81.18	417.90	77.07	568.30	55.67

The results are expressed as means ± SEM (standard error of the mean). Statistically significant differences: 14 day, ^{A} Control Group vs. ZEN II, $p < 0.001$; ^{B} ZEN I vs. ZEN II, $p < 0.001$; 42 day, ^{C} Control Group vs. ZEN II, $p < 0.0001$, ^{D} Control Group vs. ZEN III, $p < 0.01$, ^{E} ZEN I vs. ZEN II, $p < 0.0001$, ^{F} ZEN I vs. ZEN III, $p < 0.0001$.

Table 3. Changes in IL-1β content in the ileum. Experimental group and ZEN concentration: Control Group, ZEN I—5 μg/kg BW, ZEN II—10 μg/kg BW, ZEN III—15 μg/kg BW.

Group	Experimental Day					
	14		28		42	
	mean	SEM	mean	SEM	mean	SEM
Control	113.70	9.32	114.60	9.12	99.01 B,C	8.78
ZEN I	93.11 A	10.43	91.40	9.74	75.80 D,E	9.87
ZEN II	167.30	22.20	86.21	9.89	193.60	37.13
ZEN III	154.80	22.13	103.60	27.92	187.20	21.79

The results are expressed as means ± SEM (standard error of the mean). Statistically significant differences: 14 day, A ZEN I vs. ZEN II, $p < 0.05$; 42 days, B Control Group vs. ZEN II, $p < 0.01$, C Control Group vs. ZEN III, $p < 0.01$, D ZEN I vs. ZEN II, $p < 0.001$, E ZEN I vs. ZEN III, $p < 0.001$.

Table 4. Changes in IL-2 content in the ileum. Experimental group and ZEN concentration: Control Group, ZEN I—5 μg/kg BW, ZEN II—10 μg/kg BW, ZEN III—15 μg/kg BW.

Group	Experimental Day					
	14		28		42	
	mean	SEM	mean	SEM	mean	SEM
Control	6.39 A,B	1.21	11.93	1.78	10,10 E,F	1.29
ZEN I	6.56 C,D	0.88	8.97	1.30	7.43 G,H	1.46
ZEN II	28.68	2.67	22.52	2.90	55.04 I	13.55
ZEN III	24.81	4.65	22.03	4.19	31.93	1.28

The results are expressed as means ± SEM (standard error of the mean). Statistically significant differences: 14 day, A Control Group vs. ZEN II, $p < 0.01$, B Control Group vs. ZEN III, $p < 0.05$, C ZEN I vs. ZEN II, $p < 0.01$, D ZEN I vs. ZEN III, $p < 0.05$; 42 day, E Control Group vs. ZEN II, $p < 0.0001$, F Control Group vs. ZEN III, $p < 0.01$, G ZEN I vs. ZEN II, $p < 0.0001$, H ZEN I vs. ZEN III, $p < 0.01$, I ZEN II vs. ZEN III, $p < 0.01$.

Similar observations were made in the concentration of IL-1β, which increased by 53.6 and 94.59 pg/mg in group ZEN II, and by 41.10 and 88.19 pg/mg in group ZEN III on days 14 and 42, respectively (Table 3).

The secretion of IL-2 also increased in response to ZEN doses of 10 and 15 μg ZEN/kg BW. The highest increase in the concentration of IL-2 was noted on day 42 in group ZEN II where its content was 21.83 pg/mg higher than in the control gilts (Table 4).

The IL-6 secretion profile was highly similar in group ZEN I and in the control group. On day 42, the concentration of IL-6 increased significantly ($p < 0.001$) by 287.78 pg/mg in group ZEN II relative to the control group (Table 5).

Table 5. Changes in IL-6 content in the ileum. Experimental group and ZEN concentration: Control Group, ZEN I—5 μg/kg BW, ZEN II—10 μg/kg BW, ZEN III—15 μg/kg BW.

Group	Experimental Day					
	14		28		42	
	mean	SEM	mean	SEM	mean	SEM
Control	45.44	4.86	67.76	12.72	47.72 A	5.39
ZEN I	41.66	5.50	54.35	6.11	47.15 B	8.31
ZEN II	141.80	13.12	160.20	21.34	345.50 C	175.80
ZEN III	88.69	12.10	103.20	17.43	126.90	6.81

The results are expressed as means ± SEM (standard error of the mean). Statistically significant differences: 42 day, A Control Group vs. ZEN II, $p < 0.001$, B ZEN I vs. ZEN II, $p < 0.001$, C ZEN II vs. ZEN III, $p < 0.05$.

In turn, IL-8 concentration tended to decrease throughout the experiment and was proportional to the administered ZEN dose (Table 6). The content of IL-8 decreased by 589.00 pg/mg in group ZEN I on day 42, by 906.00 pg/mg in group ZEN II on day 28, and by 929.20 pg/mg in group ZEN III on day 42 relative to the control group.

Table 6. Changes in IL-8 content in the ileum. Experimental group and ZEN concentration: Control Group, ZEN I—5 μg/kg BW, ZEN II—10 μg/kg BW, ZEN III—15 μg/kg BW.

Group	Experimental Day 14		28		42	
	mean	SEM	mean	SEM	mean	SEM
Control	1390.00	79.32	1916.00 [A,B]	266.60	1619.00	177.20
ZEN I	1400.00	141.90	1840.00 [C,D]	147.60	1030.00 [E]	89.83
ZEN II	1502.00	150.60	1010.00	114.20	1700.00	321.90
ZEN III	1203.00	165.30	986.80	163.30	1267.00	64.91

The results are expressed as means ± SEM (standard error of the mean). Statistically significant differences: 28 day, [A] Control Group vs. ZEN II, $p < 0.01$, [B] Control Group vs. ZEN III, $p < 0.01$, [C] ZEN I vs. ZEN II, $p < 0.01$, [D] ZEN I vs. ZEN III, $p < 0.01$; 42 day, [E] ZEN I vs. ZEN II, $p < 0.05$.

The concentration of IL-12/23p40, which is produced by, among others, macrophages, increased significantly ($p < 0.001$) after 42 days of administration to ZEN (Table 7), and it was 701.40 pg/mg higher than in the control group.

Table 7. Changes in IL-12/23p40 content in the ileum. Experimental group and ZEN concentration: Control Group, ZEN I—5 μg/kg BW, ZEN II—10 μg/kg BW, ZEN III—15 μg/kg BW.

Group	Experimental Day 14		28		42	
	mean	SEM	mean	SEM	mean	SEM
Control	873.60	76.30	918.90	67.70	827.60 [A]	38.05
ZEN I	653.70	117.30	823.80	87.98	539.50 [B]	42.28
ZEN II	762.90	89.15	652.00	68.79	1529.00 [C]	323.40
ZEN III	753.50	36.76	590.90	121.10	763.80	99.28

The results are expressed as means ± SEM (standard error of the mean). Statistically significant differences: 42 day, [A] Control Group vs. ZEN II, $p < 0.001$, [B] ZEN I vs. ZEN II, $p < 0.0001$, [C] ZEN II vs. ZEN III, $p < 0.001$.

Similarly to IL-2, IL-1α and IL-1β, the secretion of tumor necrosis factor alpha (TNFα) was clearly affected by ZEN. The increasing trend in the concentration of TNFα was observed in groups ZEN II and ZEN III. The content of TNFα was highest in group ZEN III on day 42 when it exceeded TNFα levels in the control group by 149.84 pg/mg ($p < 0.0001$) (Table 8).

Table 8. Changes in TNFα content in the ileum. Experimental group and ZEN concentration: Control Group, ZEN I—5 μg/kg BW, ZEN II—10 μg/kg BW, ZEN III—15 μg/kg BW.

Group	Experimental Day 14		28		42	
	mean	SEM	mean	SEM	mean	SEM
Control	38.45 [A]	6.52	100.30	15.27	46.16 [B,C]	6.79
ZEN I	100.90	32.73	98.03	21.61	49.26 [D,E]	9.38
ZEN II	132.90	17.87	111.60	9.35	196.00	41.90
ZEN III	97.99	10.89	135.50	20.92	125.20	7.20

The results are expressed as means ± SEM (standard error of the mean). Statistically significant differences: 14 day, [A] Control Group vs. ZEN II, $p < 0.01$; 42 days, [B] Control Group vs. ZEN II, $p < 0.0001$, [C] Control Group vs. ZEN III, $p < 0.05$, [D] ZEN I vs. ZEN II, $p < 0.0001$, [E] ZEN I vs. ZEN III, $p < 0.05$.

2.1.2. Anti-Inflammatory and Regulatory Cytokines

The dynamics of IL-4 secretion was similar to the changes in the concentration of the proinflammatory IFN-γ. The concentration of IL-4 ranged from 277.40 to 2894.00 pg/mg. The highest increase in IL-4 content was noted in group ZEN II where the analyzed parameter was 2616.60 pg/mg higher on day 14 and 2094.30 pg/mg higher on day 42 than in the control group (Table 9).

Table 9. Changes in IL-4 content in the ileum. Experimental group and ZEN concentration: Control Group, ZEN I—5 μg/kg BW, ZEN II—10 μg/kg BW, ZEN III—15 μg/kg BW.

Group	Experimental Day 14		28		42	
	mean	SEM	mean	SEM	mean	SEM
Control	277.40 ^{A}	56.61	572.30	130.40	470.70 ^{D}	91.95
ZEN I	356.80 ^{B}	56.00	463.80	77.36	330.90 ^{E}	56.96
ZEN II	2894.00 ^{C}	955.90	1352.00	318.90	2565.00 ^{F}	680.60
ZEN III	589.80	91.53	994.20	260.70	908.50	93.04

The results are expressed as means ± SEM (standard error of the mean). Statistically significant differences: 14 day, ^{A} Control Group vs. ZEN II, $p < 0.001$, ^{B} ZEN I vs. ZEN II, $p < 0.001$, ^{C} ZEN II vs. ZEN III, $p < 0.01$; 42 day, ^{D} Control Group vs. ZEN II, $p < 0.01$, ^{E} ZEN I vs. ZEN II, $p < 0.01$, ^{F} ZEN II vs. ZEN III, $p < 0.05$.

The changes in the IL-10 profile were similar to those noted in the content of IL-4, IL-1α, IL-1β, IFN-γ, and IL-2. During the experiment, the concentration of IL-10 was lowest in group ZEN I. In turn, IL-10 levels increased in gilts administered ZEN doses of 10 and 15 pg/kg BW. The highest concentration of IL-10 relative to the control group was observed on day 42 in gilts from group ZEN II where it reached 62.61 pg/mg (Table 10).

Table 10. Changes in IL-10 content in the ileum. Experimental group and ZEN concentration: Control Group, ZEN I—5 μg/kg BW, ZEN II—10 μg/kg BW, ZEN III—15 μg/kg BW.

Group	Experimental Day 14		28		42	
	mean	SEM	mean	SEM	mean	SEM
Control	37.40	6.45	57.49 ^{B}	8.16	35.35 ^{C}	4.89
ZEN I	23.69 ^{A}	6.44	25.87	2.70	14.46 ^{D,E}	1.47
ZEN II	52.33	6.76	39.31	2.41	77.06	15.46
ZEN III	42.51	5.36	44.76	7.42	56.36	3.66

The results are expressed as means ± SEM (standard error of the mean). Statistically significant differences: 14 day, ^{A} ZEN I vs. ZEN II, $p < 0.05$; 28 day, ^{B} Control Group vs. ZEN I, $p < 0.01$; 42 day, ^{C} Control Group vs. ZEN II, $p < 0.001$, ^{D} ZEN I vs. ZEN II, $p < 0.0001$, ^{E} ZEN I vs. ZEN III, $p < 0.001$.

Similarly to IL-8, the concentration of TGFβ decreased throughout the experiment, and significant differences were observed in groups ZEN II and III relative to the control group (Table 11).

2.2. The Effect of Zearalenone on Oxidative Stress Markers

Oxidative stress in the ileal Peyer's patches of gilts was evaluated based on the activity of catalase (CAT), peroxidase (POD) and superoxide dismutase (SOD), MDA content, and the number of free metabolic sulfhydryl groups (-SH). The changes in the values of oxidative stress markers were similar to those noted in the concentrations of IL-1α, IL-1β, IFN-γ, IL-2, IL4, IL-10 and TNFα.

Table 11. Changes in TGFβ content in the ileum. Experimental group and ZEN concentration: Control Group, ZEN I—5 μg/kg BW, ZEN II—10 μg/kg BW, ZEN III—15 μg/kg BW.

Group	Experimental Day 14		28		42	
	mean	SEM	mean	SEM	mean	SEM
Control	215.20 [A]	19.46	263.30 [B,C]	19.46	134.50	15.71
ZEN I	143.40	23.98	212.20 [D,E]	28.75	160.20	13.42
ZEN II	125.60	15.56	116.60	14.16	129.90	8.25
ZEN III	193.80	28.69	126.50	7.21	170.20	39.46

The results are expressed as means ± SEM (standard error of the mean). Statistically significant differences: 14 day, [A] Control Group vs. ZEN III, $p < 0.05$; 28 days, [B] Control Group vs. ZEN II, $p < 0.001$, [C] Control Group vs. ZEN III, $p < 0.0001$, [D] ZEN I vs. ZEN II, $p < 0.05$, [E] ZEN I vs. ZEN III, $p < 0.05$.

Peroxidase activity in the ileal Peyer's patches of gilts ranged from 47.66 to 231.30 units/mg/min (Table 12).

Table 12. Changes in peroxidase activity in the ileum. Experimental group and ZEN concentration: Control Group, ZEN I—5 μg/kg BW, ZEN II—10 μg/kg BW, ZEN III—15 μg/kg BW.

Group	Experimental Day 14		28		42	
	mean	SEM	mean	SEM	mean	SEM
Control	74.88 [A,B]	4.56	74.63 [E,F]	4.85	87.04 [I,J]	6.63
ZEN I	47.66 [C,D]	5.42	79.13 [G,H]	3.09	73.62 [K,L]	5.64
ZEN II	184.40	19.23	156.90	13.99	231.30	19.70
ZEN III	157.90	16.06	179.20	29.34	214.80	16.03

The results are expressed as means ± SEM (standard error of the mean). Statistically significant differences: 14 day, [A] Control Group vs. ZEN II, $p < 0.0001$, [B] Control Group vs. ZEN III, $p < 0.001$, [C] ZEN I vs. ZEN II, $p < 0.0001$, [D] ZEN I vs. ZEN III, $p < 0.0001$; 28 day, [E] Control Group vs. ZEN II, $p < 0.001$, [F] Control Group vs. ZEN III, $p < 0.0001$, [G] ZEN I vs. ZEN II, $p < 0.01$, [H] ZEN I vs. ZEN III, $p < 0.0001$; 42 day, [I] Control Group vs. ZEN II, $p < 0.0001$, [J] Control Group vs. ZEN III, $p < 0.0001$, [K] ZEN I vs. ZEN II, $p < 0.0001$, [L] ZEN I vs. ZEN III, $p < 0.0001$.

Catalase activity ranged from 2.32 to 12.36 H_2O_2 [mM/mg/min] (Table 13). Significant differences were observed between experimental and control gilts and analytical dates. Catalase activity was strongly suppressed in gilts administered a ZEN dose of 10 μg ZEN/kg BW. The highest CAT activity was observed in group ZEN II.

Table 13. Changes in catalase activity in the ileum. Experimental group and ZEN concentration: Control Group, ZEN I—5 μg/kg BW, ZEN II—10 μg/kg BW, ZEN III—15 μg/kg BW.

Group	Experimental Day 14		28		42	
	mean	SEM	mean	SEM	mean	SEM
Control	6.67 [A]	0.35	6.49 [E]	0.40	3.92 [H,I]	0.25
ZEN I	4.24 [B,C]	0.62	3.58 [F,G]	0.38	2.32 [J,K]	0.14
ZEN II	12.23 [D]	1.12	9.20	0.81	12.36 [L]	1.50
ZEN III	7.65	0.61	7.83	1.09	8.73	0.52

The results are expressed as means ± SEM (standard error of the mean). Statistically significant differences: 14 day, [A] Control Group vs. ZEN II, $p < 0.0001$, [B] ZEN I vs. ZEN II, $p < 0.0001$, [C] ZEN I vs. ZEN III, $p < 0.01$, [D] ZEN II vs. ZEN III, $p < 0.001$; 28 day, [E] Control Group vs. ZEN I, $p < 0.05$, [F] ZEN I vs. ZEN II, $p < 0.0001$, [G] ZEN I vs. ZEN III, $p < 0.001$; 42 day, [H] Control Group vs. ZEN II, $p < 0.0001$, [I] Control Group vs. ZEN III, $p < 0.0001$, [J] ZEN I vs. ZEN II, $p < 0.0001$, [K] ZEN I vs. ZEN III, $p < 0.01$.

The activity of SOD was highly similar to POD and CAT activity, and significant differences were observed between control and ZEN groups and analytical dates. The greatest increase in SOD activity was noted in group ZEN II where this parameter reached 17.94 H_2O_2 [mM/mg/min] on day 42 and was 7.62 H_2O_2 [mM/mg/min] higher than in the control gilts (Table 14).

Table 14. Changes in the activity of superoxide dismutase in the ileum. Experimental group and ZEN concentration: Control Group, ZEN I—5 μg/kg BW, ZEN II—10 μg/kg BW, ZEN III—15 μg/kg BW.

Group	Experimental Day 14		28		42	
	mean	SEM	mean	SEM	mean	SEM
Control	11.35 ^{A}	0.65	11.63	0.78	10.32 ^{E,F}	0.32
ZEN I	8.14 ^{B}	0.91	7.31 ^{C,D}	0.73	1.14 ^{G,H}	0.10
ZEN II	17.94	1.55	14.12	1.28	18.63	2.75
ZEN III	13.57	1.33	14.62	2.54	15.80	0.36

The results are expressed as means ± SEM (standard error of the mean). Statistically significant differences: 14 day, ^{A} Control Group vs. ZEN II, $p < 0.05$, ^{B} ZEN I vs. ZEN II, $p < 0.001$; 28 day, ^{C} ZEN I vs. ZEN II, $p < 0.05$, ^{D} ZEN I vs. ZEN III, $p < 0.01$; 42 day, ^{E} Control Group vs. ZEN I, $p < 0.001$, ^{F} Control Group vs. ZEN II, $p < 0.01$, ^{G} ZEN I vs. ZEN II, $p < 0.0001$, ^{H} ZEN I vs. ZEN III, $p < 0.0001$.

The MDA profile in group ZEN I was highly similar to that noted in the control group. The concentration of MDA was highest on day 42 in groups ZEN II and ZEN III where it reached 1.43 and 1.39 pM/mg, respectively (Table 15).

Table 15. Changes in the concentration of malondialdehyde in the ileum. Experimental group and ZEN concentration: Control Group, ZEN I—5 μg/kg BW, ZEN II—10 μg/kg BW, ZEN III—15 μg/kg BW.

Group	Experimental Day 14		28		42	
	mean	SEM	mean	SEM	mean	SEM
Control	0.28 ^{A}	0.08	0.67	0.34	0.41 ^{C,D}	0.11
ZEN I	0.25 ^{B}	0.12	0.49	0.18	0.32 ^{E,F}	0.20
ZEN II	1.26	0.12	0.65	0.06	1.43	0.50
ZEN III	0.54	0.07	0.83	0.15	1.39	0.52

The results are expressed as means ± SEM (standard error of the mean). Statistically significant differences: 14 day, ^{A} Control Group vs. ZEN II, $p < 0.05$, ^{B} ZEN I vs. ZEN II, $p < 0.05$; 42 day, ^{C} Control Group vs. ZEN II, $p < 0.05$, ^{D} Control Group vs. ZEN III, $p < 0.05$, ^{E} ZEN I vs. ZEN II, $p < 0.05$, ^{F} ZEN I vs. ZEN III, $p < 0.05$.

During the experiment, the concentration of -SH groups decreased in gilts from group ZEN I. In turn, a very high increase in the concentration of -SH groups was noted in groups ZEN II and ZEN III. On day 42, this parameter reached 0.73 μM/mg in group ZEN II and 0.63 μM/mg in group ZEN III (Table 16).

Table 16. Changes in the content of -SH groups in the ileum. Experimental group and ZEN concentration: Control Group, ZEN I—5 μg/kg BW, ZEN II—10 μg/kg BW, ZEN III—15 μg/kg BW.

Group	Experimental Day 14		28		42	
	mean	SEM	mean	SEM	mean	SEM
Control	0.27 $^{A, B}$	0.01	0.30 $^{E, F}$	0.02	0.27 $^{I, J}$	0.01
ZEN I	0.21 $^{C, D}$	0.02	0.26 $^{G, H}$	0.02	0.17 $^{K, L}$	0.01
ZEN II	0.63	0.04	0.53	0.04	0.77	0.09
ZEN III	0.51	0.04	0.54	0.07	0.63	0.02

The results are expressed as means ± SEM (standard error of the mean). Statistically significant differences: 14 day, ^{A} Control Group vs. ZEN II, $p < 0.0001$, ^{B} Control Group vs. ZEN III, $p < 0.001$, ^{C} ZEN I vs. ZEN II, $p < 0.0001$, ^{D} ZEN I vs. ZEN III, $p < 0.0001$; 28 day, ^{E} Control Group vs. ZEN II, $p < 0.001$, ^{F} Control Group vs. ZEN III, $p < 0.001$, ^{G} ZEN I vs. ZEN II, $p < 0.0001$, ^{H} ZEN I vs. ZEN III, $p < 0.0001$, 42 day, ^{I} control group vs. ZEN II, $p < 0.0001$, ^{J} Control Group vs. ZEN III, $p < 0.0001$, ^{K} ZEN I vs. ZEN II, $p < 0.0001$, ^{L} ZEN I vs. ZEN III, $p < 0.0001$.

2.3. The Effect of Zearalenone on Basal Metabolic Markers

Basal metabolism in ileal Peyer's patches was evaluated based on the concentrations of glucose, fructose, and proline. In groups ZEN II and ZEN III, the values of metabolic markers were characterized by similar change trends to the values of oxidative stress markers (POD, CAT, SOD, MDA, -SH) and cytokine concentrations (IL-2, IFN-γ, IL-1α, IL-1β, IL4, IL-10, TNFα). Significant differences were observed mainly in groups ZEN II and ZEN III (Tables 17–19).

Table 17. Changes in glucose content in the ileum. Experimental group and ZEN concentration: Control Group, ZEN I—5 µg/kg BW, ZEN II—10 µg/kg BW, ZEN III—15 µg/kg BW.

Group	Experimental Day 14		28		42	
	mean	SEM	mean	SEM	mean	SEM
Control	5.60 ^{A,B}	1.68	11.79 ^{D}	2.71	20.59 ^{E,F}	1.98
ZEN I	23.20 ^{C}	2.45	15.67	2.41	16.39 ^{G,H}	1.73
ZEN II	44.93	8.86	40.80	4.36	80.79 ^{I}	20.29
ZEN III	62.39	6.92	29.37	3.69	51.74	11.37

The results are expressed as means ± SEM (standard error of the mean). Statistically significant differences: 14 day, ^{A} Control Group vs. ZEN II, $p < 0.01$, ^{B} Control Group vs. ZEN III, $p < 0.0001$, ^{C} ZEN I vs. ZEN III, $p < 0.01$; 28 day, ^{D} Control Group vs. ZEN II, $p < 0.05$; 42 day, ^{E} Control Group vs. ZEN II, $p < 0.0001$, ^{F} Control Group vs. ZEN III, $p < 0.05$, ^{G} ZEN I vs. ZEN II, $p < 0.0001$, ^{H} ZEN I vs. ZEN III, $p < 0.01$, ^{I} ZEN II vs. ZEN III, $p < 0.05$.

Table 18. Changes in fructose content in the ileum. Experimental group and ZEN concentration: Control Group, ZEN I—5 µg/kg BW, ZEN II—10 µg/kg BW, ZEN III—15 µg/kg BW.

Group	Experimental Day 14		28		42	
	mean	SEM	mean	SEM	mean	SEM
Control	5.24 ^{A,B}	0.26	4.94 ^{F}	0.32	5.18 ^{H,I}	0.25
ZEN I	5.78 ^{C,D}	0.69	4.98 ^{G}	0.32	2.73 ^{J,K}	0.20
ZEN II	19.23 ^{E}	2.11	12.26	0.67	18.28 ^{L}	2.73
ZEN III	12.70	1.36	8.60	1.04	10.46	0.29

The results are expressed as means ± SEM (standard error of the mean). Statistically significant differences: 14 day, ^{A} Control Group vs. ZEN II, $p < 0.0001$, ^{B} Control Group vs. ZEN III, $p < 0.0001$, ^{C} ZEN I vs. ZEN II, $p < 0.0001$, ^{D} ZEN I vs. ZEN III, $p < 0.001$, ^{E} ZEN II vs. ZEN III, $p < 0.001$; 28 day, ^{F} Control Group vs. ZEN II, p < 0.0001, ^{G} ZEN I vs. ZEN II, $p < 0.001$; 42 day, ^{H} Control Group vs. ZEN II, $p < 0.0001$, ^{I} Control Group vs. ZEN III, $p < 0.01$, ^{J} ZEN I vs. ZEN II, $p < 0.0001$, ^{K} ZEN I vs. ZEN III, $p < 0.0001$, ^{L} ZEN II vs. ZEN III, $p < 0.0001$.

Table 19. Changes in proline content in the ileum. Experimental group and ZEN concentration: Control Group, ZEN I—5 µg/kg BW, ZEN II—10 µg/kg BW, ZEN III—15 µg/kg BW.

Group	Experimental Day 14		28		42	
	mean	SEM	mean	SEM	mean	SEM
Control	6.39	0.56	6.32 ^{B}	0.73	5.04 ^{E,F,G}	0.52
ZEN I	3.76 ^{A}	0.49	2.51 ^{C,D}	0.33	1.27 ^{H,I}	0.16
ZEN II	8.29	1.19	5.77	0.36	10.76	1.32
ZEN III	6.57	0.83	6.37	1.08	10.21	0.98

The results are expressed as means ± SEM (standard error of the mean). Statistically significant differences: 14 day, ^{A} ZEN vs. ZEN II, $p < 0.001$; 28 day, ^{B} Control Group vs. ZEN I, $p < 0.01$, ^{C} ZEN I vs. ZEN II, $p < 0.05$, ^{D} ZEN I vs. ZEN III, $p < 0.01$; 42 day, ^{E} Control Group vs. ZEN I, $p < 0.01$, ^{F} Control Group vs. ZEN II, $p < 0.0001$, ^{G} Control Group vs. ZEN III, $p < 0.0001$, ^{H} ZEN I vs. ZEN II, $p < 0.0001$, ^{I} ZEN I vs. ZEN III, $p < 0.0001$.

3. Discussion

In the European Union, the maximum limits for ZEN in livestock have been set at 0.25 mg ZEN/kg feed/day for adult pigs and 0.10 mg ZEN/kg feed/day for sows and piglets [24]. These limits have been introduced to protect pigs against the hyperestrogenic effects of ZEN. In morphometric analyses of porcine vulvae and uteri in animals exposed to ZEN, the NOAEL for ZEN was determined at 10.4 μg ZEN/kg BW/day for piglets and 40 μg ZEN/kg BW/day for mature female pigs [25]. When designing this study, the authors aimed to determine whether the established limits for protecting pigs against the estrogenic effects of ZEN are safe for GALT homeostasis. The theory proposed nearly 30 years ago [26] postulates that chemical processes involving oxidation-reduction reactions elicit oxidative stress and play a key role in the pathophysiology of inflammations [27]. This experiment was carry out on pre-pubertal gilts administered ZEN *per os* in doses equivalent to 50%, 100%, and 150% NOAEL values for 14, 28, and 42 days to evaluate the oxidoreductive balance, cytokine secretion, and basal metabolic markers in ileal Peyer's patches. This research concept was formulated to address the general scarcity of the relevant data in the literature and to determine whether the elimination of ZEN's estrogenic effects can also inhibit other adverse effects exerted by this mycotoxin.

Macrophages, mucosal cells, dendritic cells, NK cells, and neutrophils are responsible for innate immune defense in the digestive tract. The adaptive immune system is composed of T and B lymphocytes which, when activated, secrete cytokines and produce antibodies, respectively [28]. When the gastrointestinal mucosa is in a state of homeostasis, a strict balance is maintained between proinflammatory (IL-6, TNFα, IL-17, IL-1, IL-23, and IL-8) and anti-inflammatory cytokines (TGFβ, IL-5, IL-10, and IL-11), where neutrophils and other types of cells initiate inflammation by releasing proinflammatory interleukins to activate the adaptive immune response [29].

In the body, homeostasis is determined by the oxidant–antioxidant balance, which is most often disrupted by excessive production of ROS. Oxidation-reduction reactions are powerful chemical processes that involve oxidation and reduction of proteins and lipids, in particular unsaturated fatty acids, which generate superoxide radicals. The produced metabolites disrupt cell functions and regulatory mechanisms that are essential for maintaining homeostasis and a balance between health and disease [30].

Estradiol is a naturally occurring estrogen that delivers immunosuppressive effects when applied at high concentrations under experimental conditions. Previous research has demonstrated that low doses of mycoestrogens such as ZEN activated the porcine immune system [31,32]. The present study focused on evaluating the effect of low ZEN doses that do not exert estrogenic effects (5, 10 and 15 μg ZEN/kg BW) on GALT in the ileum's walls of gilts during weaning. Zearalenone's effects were evaluated based on cytokine secretion, the enzymatic and non-enzymatic markers of oxidative stress, and cellular metabolism in ileal Peyer's patches.

The study demonstrated that each of the tested ZEN doses increased the secretion of proinflammatory cytokines (IFN-γ, IL-12/23p40, IL-1α, IL-1β, IL-2, IL-6, TNFα) in a different manner, which corroborates the findings of other authors [32,33]. In an in vitro study of the IPEC-2 cell line and mouse peritoneal macrophages, Fan et al. [34] noted an increase in the secretion of IL-1β and IL-18 under exposure to ZEN, and similar observations were made by Yousef et al. [35] in cultures of bovine oviductal epithelial cells. The present study also demonstrated that the concentrations of regulatory and anti-inflammatory cytokines (IL-4, IL-10) tended to increase, whereas TGFβ levels tended to decrease on successive days of exposure to ZEN. This is a very important observation, because TGFβ determines the integrity of the intestinal epithelium and is responsible for its repair as well as cell migration [36]. The binding of TGFβ signal molecules to the TGFβ IIR receptor may be disrupted in non-specific inflammatory bowel diseases (e.g., Crohn's disease). In this experiment, each ZEN dose increased the concentrations of IL-2, IL-6, IL-12β, and IFN-γ in ileal Peyer's patches. However, according to the literature, cytokine expression is determined not only by ZEN concentration, but also by the type of the exposed tissue or organ. Tarnau et al. [37] reported a decrease in the expression of IL-6, IL-8, IL-1β, TNFα, and IFN-γ genes in the liver and duodenum of gilts receiving a ZEN dose of

100 ppb. In contrast, the mRNA expression of the above cytokines increased in the pancreas, kidneys, and the colon, and IFN-γ expression increased in the liver. The real-time PCR method used in this study measures RNA levels; therefore, it does not have to be closely related to the direct concentration of cytokine protein. Despite the above, the expression of cytokine genes in the cited studies was similar to that noted in this experiment. Liu et al. [38] administered ZEN at 2.77 mg/kg feed to pregnant sows and reported an increase in the expression of genes encoding proinflammatory cytokines IL-6, IL-1α, IL-1β, and TNFα, but it did not observe changes in the expression of the IL-8 gene. The cited study revealed inflammation of the small intestine in pregnant sows as well as changes in newborn piglets where immune dysfunctions and increased expression of the genes encoding proinflammatory cytokines IL-6, IL-1α, IL-1β, and TNFα were observed in the small intestine.

Changes in the expression ratio of proinflammatory to anti-inflammatory cytokines can influence the permeability of the intestinal epithelium. Zearalenone can also affect the intestinal microbiome by decreasing the counts of selected bacterial strains, including Lactobacillus sp. [39]. Cytokines such as IL-12, IL-1α, IL-1β, TNFα, and IFN-γ are the key mediators of inflammation in GALT, whereas IL-10 and TGFβ should alleviate inflammatory states. According to some authors, ZEN induces intestinal inflammations via the NLRP3 inflammasome, which is composed of a NOD-like receptor and caspase-1 precursor. The NRLP inflammasome activated by ZEN is responsible for the maturation and secretion of proinflammatory cytokines IL-1β and IL-18, which play a role in the etiology of non-specific inflammatory bowel diseases. According to the literature, ZEN induces oxidative stress and increases ROS levels in the mitochondria, which also activates the NLRP inflammasome [34].

The intestinal epithelium significantly affects intestinal immunity because enterocytes form a protective barrier that shields the body from external influences. Enterocytes secrete retinoic acid metabolites, thymic stromal lymphopoietin (TSLP), and TGFβ, which induce immune tolerance to dietary antigens. Mycotoxins increase the permeability of the intestinal epithelium by influencing cell viability [40], and they decrease the proliferation of enterocytes and intestinal epithelial stem cells in intestinal crypts. Damage to the intestinal barrier allows the passage of intraluminal antigens, including bacterial antigens such as lipopolysaccharides (LPS), which stimulate macrophages and dendritic cells. Macrophages are polarized toward the M1 subpopulation, which increases the secretion of IL-12, IL-1β, and IL-1α [41]. A similar mechanism was observed in the current study, where the concentrations of IL-12/23p40, IL-1β, and IL-1α in ileal Peyer's patches increased under exposure to ZEN doses of 10 and 15 μg ZEN/ kg BW. Elevated levels of IL-2 and IFN-γ are also indicative of the Th1/Tc1 response, which is induced by, among others, IL-12 secreted by macrophages. Macrophage metabolism also changes during inflammation because, according to the literature, M1 macrophages and dendritic cells activated by LPS undergo metabolic reprogramming [42]. This process is characterized by a decrease in the intensity of mitochondrial biochemical pathways, including the tricarboxylic acid cycle, respiratory chain, and fatty acid β-oxidation, as well as an increase in the activity of metabolic pathways in the cytoplasm, including glycolysis, the pentose phosphate pathway, and the synthesis of fatty acids and nucleic acids.

In the present study, glucose, fructose, and proline levels were elevated in gilts administered ZEN doses equivalent to 100% and 150% NOAEL values. According to the literature, mitogen-activated immune cells increase the demand for energy even 20-fold, which can be attributed to higher expression of the glucose transporter 1 (GLUT 1) gene [43]. The fact that glucose and fructose levels were higher than in the control gilts could be linked with the intensification of glycolysis and the pentose phosphate pathway, in particular in M1 macrophages and dendritic cells activated by LPS, whereas the increase in proline concentration in lymphoid tissues could be induced by activated metalloproteinases and the intensified synthesis of proline from glutamine.

Proline is a stress response marker in both plants and animals. In the current experiment, proline levels tended to increase in the GALT of gilts exposed to ZEN. According to the literature, inflammatory states lead to metabolic reprogramming as well as an increase in glucose uptake by cells and an increase in the expression of the GLUT 1 gene. An increase in glucose uptake, lactic acid production, and

NADPH oxidase activity was reported in cultures of smooth muscle cells isolated from rat mesenteric veins and in human aortic smooth muscle cells treated with IL-1β [44]. Other authors demonstrated that the activation of T lymphocytes is determined by the amount of glucose transported into the cells, the GLUT 1 transporter, and the PI3K/Akt signaling pathway. Interestingly, a lower availability of glucose has been linked with a decrease in the IFN-γ secretion and proliferation of T cells. Despite the presence of other components, glucose is essential for the proliferation of T lymphocytes and IL-2 synthesis. Interleukin 10 exerts anti-inflammatory effects e.g., by blocking the expression of glycolytic enzymes and glucose transporters GLUT 1 and GLUT 4. In the present experiment, the progressive increase in IL-10 secretion was correlated with ZEN dose. However, other studies have demonstrated that the effects of IL-10 are reversible. Even low doses of estrogenic substances can stimulate IL-10 production by monocytes and macrophages [45] and by dorsal root ganglion cells in the nervous system [45].

Proline is also regarded as a marker of metabolic reprogramming. In pigs and humans, the small intestinal epithelium is the main site of proline synthesis from glutamate, but proline synthesized from arginine was not found in the intestinal epithelium of suckling piglets. Proline synthesis from glutamate is induced after weaning, and it is also linked with synthetase of pyrroline-5-carboxylic acid and proline oxidase. These enzymes are regulated by glucocorticoids [46], but also by lactic acid, which is present in quite high concentrations in the plasma of suckling animals. Research has also demonstrated that the synthesis of arginine and citrullin from proline in the intestinal epithelium decreases in environments with high concentrations of lactic acid [47]. In ileal Peyer's patches, the synthesis of large quantities of lactic acid is linked with the metabolic reprogramming of M1 macrophages, high glycolytic activity, and high secretion of IL-12 and IL-1β [14]. M1 macrophages are produced after activation with bacterial products such as LPS during inflammations, including in the intestinal epithelium [47]. Lactic acid decreases the activity of proline oxidase, which catalyzes proline, an amino acid that participates in the biochemical transformation of arginine, a source of nitric oxide (NO). NO plays an important function in the regulation of the intestinal barrier function, and decreased proline metabolism can compromise intestinal integrity and lead to intestinal damage [47]. Interestingly, hypoargininemia was found to coexist with hyperargininemia and hyperlactacidemia in the plasma of patients with sepsis. During inflammations, proline can be released from the extracellular matrix, such as collagen, under the influence of metalloproteinases activated by stress responses and/or inflammation [48]. It should also be noted that proline and collagen synthesis can increase during inflammation, which can lead to tissue fibrosis [49]. More proline is synthesized from glutamine when the activity of the c-MYC transcription factor, also known as the oncogenic transcription factor, is intensified [50]. Proline contains an amine group with a ring structure, and unlike other amino acids, it does not undergo biochemical reactions, but is catabolized within the proline cycle [51]. This biochemical pathway occurs in the mitochondria, and its secondary effects include the production of superoxide anion radicals and an increase in the concentration of SOD. The proline cycle supports the generation of ATP when the activity of the respiratory chain decreases. The proline cycle also donates electrons to complexes III and IV of the respiratory chain, which is broken during inflammatory states caused by the metabolic reprogramming of cells [50,51]. Metabolic reprogramming also leads to oxidative stress that can be induced by ZEN.

An analysis of selected oxidative stress parameters revealed a correlation between immunological, metabolic, and oxidative stress parameters. Superoxide dismutase activity was proportional to proline content as well as CAT and POD activity. The activity of CAT, POD, and SOD increases at the beginning of inflammation, and it appears to be a defense response [52,53]. Oxidative stress is also linked with the activity of the inflammasome, IL-1α, and IL-1β, and it increases the permeability of the intestinal barrier, which affects the intestinal epithelium [54]. In reality, these processes are linked by biochemical pathways, and they are also dependent on the ZEN dose. The concentration of proinflammatory cytokines (IL-1β, TNFα, IFNγ) in the liver and spleen decreased in gilts fed diets containing 316 μg ZEN/kg feed [55]. According to the literature, the expression of CAT and glutathione peroxidase genes was elevated in oxidative stress [55]. In the current experiment, the values of metabolic, oxidative stress, and immunological parameters peaked under exposure to a ZEN dose of 10 μg ZEN/kg feed. However, in piglets administered 15 μg ZEN/kg feed, these parameters decreased or showed a decreasing trend relative to the previous dose, but, interestingly, their values were still higher than in the control gilts. In other studies, the values of immunological and oxidative stress parameters in the kidneys increased in pigs administered ZEN doses of 0.0003, 0.048, 0.098, and 0.146 g ZEN/kg feed.

In the present study, the concentrations of metabolic sulfhydryl groups (-SH) increased in ileal Peyer's patches. Metabolic -SH groups are linked with the pentose phosphate pathway and glycolysis, which are elements of metabolic reprogramming in cells during inflammation [39,56]. Cysteine residues in proteins and glutathione (GSH) are the main donors of -SH groups in cells. The reduced form of glutathione (GSSG) is synthesized by glutathione reductase. The activity of this enzyme is linked with intracellular redox states and reduced forms of nicotinamide adenine coenzymes NADH and NADPH [22]. The supply of reduced coenzymes is strictly correlated with the activity of the glycolysis pathway and the pentose phosphate pathway. Moreover, GSH also stimulates glycolysis and glutaminolysis, which are characteristic of metabolic reprogramming of active T lymphocytes via the myc transcription factor [39]. Catalase activity can be closely linked with the level of NADPH because this coenzyme is molecularly bound to CAT. For example, human CAT contains four molecules of tightly bound NADPH [39]. At the same time, metabolic changes decrease oxidative phosphorylation, which disrupts the respiratory chain and leads to the production of ROS and hydrogen peroxide. The above intensifies the peroxidation of cell membrane lipids, increases MDA concentration, and enhances the activity of cellular scavenger enzymes: CAT and POD [57]. Malondialdehyde is a reactive compound that is formed during reactions involving ROS (mainly superoxide anion radicals and hydrogen peroxide) and unsaturated fatty acids which are found in plasma membranes and endoplasmic reticulum in mitochondrial, microsomal, and nuclear membranes. Reactive oxygen species are produced during inflammatory states in the mitochondria, where metabolic reprogramming contributes to alterations in the Krebs cycle. The above leads to the accumulation of succinic acid, a decrease in the concentrations of reduced forms of flavin adenine dinucleotide (FAD) and nicotinamide adenine dinucleotide (NAD), and uncoupling of the respiratory chain. Reactive oxygen species are most dynamically formed in complexes I and III of the electron transport chain [50].

4. Conclusions

To sum up, the close links between metabolic, oxidative stress, and immunological parameters during inflammatory states testify to the presence of a single process that begins with biochemical changes in metabolism in response to stress factors. The results of the present study indicate that ZEN doses are equivalent to 50%, 100%, and 150% NOAEL values, which induce such changes in ileal Peyer's patches. It could be postulated that cell responses to low ZEN doses initiate reactions that lead to inflammation. The proinflammatory properties of ZEN and intensified oxidative stress can impair the function of the intestinal epithelium, as demonstrated by oxidative stress markers such as the biochemical changes associated with the metabolism of sugars (enhanced glycolysis) and amino acids (proline).

5. Materials and Methods

5.1. Animals and Diet

Experiments were carried out in compliance with Polish regulations setting forth the terms and conditions of animal experimentation (Opinions No. 12/2016 and 45/2016/DLZ of the Local Ethics Committee for Animal Experimentation of 27 April 2016 and 30 November 2016).

A total of 60 gilts (14.5 ± 2 kg BW) were used in the experiment. The animals were acclimated in the experimental facilities of the Faculty of Veterinary Medicine of the University of Warmia and Mazury in Olsztyn (Poland) for one week before the experiment. Gilts were penned in groups with ad libitum access to feed and water throughout the experiment. The animals were randomly divided into three experimental groups (ZEN I, ZEN II, and ZEN III; n = 15) and a control group (Control; n = 15). The experimental gilts were orally administered ZEN (Z2125-25MG, Sigma-Aldrich, USA) at a dose of 5 μg ZEN/kg BW (group ZEN I), 10 μg ZEN/kg BW (group ZEN II), and 15 μg ZEN/kg BW (group ZEN III). The appropriate analytical doses of mycotoxins were dissolved in 96% ethanol, combined with the feed carrier and placed in gel capsules (Polskie Odczynniki SA, Gliwice, Poland). Before administration, open capsules were stored to evaporate ethanol. The capsules were administered daily before morning feeding. The gilts were weighed every week to adjust the ZEN dose to the body weight of each pig.

All animals were fed pelleted feed supplied by the same producer. The gilts were fed *ad libitum* at 08:00 and 17:00 during the entire experiment. The composition of pelleted feed is presented in Table 20 [58].

Table 20. Declared composition of the complete diet.

Parameters	Composition Declared by the Manufacturer (%)
Wheat	55
Barley	22
Soybean meal	16
Wheat bran	4.0
Chalk	0.3
Zitrosan	0.2
Vitamin–mineral premix [1]	2.5

[1] Composition of the vitamin–mineral premix per kg: vitamin A—500.000 IU; iron—5000 mg; vitamin D^3—100.000 IU; zinc—5000 mg; vitamin E (alpha-tocopherol)—2000 mg; manganese—3000 mg; vitamin K—150 mg; copper ($CuSO_4{\cdot}5H_2O$)—500 mg; vitamin B_1—100 mg; cobalt—20 mg; vitamin B_2—300 mg; iodine—40 mg; vitamin B_6—150 mg; selenium—15 mg; vitamin B_{12}—1500 μg; L-lysine—9.4 g; niacin—1200 mg; DL-methionine+cystine—3.7 g; pantothenic acid—600 mg; L-threonine—2.3 g; folic acid—50 mg; tryptophan—1.1 g; biotin—7500 μg; phytase+choline—10 g; ToyoCerin probiotic+calcium—250 g; antioxidant+mineral phosphorus and released phosphorus—60 g; magnesium—5 g; sodium; calcium—51 g.

5.2. Tissue Sampling and Preparation for Analyses

The experimental material comprised segments of the ileum collected from gilts after 14, 28, and 42 days of exposure to ZEN. On each sampling date, five gilts selected randomly from each group were premedicated with azaperone in dose 4 mg/kg BW, im (Stresnil, Jansen Pharmaceutica NV, Belgium) and euthanized (sodium pentobarbital, 0.6 ml/kg BW, iv) after 15 min (Morbital, Biowet Puławy, Poland). Segments of the ileum (with a length of 5 cm each, sampled from the same site in every animal) located 2 cm away from the ileocecal valve were collected immediately after euthanasia. The samples were stored at a temperature of −80 °C until cytokine analysis.

Samples of 1 g of minced ileum were processed with 2.5 ml of the extraction dilution (137 mM NaCl, 2.7 mM KCl, 8.1 mM Na_2HPO_4, 1.5 mM KH_2PO_4), 0.5% sodium citrate (Avantor, Poland), 0.05% Tween 20 (Serva, Germany), protease inhibitors (Roche, Germany)] in a homogenizer (Omni-TipsTM Disposable, Omni International, Kennesaw, GA, USA). The homogenate was centrifuged (8600 g for 1 hour) in an Eppendorf 5804R centrifuge, and supernatant samples were stored at −80 °C until analysis.

5.3. Immunoenzymatic Determination of Cytokines and Oxidative Stress Markers

To determine the concentrations of cytokines in porcine tissues, commercial ELISA kits were used (Table 21). The tissues were homogenized in radioimmunoprecipitation assay buffer (RIPA) buffer in 4 °C and were centrifuged (30,000× *g* for 1 h). After centrifugation, the obtained tissue supernatants were aliquoted and stored at −80 °C. The supernatants were used for cytokines measurements. The ELISA test plate was measured with the TECAN Infinite M200PRO (Austria) plate reader at a wavelength of λ = 492 nm. The concentrations of cytokines in the tissues were measured by the bicinchoninic acid (BCA) method (Pierce BCA Protein Assay Kit, Thermo Scientific, Rockford, IL, USA) and expressed per milligram of protein.

Table 21. ELISA kits used for the determination of cytokine concentrations in porcine tissues. IL: interleukin, TGF: transforming growth factor, TNF: tumor necrosis factor.

Antigen	ELISA Test and Catalogue Number	Manufacturer, Country	Assay Range pg/mL
IL-1α	Porcine IL-1 alpha/IL-1F1 DuoSet ELISA, DY680	R&D systems, Minneapolis, MN, USA	93.8–6000 Intra-assay CV < 2.92% Inter-assay CV < 4.21%
IL-1β	Porcine IL-1 beta/IL-1F2 DuoSet ELISA, DY681	R&D systems, Minneapolis, MN, USA	62.5–4000 Intra-assay CV < 1.1% Inter-assay CV < 3.2%
IL-2	Porcine IL-1 alpha/IL-1F1 DuoSet ELISA, DY652	R&D systems, Minneapolis, MN, USA	46.9–3000 Intra-assay CV < 4% Inter-assay CV < 5.6%
IL-4	Porcine IL-4 DuoSet ELISA, DY654	R&D systems, Minneapolis, MN, USA	156.0–10,000 Intra-assay CV < 5% Inter-assay CV < 6.69%
IL-6	Porcine IL-6 DuoSet ELISA, DY686	R&D systems, Minneapolis, MN, USA	125.0–8000 Intra-assay CV < 1.98% Inter-assay CV < 5.76%
IL-8	Porcine IL-8/CXCL8 DuoSet ELISA, DY535	R&D systems, Minneapolis, MN, USA	125.0–8000 Intra-assay CV < 7.3% Inter-assay CV < 8%
IL-10	Porcine IL-10 DuoSet ELISA, DY693B	R&D systems, Minneapolis, MN, USA	23.4–1500 Intra-assay CV < 3% Inter-assay CV < 4.64%
IL-12/23p40	Porcine IL-12/IL-23 p40 DuoSet ELISA, DY912	R&D systems, Minneapolis, MN, USA	78.1–5000 Intra-assay CV < 3.67% Inter-assay CV < 4.25%
TNFα	Porcine TNF-alpha DuoSet ELISA	R&D systems, Minneapolis, MN, USA	31.2–2000 Intra-assay CV < 5.11% Inter-assay CV <5.24%
IFN-γ	Porcine IFN-gamma DuoSet ELISA, DY985	R&D systems, Minneapolis, MN, USA	62.5–4000 Intra-assay CV < 3.4% Inter-assay CV < 4.6%
TGFβ	TGF beta-1 Multispecies Matched Antibody Pair, CHC1683	ThermoFisher Scientific, Waltham, MA, USA	62.5–4000 Intra-assay CV < 2.9% Inter-assay CV < 5%

5.4. Determination of Malondialdehyde Levels (Thiobarbituric Acid Assay)

The level of malondialdehyde (MDA) was measured according to the method described by Weitner et al. [59] with small modifications. Briefly, the tissue supernatant with buthylohydroxytoluene (BHT, Sigma Aldrich, Saint Louis, MO, USA) was deproteinized with 20% TCA (trichloroacetic acid, Avantor, Poland) and centrifuged for 1 h; 100 µl of the supernatant was then combined with thiobarbituric acid (TBA, Sigma Aldrich, Saint Louis, MO, USA) and incubated (1 h at 95°C). The level of MDA was read from a calibration curve (TBA, MDA Standard, Cayman, Ann Arbor, MI, USA). Absorbance was read in a Perkin Elmer spectrophotometer Lambda 25 at a wavelength of λ= 520 nm (Biocompare, Baltimore, MD, USA). The level of MDA was expressed in picomoles per milligram of whole protein in the tissue supernatant.

5.5. Determination of Sulfhydryl Groups (–SH)

Thiol groups were measured according to the modified Ellman method. Briefly, 1.0 ml of 40 mM Ellman's reagent (5,5′-dithiobis-(2-nitrobenzoic acid) (DTNB) solution (Serva, Heidelberg, Germany) was added to the sample (86 mM Tris (Sigma Aldrich, Saint Louis, MO, USA), 90 mM glycine (Avantor, Gliwice, Poland), 4 mM ethylenediaminetetraacetic acid (EDTA) (Avantor, Gliwice, Poland), 8 M urea (Sigma Aldrich, Saint Louis, MO, USA), 0.5% sodium dodecyl sulfate (SDS, Serva, Heidelberg, Germany), 0.2 M Tris-HCl (Sigma Aldrich, Saint Louis, MO, USA)) with pH 8.0. Next, 200 µl of each sample was added to 1.0 ml of DTNB. The samples were incubated at room temperature for 30 min. Cysteine was used (Avantor, Gliwice, Poland) as a standard, and absorbance was measured by a Perkin Elmer spectrophotometer Lambda 25 at a wavelength of λ= 412 nm (Biocompare, Baltimore, MD, USA). The concentration of -SH groups was measured from a calibration curve based on cysteine solution in PBS. The concentration of -SH groups was expressed in micromoles per milligram of whole protein in the ileal supernatant.

5.6. Determination of Fructose and Glucose Concentrations

Fructose level was determined in the ileum by the method described by Messineo and Musarra [60] with some modifications. This method is specific for fructose and similar to sucrose and inulin measurements method without interference from glucose (aldohexoses), aldopentose, and ketopentose. Glucose concentration was evaluated by Trinder's glucose oxidase method modified by Lott and Turner [61], with further modifications. The measurements were conducted spectrophotometrically with glucose oxidase reagent (G7521, Pointe Scientific, Canton, MI, USA) with the appropriate modifications. Fructose and glucose concentrations were expressed in micrograms per milligram of whole protein in the ileal supernatant.

5.7. Determination of Proline Concentration

Proline concentration in ileal Peyer's patches was determined by the modified method for the determination of proline levels in plants [62]. Tissue homogenates in the amount of 500 mL were placed in glass test tubes, and 1.0 mL of 2.5% ninhydrin solution, 1.0 mL of 3% sulfosalicylic acid, and 1.0 mL of glacial acetic acid were added. The samples were incubated (boiling water bath for 15 min.). The glass test tubes were cooled under running water, and 2 mL of toluene was added. The tubes were shaken for 15 min and left to stand until the separation of mixture components. The toluene layer was measured spectrophotometrically at a wavelength of 520 nm. Proline concentration was read from the standard curve, and points on the curve were determined based on the sample preparation method.

5.8. Statistical Analysis

The results were processed in Excel (Microsoft, Redmond, WA, USA) and GraphPad Prism 6 (GraphPad Software, San Diego, CA, USA) applications. Mean values and standard error of the mean (SEM) were determined for all groups. Population distributions were evaluated by the Shapiro–Wilk normality test. The results were processed by two-way ANOVA with post hoc Tukey's multiple comparison test. The results were regarded as statistically significant at $p < 0.05$.

Author Contributions: The experiments were conceived and designed by K.O. and P.W. The experiments were performed by W.T., K.O. and P.W. Data were analyzed and interpreted by K.O., P.W. and W.T. The manuscript was drafted by K.O., W.T. and P.W. and critically edited by Ł.Z., J.T. and M.G. All authors have read and agreed to the published version of the manuscript.

Funding: The study was supported by the "Healthy Animal—Safe Food" Scientific Consortium of the Leading National Research Centre (KNOW) pursuant to a decision of the Ministry of Science and Higher Education No. 05-1/KNOW2/2015. The project was financially supported by the Minister of Science and Higher 613 Education under the program entitled "Regional Initiative of Excellence" for the years 2019–2022, 614 Project No. 010/RID/2018/19, amount of funding PLN 12,000,000.

Conflicts of Interest: The authors declare no conflict of interest.

Abbreviations

4-HNE: 4-Hydroxynonenal; BCA: Bicinchoninic acid; BW: Body weight; CAT: Catalase; DNA: Deoxyribonucleic acid; DTNB: Ellman's reagent (5,5'-dithiobis-(2-nitrobenzoic acid); EDTA: ethylenediaminetetraacetic acid; ELISA: enzyme-linked immunosorbent assay; ERα: Estrogen receptors alpha; ERβ: Estrogen receptors beta; FAD: Flavin adenine dinucleotide; GALT: Gut-associated lymphoid tissue; GLUT: Glucose transporter; GSH: Glutathione; GSSG: Glutathione disulfide; IL-10: Interleukin 10; IL-12: Interleukin 12; IL-1α: Interleukin 1 alpha; IL-1β: Interleukin 1 beta; IL-2: Interleukin 2; IL-4: Interleukin 4; IL-6: Interleukin 6; IL-8: Interleukin 8; KCl: Potassium chloride; KH_2PO_4: Monopotassium phosphate; LPS: Lipopolysaccharides; MDA: Malondialdehyde; Na_2HPO_4: Disodium phosphate; NaCl: Sodium chloride; NAD: Nicotinamide adenine dinucleotide; NADP: Nicotinamide adenine dinucleotide phosphate; NADPH: Reduced form of nicotinamide adenine dinucleotide phosphate; NK: Natural killer cells; NOAEL: No-observed-adverse-effect level; PBS: Phosphate-buffered saline; RIPA: Radioimmunoprecipitation assay buffer; ROS: Reactive oxygen species; SH: Sulfhydryl group; SOD: Superoxide dismutase; TBA: 2-Thiobarbituric acid; TCA: Trichloroacetic acid; TNFα: Tumor necrosis factor alpha; TGFβ: Transforming growth factor beta; Tris HCl: Trizma hydrochloride; TSLP: thymic stromal lymphopoietin; ZEN: Zearalenone; α-ZAL: α-zearalanol; α-ZEL: α-zearalenol; β-ZAL: β-zearalanol; β-ZEL: β-zearalenol.

References

1. Bracarense, A.P.; Lucioli, J.; Grenier, B.; Pacheco, G.D.; Moll, W.D.; Schatzmayr, G.; Oswald, I.P. Chronic ingestion of deoxynivalenol and fumonisin, alone or in interaction, induces morphological and immunological changes in the intestine of piglets. *Br. J. Nutr.* **2012**, *107*, 1776–1786. [CrossRef] [PubMed]
2. Brera, C.; Debegnach, F.; De Santis, B.; Di Ianni, S.; Gregori, E.; Neuhold, S.; Valitutti, F. Exposure assessment to mycotoxins in gluten-free diet for celiac patients. *Food Chem. Toxicol.* **2014**, *69*, 13–17. [CrossRef] [PubMed]
3. Gonkowski, S.; Obremski, K.; Makowska, K.; Rytel, L.; Mwaanga, E.S. Levels of zearalenone and its metabolites in sun-dried kapenta fish and water of Lake Kariba in Zambia—A preliminary study. *Sci. Total Environ.* **2018**, *637–638*, 1046–1050. [CrossRef] [PubMed]
4. Streit, E.; Schatzmayr, G.; Tassi, P.; Tzika, E.; Marin, D.; Taranu, I.; Tabuc, C.; Nicolau, A.; Aprodu, I.; Puel, O.; et al. Current situation of mycotoxin contamination and cooccurrence in animal feed–focus on Europe. *Toxins (Basel)* **2012**, *4*, 788–809. [CrossRef]
5. Vejdovszky, K.; Hahn, K.; Braun, D.; Warth, B.; Marko, D. Synergistic estrogenic effects of Fusarium and Alternaria mycotoxins in vitro. *Arch. Toxicol.* **2017**, *91*, 1447–1460. [CrossRef]
6. Cheli, F.; Pinotti, L.; Rossi, L.; Dell'Orto, V. Effect of milling procedures on mycotoxin distribution in wheat fractions: A review. *LWT-Food Sci. Technol.* **2013**, *54*, 307–314. [CrossRef]
7. Heidari, S.; Milani, J.; Nazari, S.S. Effect of the bread-making process on zearalenone levels. *Food Addit. Contam. Part A Chem. Anal. Control Expo. Risk Assess.* **2014**, *31*, 2047–2054. [CrossRef]
8. Obremski, K.; Gajęcka, M.; Zielonka, Ł.; Jakimiuk, E.; Gajęcki, M. Morphology and ultrastructure of small intestine mucosa in gilts with zearalenone mycotoxicosis. *Pol. J. Vet. Sci.* **2005**, *8*, 301–307.
9. Fink-Gremmels, J.; Malekinejad, H. Clinical effects and biochemical mechanisms associated with exposure to the mycoestrogen zearalenone. *Anim. Feed Sci. Technol.* **2007**, *137*, 326–341. [CrossRef]
10. Kanora, A.; Maes, D. The role of mycotoxins in pig reproduction: A review. *Vet. Med.* **2009**, *54*, 565–576. [CrossRef]
11. Schoevers, E.J.; Santos, R.R.; Colenbrander, B.; Fink-Gremmels, J.; Roelen, B.A. Transgenerational toxicity of zearalenone in pigs. *Reprod. Toxicol.* **2012**, *34*, 110–119. [CrossRef] [PubMed]
12. Obremski, K.; Poniatowska-Broniek, G. Zearalenone induces apoptosis and inhibits proliferation in porcine ileal Peyer's patches lymphocytes. *Pol. J. Vet. Sci.* **2015**, *18*, 153–161. [CrossRef] [PubMed]
13. Zielonka, Ł.; Waśkiewicz, A.; Beszterda, M.; Kostecki, M.; Dąbrowski, M.; Obremski, K.; Goliński, P.; Gajęcki, M. Zearalenone in the Intestinal Tissues of Immature Gilts Exposed *per os* to Mycotoxins. *Toxins (Basel)* **2015**, *18*, 3210–3223. [CrossRef] [PubMed]
14. Avantaggiato, G.; Havenaar, R.; Visconti, A. Assessing the zearalenone binding activity of adsorbent materials during passage through a dynamin in vitro gastrointestinal model. *Food Chem. Toxicol.* **2003**, *41*, 1283–1290. [CrossRef]
15. Goliński, P.; Waśkiewicz, A.; Gromadzka, K. Mycotoxins and mycotoxicoses under climatic conditions of Poland. *Pol. J. Vet. Sci.* **2009**, *12*, 581–588.
16. Jakimiuk, E.; Rybarczyk, L.; Zwierzchowski, W.; Obremski, K.; Gajęcka, M.; Zielonka, Ł.; Gajęcki, M. Effect of experimental long-term exposure to low-dose zearalenone mycotoxicosis on selected morphometric parameters of the reproductive tract in sexually-immature gilts. *Bull. Vet. Inst. Pulawy* **2010**, *54*, 25–28.

17. Gajęcka, M.; Rybarczyk, L.; Jakimiuk, E.; Zielonka, Ł.; Obremski, K.; Zwierzchowski, W.; Gajęcki, M. The effect of experimental long-term exposure to low-dose zearalenone on uterine histology in sexually immature gilts. *Exp. Toxicol. Pathol.* **2012**, *64*, 537–542.
18. Obremski, K.; Gonkowski, S.; Wojtacha, P. Zearalenone—Induced changes in lymphoid tissue and mucosal nerve fibers in the porcine ileum. *Pol. J. Vet. Sci.* **2015**, *18*, 357–365. [CrossRef]
19. Zielonka, Ł.; Gajęcka, M.; Lisieska-Żołnierczyk, S.; Dąbrowski, M.; Gajęcki, M.T. The Effect of Different Doses of Zearalenone in Feed on the Bioavailability of Zearalenone and Alpha-Zearalenol, and the Concentrations of Estradiol and Testosterone in the Peripheral Blood of Pre-Pubertal Gilts. *Toxins (Basel)* **2020**, *12*, 144. [CrossRef]
20. Liu, F.; Cottrell, J.J.; Furness, J.B.; Rivera, L.R.; Kelly, F.W.; Wijesiriwardana, U.; Pustovit, R.V.; Fothergill, L.J.; Bravo, D.M.; Celi, P.; et al. Selenium and vitamin E together improve intestinal epithelial barrier function and alleviate oxidative stress in heat-stressed pigs. *Exp. Physiol.* **2016**, *101*, 801–810. [CrossRef]
21. Devasagayam, T.P.; Boloor, K.K.; Ramasarma, T. Methods for estimating lipid peroxidation: An analysis of merits and demerits. *Indian J. Biochem. Biophys.* **2003**, *40*, 300–308. [PubMed]
22. Montagne, L.; Boudry, G.; Favier, C.; Huerou-Luron, I.; Lalles, J.P.; Seve, B. Main intestinal markers associated with the changes in gut architecture and function in piglets after weaning. *Br. J. Nutr.* **2007**, *97*, 45–57. [CrossRef] [PubMed]
23. Binns, R.M.; Pabst, R. Lymphoid tissue structure and lymphocyte trafficking in the pig. *Vet. Immunol. Immunopathol.* **1994**, *43*, 79–87. [CrossRef]
24. European Commission. COMMISSION RECOMMENDATION 2006/576/EC of 17 August 2006 on the 402 presence of deoxynivalenol, zearalenone, ochratoxin A, T-2 and HT-2 and fumonisins in products intended 403 for animal feeding. *Off. J. Eur. Union* **2006**, *2006*, 7–9.
25. EFSA CONTAM Panel (EFSA Panel on Contaminants in the Food Chain). Scientific Opinion on the risks for human and animal health related to the presence of modified forms of certain mycotoxins in food and feed. *EFSA J.* **2014**, *12*, 3916. [CrossRef]
26. Sies, H. Biochemistry of oxidatives stress. *Angew. Chem. Int. Ed. Engl.* **1986**, *25*, 1058–1071. [CrossRef]
27. Nathan, C.; Cunningham-Bussel, A. Beyond oxidative stress: An immunologist's guide to reactive oxygen species. *Nat. Rev. Immunol.* **2013**, *13*, 349–361. [CrossRef]
28. Obremski, K.; Wojtacha, P.; Podlasz, P.; Żmigrodzka, M. The influence of experimental administration of low zearalenone doses on the expression of Th1 and Th2 cytokines and on selected subpopulations of lymphocytes in intestinal lymph nodes. *Pol. J. Vet. Sci.* **2015**, *18*, 489–497. [CrossRef]
29. Wallace, K.L.; Zheng, L.B.; Kanazawa, Y.; Shih, D.Q. Immunopathology of inflammatory bowel disease. *World J. Gastroenterol.* **2014**, *20*, 6–21. [CrossRef]
30. Tiidus, P.M. Radical species in inflammation and overtraining. *Can. J. Physiol. Pharmacol.* **1998**, *76*, 533–538. [CrossRef]
31. Straub, R.H. The complex role of estrogens in inflammation. *Endocr. Rev.* **2007**, *28*, 521–574. [CrossRef] [PubMed]
32. Obremski, K. The effect of in vivo exposure to zearalenone on cytokine secretion by Th1 and Th2 lymphocytes in porcine Peyer's patches after in vitro stimulation with LPS. *Pol. J. Vet. Sci.* **2014**, *17*, 625–632. [CrossRef] [PubMed]
33. Pierron, A.; Alassane-Kpembi, I.; Oswald, I.P. Impact of mycotoxin on immune response and consequences for pig health. *Anim. Nutr.* **2016**, *2*, 63–68. [CrossRef]
34. Fan, W.; Lv, Y.; Ren, S.; Shao, M.; Shen, T.; Huang, K.; Zhou, J.; Yan, L.; Song, S. Zearalenone (ZEA)-induced intestinal inflammation is mediated by the NLRP3 inflammasome. *Chemosphere* **2018**, *190*, 272–279. [CrossRef] [PubMed]
35. Yousef, M.S.; Takagi, M.; Talukder, A.K.; Marey, M.A.; Kowsar, R.; Abdel-Razek, A.K.; Shimizu, T.; Fink-Gremmels, J.; Miyamoto, A. Zearalenone (ZEN) disrupts the anti-inflammatory response of bovine oviductal epithelial cells to sperm in vitro. *Reprod. Toxicol.* **2017**, *74*, 158–163. [CrossRef] [PubMed]
36. Beck, P.L.; Rosenberg, I.M.; Xavier, R.J.; Koh, T.; Wong, J.F.; Podolsky, D.K. Transforming growth factor-beta mediates intestinal healing and susceptibility to injury in vitro and in vivo through epithelial cells. *Am. J. Pathol.* **2003**, *162*, 597–608. [CrossRef]
37. Tarnau, I.; Gras, M.A.; Pistol, G.C.; Motiu, M.; Marin, D.; Stancu, M. Investigation of zearalenone tolerance limit in the feedstuffs for weaned pigs. *Sci. Pap. Anim. Sci. Ser.* **2016**, *65*, 14–18.

38. Liu, M.; Zhu, D.; Guo, T.; Zhang, Y.; Shi, B.; Shan, A.; Chen, Z. Toxicity of zearalenone on the intestines of pregnant sows and their offspring and alleviation with modified halloysite nanotubes. *J. Sci. Food Agric.* **2018**, *98*, 698–706. [CrossRef]
39. Mak, T.W.; Grusdat, M.; Duncan, G.S.; Dostert, C.; Nonnenmacher, Y.; Cox, M.; Binsfeld, C.; Hao, Z.; Brüstle, A.; Itsumi, M.; et al. Glutathione primes T cell metabolism for inflammation. *Immunity* **2017**, *46*, 675–689. [CrossRef]
40. Hueza, I.M.; Raspantini, P.C.; Raspantini, L.E.; Latorre, A.O.; Górniak, S.L. Zearalenone, an estrogenic mycotoxin, is an immunotoxic compound. *Toxins (Basel)* **2014**, *6*, 1080–1095. [CrossRef]
41. Atri, C.; Guerfali, F.Z.; Laouini, D. Role of Human Macrophage Polarization in Inflammation during Infectious Diseases. *Int. J. Mol. Sci.* **2018**, *19*, 1801. [CrossRef] [PubMed]
42. Suzuki, H.; Hisamatsu, T.; Chiba, S.; Mori, K.; Kitazume, M.T.; Shimamura, K.; Nakamoto, N.; Matsuoka, K.; Ebinuma, H.; Naganuma, M.; et al. Glycolytic pathway affects differentiation of human monocytes to regulatory macrophages. *Immunol. Lett.* **2016**, *176*, 18–27. [CrossRef] [PubMed]
43. Kominsky, D.J.; Campbell, E.L.; Colgan, S.P. Metabolic shifts in immunity and inflammation. *J. Immunol.* **2010**, *184*, 4062–4068. [CrossRef] [PubMed]
44. Peiró, C.; Romacho, T.; Azcutia, V.; Villalobos, L.; Fernández, E.; Bolaños, J.P.; Moncada, S.; Sánchez-Ferrer, C.F. Inflammation, glucose, and vascular cell damage: The role of the pentose phosphate pathway. *Cardiovasc. Diabetol.* **2016**, *15*, 82. [CrossRef]
45. Shivers, K.Y.; Amador, N.; Abrams, L.; Hunter, D.; Jenab, S.; Quiñones-Jenab, V. Estrogen alters baseline and inflammatory-induced cytokine levels independent from hypothalamic–pituitary–adrenal axis activity. *Cytokine* **2015**, *72*, 121–129. [CrossRef]
46. Wu, G. Intestinal mucosal amino acid catabolism. *J. Nutr.* **1998**, *128*, 1249–1252. [CrossRef]
47. Dillon, E.L.; Knabe, D.A.; Wu, G. Lactate inhibits citrulline and arginine synthesis from proline in pig enterocytes. *Am. J. Physiol.* **1999**, *276*, G1079–G1086. [CrossRef]
48. Liu, Y.; Borchert, G.L.; Donald, S.P.; Diwan, B.A.; Anver, M.; Phang, J.M. Proline oxidase functions as a mitochondrial tumor suppressor in human cancers. *Cancer Res.* **2009**, *69*, 6414–6422. [CrossRef]
49. Verma, S.; Slutsky, A.S. Idiopathic pulmonary fibrosis—New insights. *N. Engl. J. Med.* **2007**, *356*, 1370–1372. [CrossRef]
50. Liu, W.; Le, A.; Hancock, C.; Lane, A.N.; Dang, C.V.; Fan, T.W.; Phang, J.M. Reprogramming of proline and glutamine metabolism contributes to the proliferative and metabolic responses regulated by oncogenic transcription factor c-MYC. *Proc. Natl. Acad. Sci. USA* **2012**, *109*, 8983–8988. [CrossRef]
51. Phang, J.M.; Pandhare, J.; Liu, Y. The metabolism of proline as microenvironmental stress substrate. *J. Nutr.* **2008**, *138*, 2008S–2015S. [CrossRef] [PubMed]
52. Mittal, M.; Siddiqui, M.R.; Tran, K.; Reddy, S.P.; Malik, A.B. Reactive Oxygen Species in Inflammation and Tissue Injury. *Antioxid. Redox Signal.* **2014**, *20*, 1126–1167. [CrossRef] [PubMed]
53. Lubrano, V.; Balzan, S. Enzymatic antioxidant system in vascular inflammation and coronary artery disease. *World J. Exp. Med.* **2015**, *20*, 218–224. [CrossRef] [PubMed]
54. Saitoh, T.; Fujita, N.; Jang, M.H.; Uematsu, S.; Yang, B.G.; Satoh, T.; Omori, H.; Noda, T.; Yamamoto, N.; Komatsu, M.; et al. Loss of the autophagy protein Atg16L1 enhances endotoxininduced IL-1beta production. *Nature* **2008**, *456*, 264–268. [CrossRef]
55. Marin, D.E.; Pistol, G.C.; Neagoe, I.V.; Calin, L.; Taranu, I. Effects of zearalenone on oxidative stress and inflammation in weanling piglets. *Food. Chem. Toxicol.* **2013**, *58*, 408–415. [CrossRef]
56. Pilz, J.; Meineke, I.; Gleiter, C.H. Measurement of free and bound malondialdehyde in plasma by high-performance liquid chromatography as the 2, 4-dinitrophenylhydrazine derivative. *J. Chromatogr. B Biomed. Sci.* **2000**, *742*, 315–325. [CrossRef]
57. Qin, X.; Cao, M.; Lai, F.; Yang, F.; Ge, W.; Zhang, X.; Cheng, S.; Sun, X.; Qin, G.; Shen, W.; et al. Oxidative stress induced by zearalenone in porcine granulosa cells and its rescue by curcumin in vitro. *PLoS ONE* **2015**, *6*, e0127551. [CrossRef]
58. Cieplińska, K.; Gajęcka, M.; Dąbrowski, M.; Rykaczewska, A.; Zielonka, Ł.; Lisieska-Żołnierczyk, S.; Bulińska, M.; Gajęcki, M.T. Time-dependent changes in the intestinal microbiome of gilts exposed to low zearalenone doses. *Toxins* **2019**, *11*, 296. [CrossRef]

59. Weitner, T.; Inić, S.; Jablan, J.; Gabričević, M.; Domijan, A. Spectrophotometric determination of malondialdehyde in urine suitable for epidemiological studies. *Croat. Chem. Acta* **2016**, *89*, 133–139. [CrossRef]
60. Messineo, L.; Musarra, E. A sensitive spectrophotometric method for the determination of free or bound tryptophan. *Int. J. Biochem.* **1972**, *3*, 700–704. [CrossRef]
61. Lott, J.A.; Turner, K. Evaluation of Trinder's glucose oxidase method for measuring glucose in serum and urine. *Clin. Chem.* **1975**, *21*, 1754–1760. [CrossRef] [PubMed]
62. Abrahám, E.; Hourton-Cabassa, C.; Erdei, L.; Szabados, L. Methods for determination of proline in plants. *Methods Mol. Biol.* **2010**, *639*, 317–331. [PubMed]

© 2020 by the authors. Licensee MDPI, Basel, Switzerland. This article is an open access article distributed under the terms and conditions of the Creative Commons Attribution (CC BY) license (http://creativecommons.org/licenses/by/4.0/).

Article

The Effect of Different Doses of Zearalenone in Feed on the Bioavailability of Zearalenone and Alpha-Zearalenol, and the Concentrations of Estradiol and Testosterone in the Peripheral Blood of Pre-Pubertal Gilts

Łukasz Zielonka [1], Magdalena Gajęcka [1,*], Sylwia Lisieska-Żołnierczyk [2], Michał Dąbrowski [1] and Maciej T. Gajęcki [1]

1 Department of Veterinary Prevention and Feed Hygiene, Faculty of Veterinary Medicine, University of Warmia and Mazury in Olsztyn, Oczapowskiego 13/29, 10-718 Olsztyn, Poland; lukaszz@uwm.edu.pl (Ł.Z.); michal.dabrowski@uwm.edu.pl (M.D.); gajecki@uwm.edu.pl (M.T.G.)

2 Independent Public Health Care Center of the Ministry of the Interior and Administration, and the Warmia and Mazury Oncology Center in Olsztyn, Wojska Polskiego 37, 10-228 Olsztyn, Poland; lisieska@wp.pl

* Correspondence: mgaja@uwm.edu.pl; Tel.: +48-89-523-37-73; Fax: +48-89-523-3618

Received: 31 January 2020; Accepted: 24 February 2020; Published: 26 February 2020

Abstract: The objective of this study was to determine the effect of long-term (48 days), *per os* administration of specific zearalenone (ZEN) doses (20 and 40 μg ZEN/kg BW in experimental groups EI and EII, which were equivalent to 200% and 400% of the upper range limit of the no-observed-adverse-effect-level (NOAEL), respectively) on the bioavailability of ZEN and the rate of changes in estradiol and testosterone concentrations in the peripheral blood of pre-pubertal gilts. ZEN and α-ZEL levels were similar until day 28. After day 28, α-ZEL concentrations increased significantly in group EI, whereas a significant rise in ZEN levels was noted in group EII. The presence of estradiol in peripheral blood plasma was not observed until day 20 of the experiment. Spontaneous secretion of estradiol was minimal, and it was determined at very low levels of up to 10 pg/mL in EI and EII groups. Testosterone concentrations ranged from 4 to 9 ng/mL in all groups. A decrease in the concentrations of both analyzed hormones was reported in the last stage of the experiment. The results of the experiment indicate that: (i) The bioavailability of ZEN in peripheral blood has low diagnostic value, (ii) exposure to low doses of ZEN induces minor changes in the concentrations of the analyzed hormones, which could lead to situational supraphysiological hormone levels and changes in endogenous hormonal balance.

Keywords: zearalenone; bioavailability; estradiol; testosterone; blood concentration; pre-pubertal gilts

Key Contribution: The exposure to ZEN doses equivalent to 200% and 400% of the upper range limit of NOAEL had no significant effect on the bioavailability of ZEN and α-ZEL in the blood serum of pre-pubertal gilts; but it induced supraphysiological hormone levels and improved the bioavailability of "free ZEN" in group EI in line with the hormesis principle.

---

1. Introduction

The majority of environmental estrogens (not all of them have toxic or contaminating effects) are endocrine disrupters (EDs) [1] which are commonly encountered in soil, air, water, food products, and animal feeds [2,3]. Phytoestrogens (e.g., genistein and coumestrol) and mycoestrogens (including ZEN produced by moulds) [4,5] are natural EDs found in the environment.

Zearalenone is a resorcylic acid lactone chemically described as 6-(10-hydroxy-6-oxo-*E*-1-undecenyl)-β-resorcylic acid lactone ($C_{18}H_{22}O_5$, molar mass: 318.36 g/mol, CAS: 17924-92-4). Zearalenone is a white crystalline compound with melting temperature of 164–165 °C. This mycotoxin is insoluble in water, but it is soluble in alkaline solutions and organic solvents. Zearalenone remains stable during storage, grinding and processing, including thermal processing [4].

Endocrine disruptors also influence the transport and distribution [6] of endogenous hormones, leading to changes in their concentrations and free hormone levels [7,8]. The effect of endogenous hormones and xenoestrogens on ion channels is an example of such activity. In general, EDs including environmental mycotoxins, such as ZEN and its metabolites, disrupt the activity of Ca^{2+} channels and/or Ca^{2+} signaling in certain types of cells [9]. These findings indicate that endogenous hormones and EDs exhibit hormonal activity by binding to nuclear hormone receptors. By acting outside the normal hormonal milieu, they are capable of producing unexpected results [10–12].

Zearalenone and its metabolite α-zearalenol (α-ZEL) have a similar chemical structure to estrogen, but unlike steroids, they do not originate from sterane structures [6]. Endocrine disruptors, including ZEN, are involved in several mechanisms [7,13] that can influence hormonal systems [14] and produce adverse results, in particular in pubertal gilts: (i) They compete with endogenous estrogens for estrogens receptors (ERs) or androgen receptors (ARs) binding sites, which leads to changes in *m*RNA expression and protein synthesis and reduces the effectiveness of endogenous steroid hormones, and they are responsible for the transport of ZEN inside cells [6,12,15]; (ii) they can bind with the receptor without activating it, and the presence of a substance on the receptor prevents natural hormone binding (antagonistic effect) [11,16]; (iii) they can bind with blood transport proteins to lower the concentrations of natural hormones in blood [17]; (iv) by disrupting metabolic processes in the body, EDs may affect the degree of synthesis/breakdown and the release of natural hormones [6,18,19].

The functions of the endocrine system are modified when physiologically active ligands (hormones) such as estradiol (E_2) and testosterone (T) are present at low concentrations [20,21]. Endogenous hormones and EDs act via receptors, but their activity is determined by, among others, the applied dose [22]. Lower doses of EDs influence the production, metabolism, absorption, and secretion of hormones because minor changes in hormone concentrations have far-reaching biological consequences [23]. Endogenous hormones are active in blood at very low concentrations for several reasons: (i) Receptors have high affinity for specific hormones, and they are capable of accepting a sufficient number of molecules to produce the most effective response [11]; (ii) a nonlinear correlation exists between hormone concentrations and the number of saturated receptors [22]; (iii) a nonlinear correlation also exists between the number of saturated receptors and the magnitude of a biological effect [7].

The objective of this study was to determine the effect of 48-day administration of ZEN doses of 20 and 40 μg ZEN/kg BW (equivalent to 200% and 400% of the highest NOAEL value, respectively), on the bioavailability of ZEN and the concentrations of E_2 and T in the peripheral blood of pre-pubertal gilts.

2. Results

2.1. Bioavailability of ZEN and Its Metabolite

The blood plasma bioavailability [24] of ZEN and α-ZEL on different days of the experiment is given in Figure 1. ZEN and its metabolite were determined in both experimental groups. Their presence was not observed in control group animals (values below limits of detection—LOD—regarded as equal to 0).

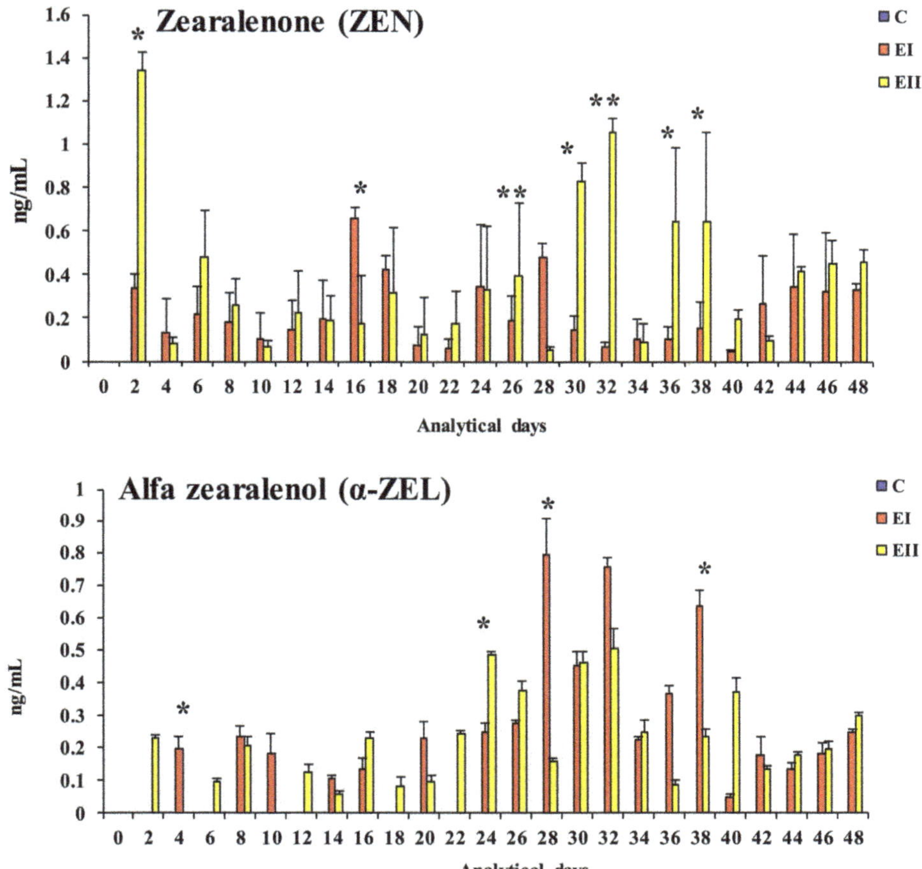

Figure 1. Arithmetic means ($\bar{x}$) and standard deviation (SD) for the bioavailability of zearalenone (ZEN) and α-zearalenol (α-ZEL) (ng/mL) in the peripheral blood of pre-pubertal gilts on different analytical days and in the experimental groups (group EI 20 μg ZEN/kg BW; group EII 40 μg ZEN/kg BW). Limits of detection (LOD) > values below the limit of detection were regarded as equal to 0. Statistically significant differences were determined at * $p \leq 0.05$ and ** $p \leq 0.01$.

Significant differences (* $p < 0.05$) were observed between groups EII and EI: (i) In the bioavailability of ZEN on days 2, 16, 30, 36, and 38 and in their mean values for the entire experimental period (EI < EII on all sampling dates, excluding day 16); (ii) in the bioavailability of α-ZEL on days 4, 24, 28, and 38 (EI < EII on the first two sampling dates, and EI > EII on the remaining days); (iii) in the bioavailability of ZON (see Figure 2) on days 22, 30, 32, and 40, and in their mean values during the entire experiment (EI < EII on all sampling dates). Highly significant differences (** $p < 0.01$) were noted between groups EI and EII: (i) In the bioavailability of ZEN on days 28 (EI > EII) and 32 (EI < EII) and in ZON concentrations on days 2 (EI < EII) and 28 (EI > EII).

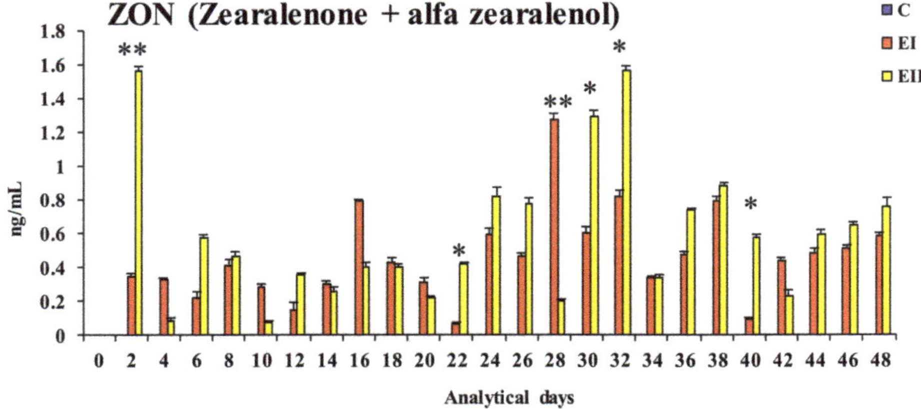

Figure 2. Arithmetic means ($\bar{x}$) and standard deviation (SD) for the bioavailability of ZON (ZEN + α-ZEL) (ng/mL) in the peripheral blood of pre-pubertal gilts on different sampling dates and in the experimental groups (group EI 20 µg ZEN/kg BW; group EII 40 µg ZEN/kg BW). Limits of detection (LOD) > values below the limit of detection were regarded as equal to 0. Statistically significant differences were determined at * $p \leq 0.05$ and ** $p \leq 0.01$.

2.2. Estradiol Concentrations

On the first ten sampling dates, the concentrations of E_2 were below the limit of sensitivity in all experimental groups (see Figure 3). This was also noted in group EI on day 20. Significant differences at $p < 0.05$ were noted between groups EI and C on days 24 and 36 (EI > C), and between groups EII and C on day 26 (EII > C). Highly significant differences at $p < 0.01$ were observed between groups EI and C on day 38 (EI > C).

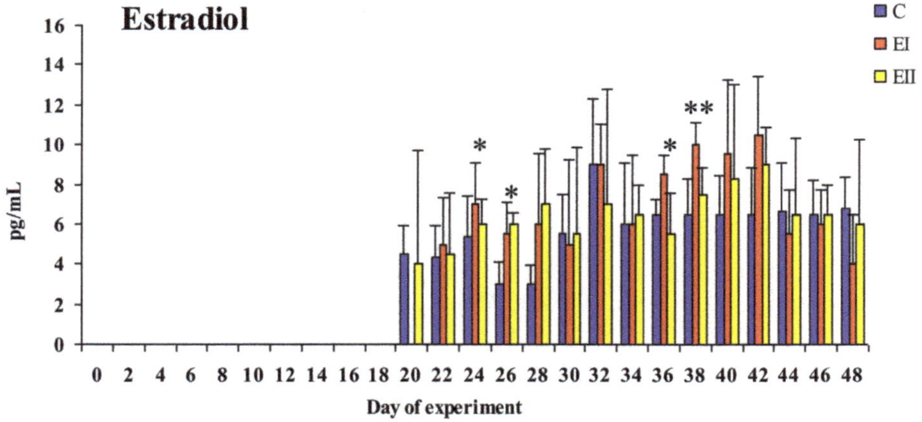

Figure 3. Arithmetic means ($\bar{x}$) and standard deviation (SD) for the concentrations of estradiol (pg/mL) in the peripheral blood of pre-pubertal gilts on different sampling dates and in the experimental groups (group EI 20 µg ZEN/kg BW; group EII 40 µg ZEN/kg BW). Limits of detection (LOD) > values below the limit of detection were regarded as equal to 0. Statistically significant differences were determined at * $p \leq 0.05$ and ** $p \leq 0.01$.

Between experimental days 36 and 42, a clear increase in E_2 levels was noted in both E groups in comparison with group C (see Figure 3). The concentrations of E_2 in the analyzed pre-pubertal gilts ranged from 3.0 to 10.5 pg/mL during the entire experiment. The mean increase in the values of E_2 was similar during the experiment, but considerably higher in both E groups until day 42. The concentrations of E_2 in experimental groups decreased towards the end of the experiment in comparison with control. The mean values of E_2 in peripheral blood were determined at 3.46 pg/mL in group C, 3.90 pg/mL in group EI and 3.83 pg/mL in group EII on 24 sampling dates, and at 5.77 pg/mL, 6.33 pg/mL, and 6.22 pg/mL, respectively, on the last 15 sampling dates (see Figure 3).

2.3. Testosterone Concentrations

Testosterone concentrations in blood plasma samples from pre-pubertal gilts are presented in Figure 4. During the entire experiment, T concentrations in the analyzed pre-pubertal gilts ranged from 4.0 to 9.0 ng/mL. Testosterone concentrations were higher in group C on 15 out of 25 sampling dates (60%). The changes in T concentrations were highly similar during the experiment. The mean concentration of T in peripheral blood during the entire experiment (25 sampling dates) was determined at 6.18 ng/mL in group C, 5.98 ng/mL in group EI, and 6.35 ng/mL in group EII.

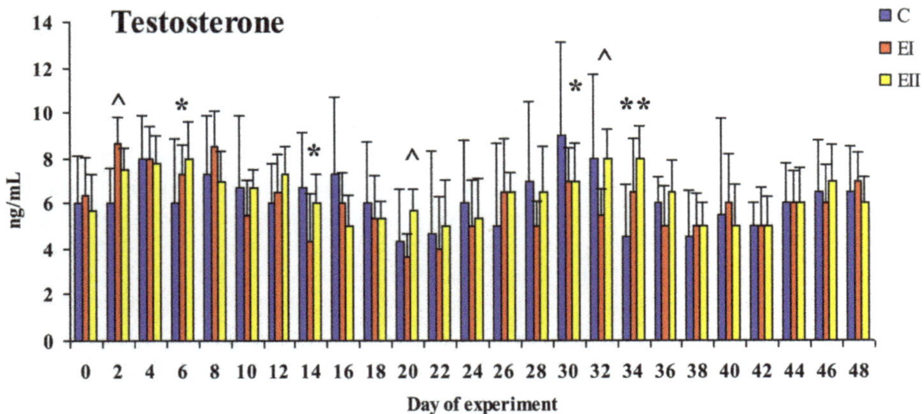

Figure 4. Arithmetic means ($\overline{x}$) and standard deviation (SD) for the concentrations of testosterone (ng/mL) in the peripheral blood of pre-pubertal gilts on different sampling dates and in the experimental groups (group EI 20 μg ZEN/kg BW; group EII 40 μg ZEN/kg BW). Statistically significant differences were determined at * or ^ $p \leq 0.05$ and ** $p \leq 0.01$; *—EI or EII compared with C; ^—EII compared with EI.

Statistically significant differences at $p < 0.05$ were reported between groups EI and C on days 14 and 30 (C > EI), between groups EII and C on days 6 (C < EII) and 30 (C > EII), and between groups EII and EI on days 2 (EI > EII), 20 (EI < EII), and 32 (EI < EII). Significant differences at $p < 0.01$ were observed between groups EII and C on day 34 (C < EII). The differences were statistically significant on days 6 and 34. A decrease in T concentrations in the experimental groups, relative to the control group, or no changes in T levels were observed more frequently.

3. Discussion

According to Shephard [25], animals are constantly exposed to mycotoxins which are secondary products of mould metabolism. Their elimination depends on the animal's age and the bioavailability of toxins in the blood plasma. This suggests that mycotoxins are always present in the natural environment, but at LOD doses. The bioavailability of ZEN and α-ZEL in the blood plasma is determined by the rate and the form of their biotransformation to water-soluble substances which are broken down by enzymes in the liver or other tissues as part of the detoxification process. According to some authors [26,27], biotransformation can generate products which are more toxic (bioactivation) than the parent mycotoxin. One of such compounds is ZEN which is transformed to α-ZEL in pre-pubertal gilts, and the consequences of this process are visible in tissues sensitive to that metabolite [13]. Tiemann et al. [28] observed that a decrease in steroid synthesis in cultures of porcine ovarian follicular granulosa cells resulting from the inhibition of P450scc activity is inversely proportional to the α-ZEL dose.

3.1. Zearalenone and Its Metabolite

In this study, the bioavailability of ZEN and α-ZEL in the blood plasma indicates that ZEN levels remained low in the peripheral blood of the experimental pre-pubertal gilts (groups EI and EII) in the first 28 days of the experiment. This can probably be attributed to enhanced biotransformation of ZEN and the distribution (carryover) of ZEN and α-ZEL to estrogen-sensitive tissues such as enterocytes, in particular in the duodenum and jejunum [19] in the initial period of exposure [29] and, subsequently, their bioavailability in peripheral blood.

The decrease in the values of ZEN in peripheral blood (see Figure 5) in both groups in the first two weeks of exposure provides evidence for the above, and it could indicate that: (i) All forms of mycoestrogen are intercepted by enterocytes (at absorptive level) or (ii) estrogen-sensitive tissues have a high demand for estrogen (at post-absorptive level) in pre-pubertal animals. In this experiment, the bioavailability of ZEN in peripheral blood 30 min after *per os* administration (maximum ZEN concentration in the blood [11]), was very low in comparison with the values reported by Dänicke and Winkler [30] and the values noted in intestinal tissues by Zielonka et al. [19] and Zheng et al. [31], which makes the resulting data difficult to interpret. It should also be noted that those values are inversely proportional to the ZEN dose (see Figure 5). Therefore, exposure to a lower ZEN dose leads to a hormetic dose response [32] by influencing the concentrations of the parent compound and its metabolite in peripheral blood, as well as the synthesis and secretion of sex steroid hormones (see Figures 3 and 4) [31,33,34]. The above state limits bioavailability. The analyzed values reflect on liver metabolism, i.e., the unmetabolised portion of the active substance (e.g., ZEN) and the produced and unutilised α-ZEL metabolite. The resulting "free ZEN" can be utilized for specific purposes [17], for example, for competing with endogenous E_2 or coregulating its supply. The above is illustrated by the fact that both doses lead to a significant increase in the concentrations that induce hyperestrogenism although rather supraphysiological hormone levels (increase in the concentrations of E_2 in both E groups and increase in T levels in group EII relative to group C) [34,35]. This observation points to a counter response to the presence of specific ZEN doses in pre-pubertal gilts. The presented results can be extrapolated [19,33,34] to suggest that "free ZEN" is captured by estrogen receptors (ERs) in the digestive tract [36], and that it stimulates qualitative changes (activation?) of ERs. The analyzed scenario testifies to the multidirectional effects of ZEN and its metabolite [33]. Therefore, it could be postulated that a dose of 20 μg ZEN/kg is more effectively utilized by pre-pubertal gilts. The above implies that biotransformation processes proceed in an identical manner, but the bioavailability of ZEN is higher in group EII and the bioavailability of α-ZEL is higher in group EI, whereas the parent compound (ZEN) and its metabolite, α-ZEL, are utilized more effectively in group EI (due to lower supply).

The values of ZON (see Figure 2), i.e., the total concentration of the parent compound and its metabolite, confirm the above suggestions. The values of ZON were higher in group EII on 15 sampling dates (60%). These findings indicate that the bioavailability of ZEN was higher in group EII, but they do not prove that ZEN was more effectively utilized for physiological processes. Low values of ZON at the beginning of the experiment can probably be attributed to the "saturation" of digestive tract tissues, in particular in regions with a higher concentration of ERs [19,36], and the low concentrations of endogenous hormones in animals during sexual maturation [5,8,34,37]. In other words, the involvement of the analyzed hormones in the stimulation of steroidogenesis and the conversion of T to E_2 ultimately led to supraphysiological hormone levels, in particular in group EII.

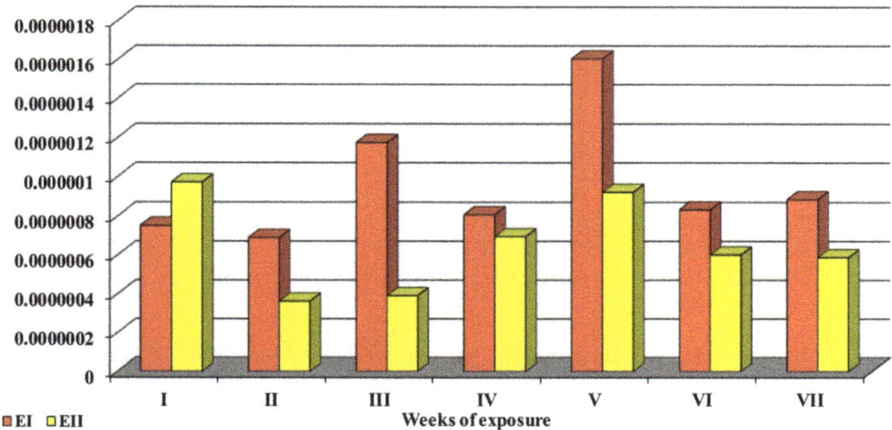

Figure 5. The bioavailability factor. The mean values ($\bar{x}$) of the ratio of ZON (ZEN + α-ZEL) bioavailability in peripheral blood (ppb) to ZEN concentration in the diet (ppb), expressed as the total dose administered in each week of the experiment.

3.2. Concentrations of Estradiol and Testosterone

Zearalenone is an ED [22], but its presence in the diet of experimental animals did not affect the production of endogenous hormones such as E_2 and T relative to the control group [38]. The presence of E_2 (see Figure 3) was not observed on the first nine sampling dates (19 days) of the experiment. On successive sampling dates (15 dates, 53.33%), E_2 values were higher in both E groups than in group C. The concentrations of T (see Figure 4) were lower on 60% sampling dates in both experimental groups than in the control group, which provides indirect evidence for the acceleration of metabolic processes or the conversion of T to E_2 [39] under the suppressive influence of ZEN [34,40–42]. The concentrations of E_2 after 20 days of exposure and serum T levels throughout the experiment were directly proportional to the ZEN dose. According to Vandenberg et al. [22], ZEN and/or α-ZEL exert a "complementary" effect on the concentrations of endogenous hormones (see Figures 3 and 4). The above is reflected by an increase in E_2 levels from day 20 to day 42 of the experiment and the decreasing trend in the concentrations of T between day 2 and the end of the experiment, which could alleviate disruptions in hormonal homeostasis [40,41]. The observed T values could also be attributed to the fact that T regulates sexual differentiation, muscle and bone mass, and erythropoietic and metabolic processes which play a very important role in the development of pre-pubertal gilts. These findings suggest that the physiological demands of young gilts considerably exceed the supply of endogenous E_2 and T. "Free ZEN" compensates for (or relieves) the resulting endocrinological dysfunction, e.g., by changing the expression of HSD enzymes [34].

3.3. Conclusions

The results of endocrinological analyses of selected estrogens (E_2) and androgens (T) in pre-pubertal gilts, administered at specific doses (200% and 400% of the highest NOAEL value) of ZEN (20 or 40 μg ZEN/kg BW) for 48 days suggest that: (i) The bioavailability of ZEN in peripheral blood is very low and highly varied throughout exposure, and that its diagnostic values are difficult to determine; (ii) experimentally-induced hyperestrogenism or, in other words, "supraphysiological hormone levels" contributed to a minor increase in total concentrations of E_2 (which could intensify proliferative processes) and influenced the levels of T (T values were lower in the experimental groups on 60% of sampling dates); (iii) the presented results can be extrapolated to suggest that the investigated ZEN doses caused different responses—the lower dose probably generated stimulatory/adaptive effects, whereas the higher dose probably inhibited life processes in the studied animals.

By extrapolation [34], it can be stated that the analyzed ZEN values affect the levels of E_2 and T in pre-pubertal gilts, thus reducing the risk of fighting for establishing the dominance hierarchy in the herd, contributing to changes in animal behaviour, slowing down sexual maturation, and boosting metabolism. Such observations could be important for pig breeders and veterinarians supervising pig farming and commercial feed production.

4. Materials and Methods

All of the experimental procedures involving animals were carried out in compliance with Polish legal regulations determining the terms and methods for performing experiments on animals (Opinion No. 14/2005/N of the Local Ethics Committee for Animal Experimentation at the University of Warmia and Mazury in Olsztyn, Poland in of 26 January, 2005).

4.1. Experimental Animals

This study was conducted at the Department of Veterinary Prevention and Feed Hygiene, Faculty of Veterinary Medicine, University of Warmia and Mazury in Olsztyn, Poland, on 18 clinically healthy, two-month-old pre-pubertal gilts. The mean initial body weight was 30 ± 2 kg. The pre-pubertal gilts were housed in individual cages with *ad libitum* access to water. Prior to the experiment, a cannula was placed in the cranial vena cava to minimize stress during blood sampling. The animals were administered with standard feed which was tested for the following mycotoxins: Aflatoxin, ochratoxin, ZEN, α-ZEL, and deoxynivalenol. Mycotoxin levels in the feed were estimated by common separation techniques with the use of immunoaffinity columns and HPLC (Hewlett Packard, type 1050 and 1100, Santa Clara, CA, USA) with fluorescent and/or UV detection techniques [34]. The obtained values were below the sensitivity of the tests.

4.1.1. Experimental Design

The animals were divided into two experimental groups and a control group of six individuals each. Group EI gilts were orally administered 20 µg ZEN/kg BW (200% more than the highest NOAEL value) for 48 days, group EII gilts were orally administered 40 µg ZEN/kg BW (400% more than the highest NOAEL value) *per os* for 48 days, and group C animals were orally administered placebo for 48 days (Table 1). The placebo was a gelatine capsule filled with a control medium onto which ZEN was applied in experimental groups.

Table 1. Daily feed intake in a restricted feeding regimen (kg/day) and mean dietary concentrations of ZEN per kg of feed (µg/kg feed).

Weeks of Exposure	Group EI			Group EII			Group C
	Feed Intake		Total ZEN Dose	Feed Intake		Total ZEN Dose	Feed Intake
	kg/day	µg/kg BW	µg/kg Feed	kg/day	µg/kg BW	µg/kg Feed	kg/day
I	1.7	664.4	390.82	1.75	1328.8	759.314	1.7
II	1.8	737.4	409.66	1.9	1474.8	776.210	1.85
III	1.9	819	431.05	1.9	1638	862.105	1.9
IV	2.0	912.2	456.1	1.9	1818.8	957.263	2.0
V	2.15	1014	471.62	2.2	2017.2	916.909	2.1
VI	2.1	1121.8	534.19	2.25	2243.6	997.155	2.2
VII	2.1	1232.4	586.85	2.2	2476.4	1125.636	2.2

4.1.2. Reagents

Sample weight of ZEN (Z-0167, Sigma Chemical Co., Steinheim, Germany) were administered *per os* daily in gelatine capsules before the morning feeding according to the weight of gilts. ZEN samples were diluted in 300 µL 96% ethyl alcohol (96% ethyl alcohol, SWW 2442-90, Polskie Odczynniki Chemiczne SA, Gliwice, Poland) to produce ZEN doses of 20 and 40 µg/kg BW. The resulting solutions were added to the feed, placed in gelatine capsules, and stored at room temperature for 12 h to evaporate the solvent.

4.2. Mycotoxicological Analyses

4.2.1. Blood Sampling to Mycotoxicological Analyses

Blood was sampled on the first day of the experiment after the administration of ZEN, and ZON levels (average total ZEN and α-ZEL concentrations) were estimated. Blood samples for determinations of ZEN and α-ZEL concentrations were collected every 48 h, 1 h after mycotoxin administration. The samples were directly transferred to chilled centrifuge tubes containing heparin and were centrifuged at 3000 rpm for 20 min at the temperature of 4 °C. The resulting plasma was transferred to 3 mL polypropylene Eppendorf tubes 3810X, it was frozen and stored at −20 °C until analysis. Blood loss was compensated through the administration of multi-electrolyte solutions in amounts corresponding to the volume of collected samples.

4.2.2. Analysis of ZEN and Its Metabolite in Blood Plasma

The presence of ZEN (see Figure 1) and α-ZEL (see Figure 1) in the blood plasma was determined by various separation methods with the use of immunoaffinity columns (Zearala-Test^{TM} Zearalenone Testing System, G1012, VICAM, Watertown, MA, USA) and HPLC with fluorescent detection.

4.2.3. HPLC Analysis

The presence of ZEN and α-ZEL in blood plasma was determined using immunoaffinity columns (Zearala-TestTM Zearalenone Testing System, G1012, VICAM, Watertown, MA, USA) and Agilent 1260 series liquid chromatography (LC) coupled with a mass spectrometer (MS) (Agilent 6460, Agilent Technologies, Inc., Santa Clara, CA, USA). The prepared sample was identified with the use of a chromatographic column (Atlantis T3 3 µm 3.0 × 150 mm Column, Waters, AN Etten-Leur, Dublin, Ireland). The mobile phase consisted of 70% acetonitrile (LiChrosolvTM, No. 984 730 109, Merck-Hitachi, Mannheim, Germany), 20% methanol (LiChrosolvTM, No. 1.06 007, Merck-Hitachi, Mannheim, Germany), and 10% deionized water (Milipore Water Purification System, Millipore S.A. Molsheim, France) with an addition of 0.2% CH3 COOH. The flow rate was 0.4 mL/min., and the temperature of the oven column was 40 °C. The chromatographic analysis was completed in 4 min. Mycotoxin concentrations were determined according to the external standard and were expressed in ppb (ng/mL). The limits of detection (LODs) for individual mycotoxins were determined as the concentrations at which the signal-to-noise ratio decreased to three. The LOQ was estimated as the triple LOD value. Quantification limits (LOQs) were determined at 0.100 ng/mL for ZEN and α-ZEL. The correlation coefficient (r) for the calibration curve was 0.9996 and 0.9989 for ZEN and α-ZEL, respectively.

4.2.4. Determination of Mycotoxin Bioavailability in Blood Plasma

The data were recorded and integrated with the use of the Computer Integrator POL-LAB application and CHROMAX for Windows v. 2000 software for processing chromatographic data (Pol-Lab Artur Dzieniszewski, Warszawa, Poland). Mycotoxin concentrations were determined according to the external standard method and given in terms of ng/mL.

4.3. Determination of Hormone (Estradiol and Testosterone) Concentrations in Blood Plasma

Blood was sampled every 48 h (25 times), 1 h after mycotoxin administration. The samples were transferred to chilled centrifuge tubes and were centrifuged at 3000 rpm for 20 min at the temperature of 4 °C. The resulting plasma was transferred to plastic polypropylene Eppendorf tubes, it was frozen and stored at −18 °C until analysis of selected steroid hormone levels.

Estradiol (see Figure 3) and T (see Figure 4) concentrations in the blood plasma were estimated by RIA procedures described for E_2 by Hotchkiss et al. [43] and for T—by Kotwica and Williams [44]. Estradiol and T antibodies (obtained from the Institute of Animal Physiology, University of Warmia and Mazury in Olsztyn, Poland) were characterized by Szafrańska et al. [45]. The sensitivity of E_2 and T assays was 5 pg and 2.5 ng (per tube), respectively. Intra-assay and inter-assay CVs for E_2 and T were determined at 6.9% and 11.8% and 7.6% and 12.5%, respectively.

4.4. Bioavailability Factor

The daily dose of ZEN (20 or 40 μg/kg BW) was administered to each animal individually, which corresponds to exposure to 928.74–1856.8 μg ZEN/kg of a complete diet, depending on daily feed intake (Table 1).

The bioavailability factor (see Figure 5) was calculated as the ratio of ZEN + α-ZEL = ZOL bioavailability in peripheral blood (expressed as ppb) to ZEN concentration in the diet (expressed as ppb), expressed as the overall dose administered in each week of the experiment, according to a previously described method.

4.5. Statistical Analysis

Plasma concentrations of ZEN (see Figure 1), α-ZEL (see Figure 1), and ZON (see Figure 2) as well as E_2 (see Figure 3) and T (see Figure 4) levels in the blood plasma of the examined gilts were presented as arithmetic means ($\overline{x}$) and SD (±). Statistical calculations were performed using the Statistica application (Statsoft Inc., Tulsa, OK, USA). Due to the applied ZEN dose and the length of application, arithmetic means were compared by one-way analysis of variance for systems with repeatable measurements. The homogeneity of variances in the compared groups (required in analyses of variance) was verified by the Brown–Forsythe test. Differences between groups were analyzed by Tukey's honestly significant difference test ($p < 0.05$ or $p < 0.01$).

Author Contributions: Developed the concept of the article, reviewed the literature, and wrote the article, Ł.Z.; reviewed the literature and prepared the manuscript for submission, M.T.G.; performed laboratory tests, M.G., S.L.-Ż. and M.D.; critically revised the paper, M.T.G. All authors have read and agreed to the published version of the manuscript.

Funding: The experiment was supported by research grant N N311 608938 from the National Science Centre Poland. The project was financially co-supported by the Minister of Science and Higher 613 Education under the program entitled "Regional Initiative of Excellence" for the years 2019-2022, 614 Project No. 010/RID/2018/19, amount of funding PLN 12,000,000.

Conflicts of Interest: The authors declare no conflict of interest.

Abbreviations

ARs: Androgen receptors; BW: Body weight; C: Control group; CAS: Chemical Abstracts Service; EI: Experimental group I; EII: Experimental group II; E_2: Estradiol; Eds: Endocrine disruptors; ERs: Estrogen receptors; ERα: Estrogen receptors alpha; ERβ: Estrogen receptors beta; HSDs: Hydroxysteroid dehydrogenases; LOD: Values below limits of detection; NOAEL: No-observed-adverse-effect level; PCNA: Proliferating cell nuclear antigen; SD: Standard deviation; Ts: Testosterone; ZEN: Zearalenone; α-ZEL: α-zearalenol; ZON: Total ZEN and α-ZEL.

References

1. Hampl, R.; Kubatova, J.; Starka, L. Steroids and endocrine disruptors -History, recent state of art and open questions. *J. Steroid Biochem. Mol. Biol.* **2016**, *155*, 217–223. [CrossRef]
2. Yurino, H.; Ishikawa, S.; Sato, T.; Akadegawa, K.; Ito, T.; Ueha, S.; Inadera, H.; Matsushima, K. Endocrine disrupters (environmental estrogens) enhance autoantibody production by B1 cells. *Toxicol. Sci.* **2004**, *81*, 139–147. [CrossRef]
3. Crain, D.A.; Janssen, S.J.; Edwards, T.M.; Heindel, J.; Ho, S.; Hunt, P.; Iguchi, T.; Juul, A.; McLachlan, J.A.; Schwartz, J.; et al. Female reproductive disorders: The roles of endocrinedisrupting compounds and developmental timing. *Fertil. Steril.* **2008**, *90*, 911–940. [CrossRef] [PubMed]

4. Gajęcki, M.; Gajęcka, M.; Jakimiuk, E.; Zielonka, Ł.; Obremski, K. Zearalenone—Undesirable substance. In *Mycotoxins in Food, Feed and Bioweapons*; Mahendra, R., Ajit, V., Eds.; Springer: Berlin/Heidelberg, Germany, 2010; pp. 131–144. [CrossRef]
5. Kowalska, K.; Habrowska-Górczyńska, D.E.; Piastowska-Ciesielska, A.W. Zearalenone as an endocrine disruptor in humans. *Environ. Toxicol. Phar.* **2016**, *48*, 141–149. [CrossRef]
6. Lathe, R.; Kotelevtsev, Y.; Mason, J.I. Steroid promiscuity: Diversity of enzyme action. *J. Steroid Biochem.* **2015**, *151*, 1–2. [CrossRef] [PubMed]
7. Barton, M. Position paper: The membrane estrogen receptor GPER—Clues and questions. *Steroids* **2012**, *77*, 935–942. [CrossRef] [PubMed]
8. Fürst, R.W.; Pistek, V.L.; Kliem, H.; Skurk, T.; Hauner, H.; Meyer, H.H.D.; Ulbrich, S.E. Maternal low-dose estradiol-17β exposure during pregnancy impairs postnatal progeny weight development and body composition. *Toxicol. Appl. Pharm.* **2012**, *263*, 338–344. [CrossRef] [PubMed]
9. Gajęcka, M.; Przybylska-Gornowicz, B. The low doses effect of experimental zearalenone (ZEN) intoxication on the presence of Ca^{2+} in selected ovarian cells from pre-pubertal bitches. *Pol. J. Vet. Sci.* **2012**, *15*, 711–720. [CrossRef]
10. Tarasiuk, M. The Effect of Low Doses of Zearalenone and Deoxynivalenol on the Jejunal Mucosa, Metabolic Profile and Body Weight of Pre-Pubertal Gilts. Ph.D. Thesis, Faculty of Veterinary Medicine, University of Warmia and Mazury, Olsztyn, Poland, 6 November 2015.
11. Barton, M. Not lost in translation: Emerging clinical importance of the G protein-coupled estrogen receptor GPER. *Rev. Artic. Steroids* **2016**, *111*, 37–45. [CrossRef]
12. Adibnia, E.; Razi, M.; Malekinejad, H. Zearalenone and 17 β-estradiol induced damages in male rats reproduction potential; evidence for ERα and ERβ receptors expression and steroidogenesis. *Toxicon* **2016**, *120*, 133–146. [CrossRef]
13. Frizzell, C.; Ndossi, D.; Verhaegen, S.; Dahl, E.; Eriksen, G.; Sřrlie, M.; Ropstad, E.; Muller, M.; Elliott, C.T.; Connolly, L. Endocrine disrupting effects of zearalenone, alpha- and beta-zearalenol at the level of nuclear receptor binding and steroidogenesis. *Toxicol. Lett.* **2011**, *206*, 210–217. [CrossRef] [PubMed]
14. Yang, R.; Wang, Y.M.; Zhang, L.; Zhao, Z.M.; Zhao, J.; Peng, S.Q. Prepubertal exposure to an oestrogenic mycotoxin zearalenone induces central precocious puberty in immature female rats through the mechanism of premature activation of hypothalamic kisspeptin-GPR54 signaling. *Mol. Cell. Endocrinol.* **2016**, *437*, 62–74. [CrossRef] [PubMed]
15. Pizzo, F.; Caloni, F.; Schreiber, N.B.; Cortinovis, C.; Spicer, L.J. In vitro effects of deoxynivalenol and zearalenone major metabolites alone and combined, on cell proliferation, steroid production and gene expression in bovine small-follicle granulose cells. *Toxicon* **2016**, *109*, 70–83. [CrossRef] [PubMed]
16. Zielonka, Ł.; Gajęcka, M.; Rozicka, A.; Dąbrowski, M.; Żmudzki, J.; Gajęcki, M. The Effect of Environmental Mycotoxins on Selected Ovarian Tissue Fragments of Multiparous Female Wild Boars at the Beginning of Astronomical Winter. *Toxicon* **2014**, *89*, 26–31. [CrossRef] [PubMed]
17. Gajęcka, M.; Tarasiuk, M.; Zielonka, Ł.; Dąbrowski, M.; Nicpoń, J.; Baranowski, M.; Gajęcki, M.T. Changes in the metabolic profile and body weight of pre-pubertal gilts during prolonged monotonic exposure to low doses of zearalenone and deoxynivalenol. *Toxicon* **2017**, *125*, 32–43. [CrossRef]
18. Zielonka, Ł.; Gajęcka, M.; Żmudzki, J.; Gajęcki, M. The effect of selected environmental *Fusarium* mycotoxins on the ovaries in the female wild boar (*Sus scrofa*). *Pol. J. Vet. Sci.* **2015**, *18*, 391–399. [CrossRef]
19. Zielonka, Ł.; Waśkiewicz, A.; Beszterda, M.; Kostecki, M.; Dąbrowski, M.; Obremski, K.; Goliński, P.; Gajęcki, M. Zearalenone in the Intestinal Tissues of Immature Gilts Exposed *per os* to Mycotoxins. *Toxins* **2015**, *7*, 3210–3223. [CrossRef]
20. Eick, G.N.; Thornton, J.W. Evolution of steroid receptors from an estrogen-sensitive ancestral receptor. *Mol. Cell. Endocrinol.* **2011**, *334*, 31–38. [CrossRef]
21. Jiang, S.Z.; Yang, Z.B.; Yang, W.R.; Wang, S.J.; Liu, F.X.; Johnston, L.A.; Chi, F.; Wang, Y. Effect of purified zearalenone with or without modified montmorillonite on nutrient availability, genital organs and serum hormones in post-weaning piglets. *Livest. Sci.* **2012**, *144*, 110–118. [CrossRef]
22. Vandenberg, L.N.; Colborn, T.; Hayes, T.B.; Heindel, J.J.; Jacobs, D.R., Jr.; Lee, D.-H.; Shioda, T.; Soto, A.M.; vom Saal, F.S.; Welshons, W.V.; et al. Hormones and Endocrine-Disrupting Chemicals: Low-Dose Effects and Nonmonotonic Dose Responses. *Endocr. Rev.* **2012**, *33*, 378–455. [CrossRef]

23. Hayes, T.B.; Anderson, L.L.; Beasley, V.R.; de Solla, S.R.; Iguchi, T.; Ingraham, H.; Kestemont, P.; Kniewald, J.; Kniewald, Z.; Langlois, V.S.; et al. Demasculinization and feminization of male gonad by atrazine: Consistent effects across vertebrate classes. *J. Steroid Biochem. Mol. Biol.* **2011**, *127*, 64–73. [CrossRef] [PubMed]
24. Ma, Z.; Wang, N.; He, H.; Tang, X. Pharmaceutical strategies of improving oral systemic bioavailability of curcumin for clinical application. *J. Control Release* **2019**, *316*, 359–380. [CrossRef] [PubMed]
25. Shephard, G.S. Impact of mycotoxins on human health in developing countries. *Food Addit. Contam.* **2008**, *25*, 146–151. [CrossRef] [PubMed]
26. Liska, D.A.; Lyon, M.; Jones, D.S. Detoxification and biotransformational imbalances. *Explor. J. Sci. Heal.* **2006**, *2*, 122–140. [CrossRef] [PubMed]
27. Sergent, T.; Ribonnet, L.; Kolosova, A.; Garsou, S.; Schau, T.A.; De Saeger, S.; Van Peteghem, C.; Larondelle, Y.; Pussemier, L.; Schneider, Y.J. Molecular and cellular effects of food contaminants and secondary plant components and their plausible interactions at the intestinal level. *Food Chem. Toxicol.* **2008**, *46*, 813–841. [CrossRef]
28. Tiemann, U.; Tomek, W.; Schneider, F.; Vanselow, J. Effects of the mycotoxins α- and β-zearalenol on regulation of progesterone synthesis in cultured granulose cells from porcine ovaries. *Reprod. Toxicol.* **2003**, *17*, 673–681. [CrossRef]
29. Oliver, W.T.; Miles, J.R.; Diaz, D.E.; Dibner, J.J.; Rottinghaus, G.E.; Harrell, R.J. Zearalenone enhances reproductive tract development, but does not alter skeletal muscle signaling in prepubertal gilts. *Anim. Feed Sci. Tech.* **2012**, *174*, 79–85. [CrossRef]
30. Dänicke, S.; Winkler, J. Invited review: Diagnosis of zearalenone (ZEN) exposure of farm animals and transfer of its residues into edible tissues (carry over). *Food Chem. Toxicol.* **2015**, *84*, 225–249. [CrossRef]
31. Zheng, W.; Feng, N.; Wang, Y.; Noll, L.; Xu, S.; Liu, X.; Lu, N.; Zou, H.; Gu, J.; Yuan, Y.; et al. Effects of zearalenone and its derivatives on the synthesis and secretion of mammalian sex steroid hormones: A review. *Food Chem. Toxicol.* **2019**, *126*, 262–276. [CrossRef]
32. Calabrese, E.J. Hormesis: Path and Progression to Significance. *Int. J. Mol. Sci.* **2018**, *19*, 2871. [CrossRef]
33. Gajęcka, M.; Zielonka, Ł.; Gajęcki, M. Activity of zearalenone in the porcine intestinal tract. *Molecules* **2017**, *22*, 18. [CrossRef] [PubMed]
34. Rykaczewska, A.; Gajęcka, M.; Onyszek, E.; Cieplińska, K.; Dąbrowski, M.; Lisieska-Żołnierczyk, S.; Bulińska, M.; Babuchowski, A.; Gajęcki, M.T.; Zielonka, Ł. Imbalance in the Blood Concentrations of Selected Steroids in Prepubertal Gilts Depending on the Time of Exposure to Low Doses of Zearalenone. *Toxins* **2019**, *11*, 561. [CrossRef] [PubMed]
35. Lawrenz, B.; Melado, L.; Fatemi, H. Premature progesterone rise in ART-cycles. *Reprod. Biol.* **2018**, *18*, 1–4. [CrossRef] [PubMed]
36. Gajęcka, M.; Dabrowski, M.; Otrocka-Domagała, I.; Brzuzan, P.; Rykaczewska, A.; Cieplińska, K.; Barasińska, M.; Gajęcki, M.T.; Zielonka, Ł. Correlations between exposure to deoxynivalenol and zearalenone and the immunohistochemical expression of estrogen receptors in the intestinal epithelium and the mRNA expression of selected colonic enzymes in pre-pubertal gilts. *Toxicon* **2020**, *173*, 75–93. [CrossRef] [PubMed]
37. Gajęcka, M.; Zielonka, Ł.; Dąbrowski, M.; Mróz, M.; Gajęcki, M. The effect of low doses of zearalenone and its metabolites on progesterone and 17β-estradiol concentrations in blood of pre-pubertal female Beagle dogs. *Toxicon* **2013**, *76*, 260–269. [CrossRef]
38. Dunbar, B.; Patel, M.; Fahey, J.; Wira, C. Endocrine control of mucosal immunity in the female reproductive tract: Impact of environmental disruptors. *Mol. Cell. Endocrinol.* **2012**, *354*, 85–93. [CrossRef]
39. Bhasin, S.; Jasuja, R. Reproductive and Nonreproductive Actions of Testosterone. *Enc. Endocr. Dis.* **2019**, *2*, 721–734. [CrossRef]
40. Freeman, B.M.; Fisher, R.K.; Kirkpatrick, S.S.; Klein, F.A.; Freeman, M.B.; Mountain, D.J.H.; Grandas, O.H. Testosterone replacement attenuates intimal hyperplasia development in an androgen deficient model of vascular injury. *J. Surg. Res.* **2017**, *207*, 53–62. [CrossRef]
41. Ketay, S.; Welker, K.M.; Slatcher, R.B. The roles of testosterone and cortisol in friendship formation. *Psychoneuroendocrinology* **2017**, *76*, 88–96. [CrossRef]
42. Jiang, X.; Lu, N.; Xue, Y.; Liu, S.; Lei, H.; Tu, W.; Lu, Y.; Xia, D. Crude fiber modulates the fecal microbiome and steroid hormones in pregnant Meishan sows. *Gen. Comp. Endocr.* **2019**, *277*, 141–147. [CrossRef]

43. Hotchkiss, J.; Atkinson, L.E.; Knobil, E. Time course of serum estrogen and luteinizing hormone (LH) concentrations during menstrual cycle of rhesus monkey. *Endocrinology* **1971**, *89*, 177–183. [CrossRef] [PubMed]
44. Kotwica, J.; Williams, G.L. Relationship of plasma testosterone concentrations to pituitary-ovarian hormone secretions during bovine estrus cycle and the effects of testosterone propionate administered during luteal regression. *Biol. Reprod.* **1982**, *27*, 790–801. [CrossRef] [PubMed]
45. Szafrańska, B.; Zięcik, A.J.; Okrasa, S. Primary antisera against selected steroids or proteins and secondary antisera against-globulins—An available tool for studies of reproductive processes. *Reprod. Biol.* **2002**, *5*, 187–203.

© 2020 by the authors. Licensee MDPI, Basel, Switzerland. This article is an open access article distributed under the terms and conditions of the Creative Commons Attribution (CC BY) license (http://creativecommons.org/licenses/by/4.0/).

MDPI

Review

A Review of the Impact of Mycotoxins on Dairy Cattle Health: Challenges for Food Safety and Dairy Production in Sub-Saharan Africa

David Chebutia Kemboi [1,2], Gunther Antonissen [3,4], Phillis E. Ochieng [3,5], Siska Croubels [3], Sheila Okoth [6], Erastus K. Kangethe [7], Johannes Faas [8], Johanna F. Lindahl [9,10,11,*] and James K. Gathumbi [1,*]

1 Department of Pathology, Parasitology and Microbiology, Faculty of Veterinary Medicine, University of Nairobi, PO Box 29053, 00100 Nairobi, Kenya; kemboidc@gmail.com
2 Department of Animal Science, Chuka University, P.O Box 109-00625 Chuka, Kenya
3 Department of Pharmacology, Toxicology and Biochemistry, Faculty of Veterinary Medicine, Ghent University, Salisburylaan 133, 9820 Merelbeke, Belgium; Gunther.Antonissen@ugent.be (G.A.); phillisemelda.ochieng@ugent.be (P.E.O.); Siska.Croubels@ugent.be (S.C.)
4 Department of Pathology, Bacteriology and Avian Diseases, Faculty of Veterinary Medicine, Ghent University, Salisburylaan 133, 9820 Merelbeke, Belgium
5 Department of Food Sciences, University of Liège, Faculty of Veterinary Medicine, Avenue de Cureghem 10, 4000 Liège, Belgium
6 School of Biological Sciences, University of Nairobi, P.O Box 30197-00100 Nairobi, Kenya; dorisokoth@yahoo.com
7 P.O Box 34405, 00100 Nairobi, Kenya; mburiajudith@gmail.com
8 BIOMIN Research Center, Technopark 1, 3430 Tulln, Austria; johannes.faas@biomin.net
9 Department of Biosciences, International Livestock Research Institute (ILRI), P.O Box 30709, 00100 Nairobi, Kenya
10 Department of Medical Biochemistry and Microbiology, Uppsala University, P.O Box 582, 751 23 Uppsala, Sweden
11 Department of Clinical Sciences, Swedish University of Agricultural Sciences, P.O Box 7054, 750 07 Uppsala, Sweden
* Correspondence: J.Lindahl@cgiar.org (J.F.L.); jkgathumbi@uonbi.ac.ke (J.K.G.)

Received: 28 February 2020; Accepted: 29 March 2020; Published: 2 April 2020

Abstract: Mycotoxins are secondary metabolites of fungi that contaminate food and feed and have a significant negative impact on human and animal health and productivity. The tropical condition in Sub-Saharan Africa (SSA) together with poor storage of feed promotes fungal growth and subsequent mycotoxin production. Aflatoxins (AF) produced by *Aspergillus* species, fumonisins (FUM), zearalenone (ZEN), T-2 toxin (T-2), and deoxynivalenol (DON) produced by *Fusarium* species, and ochratoxin A (OTA) produced by *Penicillium* and *Aspergillus* species are well-known mycotoxins of agricultural importance. Consumption of feed contaminated with these toxins may cause mycotoxicoses in animals, characterized by a range of clinical signs depending on the toxin, and losses in the animal industry. In SSA, contamination of dairy feed with mycotoxins has been frequently reported, which poses a serious constraint to animal health and productivity, and is also a hazard to human health since some mycotoxins and their metabolites are excreted in milk, especially aflatoxin M1. This review describes the major mycotoxins, their occurrence, and impact in dairy cattle diets in SSA highlighting the problems related to animal health, productivity, and food safety and the up-to-date post-harvest mitigation strategies for the prevention and reduction of contamination of dairy feed.

Keywords: mycotoxins; dairy; aflatoxin; Sub-Saharan Africa; aflatoxin M1

Key Contribution: This paper looks at the prevalence and effects of mycotoxins in the dairy sector in Sub-Saharan Africa and its impact on animal health and food safety. Mycotoxin occurrence in dairy feed is widespread in Sub-Saharan Africa and affects animal health and productivity as well as food safety. Therefore, there is a need for enhanced regulation as well as the use of mitigation strategies to promote animal health and productivity and food safety.

1. Introduction

The livestock sector accounts for about 18% of the gross domestic product (GDP) in Sub-Saharan Africa (SSA). In 2017, there were 35.3 million tons of milk produced by 71.1 million dairy cattle in Africa, of which 66.5% is located in SSA [1]. However, Africa is producing only 5.1% of the world milk and the total yield per cow per year is also low [1]. With a population of about 1 billion people, projected to rise to over 1.2 billion by 2025 [1], there is an urgent need for improved productivity per animal since the increased production of milk over the years was rather the result of an increased number of animals than increased individual animal productivity. This may be attributed to the traditional (pastoralism) system that is mostly practiced in SSA. However, due to human population growth, increased milk consumption per capita, and land shortage and increasing interest in production, semi-intensive and intensive dairy farming systems are increasingly being adopted [2]. In these semi-intensive and intensive systems, mostly localized in rural and peri-urban regions, the animals either graze or are fed on planted fodder or crop residues supplemented with concentrates [3]. Cereal grains are the major ingredients of most of the concentrates and these are often of substandard grade, mostly due to fungal growth and discoloration and thus considered unfit for human consumption, which predisposes these feeds to mycotoxin contamination [4].

Mycotoxins are secondary metabolites of fungi that contaminate food and feed and have a significant negative impact on human and animal health including animal productivity. *Aspergillus, Fusarium,* and *Penicillium* are the major mycotoxin-producing fungi. These toxigenic fungi are classified into two groups; field fungi that contaminate crops and produce toxins while still on the field such as *Fusarium* species, and storage fungi, such as *Aspergillus* and *Penicillium* species, that mainly produce toxins after harvesting during storage. Production of mycotoxins is related to environmental conditions, stress to the plant, and damage to the grains caused by rodents and pests and abiotic factors such as pH of feed and moisture content [5,6]. The tropical condition in SSA together with poor storage of feed promotes fungal growth and subsequent mycotoxin production [6,7].

Aflatoxins (AF) produced by *Aspergillus* species, fumonisins (FUM), zearalenone (ZEN), T-2 toxin (T-2), and deoxynivalenol (DON) produced by *Fusarium* species, and ochratoxin A (OTA) produced by *Aspergillus* and *Penicillium* species are well-known mycotoxins of major agricultural importance and have been found concurrently occurring in feeds, with AF, which is a class 1 carcinogen to human beings, the most prevalent and also the most studied in SSA [8–11].

Consumption of feed contaminated with these toxins may cause mycotoxicoses in animals, characterized by a range of clinical signs depending on the toxin, and may cause losses in the animal industry. In SSA, contamination of dairy feed with mycotoxins has been frequently reported, which poses a serious constraint to animal health and productivity and is also a hazard to human health. This is because some mycotoxins and their metabolites are excreted in milk, especially aflatoxin M1 (AFM1). This review describes the major mycotoxins and their occurrence and impact in dairy cattle diets in SSA, highlighting the problems related to animal health and productivity, food safety, and provides the up-to-date post-harvest mitigation strategies for the prevention and reduction of contamination of dairy feed.

2. Legislation of Mycotoxins in Africa

As per 2003, 15 countries in Africa had mycotoxin regulations covering 59% of the inhabitants of Africa. These regulations only are concerned with AF except for South Africa where guidance levels exist for ZEN, FUM, and DON in dairy feeds [10,12]. Regionally, East Africa Community (EAC) harmonized and set regulatory limits for AFB1 at 5 µg/kg in dairy feed and 0.5 µg/kg for AFM1 in milk. In addition, Rwanda through the Rwanda Standards Board (RSB) has established a regulatory limit of 5 µg/kg for AFB1 in cattle feed supplements [13]. No other regional regulatory limits have been established and implemented in Africa. In West Africa, Nigeria through the National Agency for Food and Drug Administration and Control has set a regulatory limit of 5 µg/kg and 0.5 µg/kg for AFB1 in dairy feed and AFM1 in milk, respectively [14]. Cote d' Ivoire has set a limit of 10 µg/kg for total AF in complete feed, while Senegal has a limit of 50 µg/kg for AFB1 for animal feeds from peanuts [12]. In Southern Africa, Republic of South Africa has a regulatory limit for AF and guidance levels for other mycotoxins in dairy feed, i.e., at 5 µg/kg for AFB1, 50,000 µg/kg for FUM, 500 µg/kg for ZEN, and 3000 µg/kg for DON. The regulatory limit for AFM1 in milk is set at 0.05 µg/kg [10]. Mozambique has set regulatory limits for total AF at 10 µg/kg in dairy feed, while Zimbabwe has not set the limit for dairy feed but use a 10 µg/kg limit for poultry feed [12]. Worldwide, the World Health Organization/Food and Agriculture Organization of the United Nations (WHO/FAO) through the Codex Alimentarius Commission (CODEX) have set up a regulatory limit for AFB1 in dairy feeds at 5 µg/kg and for AFM1 in milk at 0.5 µg/kg [15]. The European Union (EU) and the United States of America through the United States Food and Drug Agency (USFDA) have also established a regulatory limit for AF and guidance limits for other mycotoxins, as shown in Table 1.

Sirma et al. [16] noted that countries with mycotoxin problems including most SSA countries tend to have laxer enforcement of the regulations set and this may be due to the countries setting limits that are beyond their capacity to implement. Some developed regions such as the EU, USA, and Canada have set up regulations that allow the contaminated feed to be used in less susceptible species. The EAC through Policy Brief on Aflatoxin Prevention and Control (Policy Brief No. 8, 2018) has recommended setting up a policy for cascading for direct utilization of AF-contaminated food based on the level of contamination and negative effects to the animal species, however, this has not been established yet in EAC and other SSA countries [16,17]. Local influencing factors such as enforcement capacity, levels of contamination with mycotoxins, and existing regional standards and the use of these commodities need to be considered in order to improve acceptability and uptake of these regulations [16].

Table 1. Regulatory and guidance levels of mycotoxins in dairy feed and milk.

Country/Region	Regulatory Limit (µg/kg)			Guidance Values (µg/kg)				
	Total AF	AFB1	Milk AFM1	DON	FUM	OTA	ZEN	Reference
Central Africa region	-	-	-	-	-	-	-	-
East Africa Community	10	5	0.5	-	-	-	-	[16]
West Africa region	-	-	-	-	-	-	-	-
South Africa	10	5	0.05	3000	50,000	-	500	[10]
Rwanda	10	5	-	-	-	-	-	[13]
Nigeria	-	5	0.5	-	-	-	-	[14]
Senegal	-	50	-	-	-	-	-	[12]
Cote d' Ivorie	10	-	-	-	-	-	-	[12]
Mozambique	10	-	-	-	-	-	-	[12]
CODEX	-	5	0.5	-	-	-	-	[15]
European Union	-	5	0.05	5000	50,000	-	500	[18]
USA	20	-	0.5	-	30,000	-	-	[19]

AF—Aflatoxins, AFB1—Aflatoxin B1, AFM1—Aflatoxin M1, CODEX—Codex Alimentarius DON—Deoxynivalenol, EU—European Union, FUM—Fumonisins, OTA—Ochratoxin A, USA—The United States of America. - Not detected.

3. Incidence of Mycotoxins in Dairy Feed in Sub-Saharan Africa

In SSA, dairy cattle are fed on fodder that includes grazed grass, hay, legumes, and silage and supplemented with concentrates. These concentrates are often compounded feeds, grain millings, and oilseed cakes. The compounded feed is made from mixing raw materials such as cereals including maize, small grains, oil cakes such as cotton seed cake, sunflower cake, soy meal cake, copra, noug seed, and fish meal [10,20,21]. Taking into account the feed and food shortage in SSA, spoiled and moldy maize that has been considered unfit for human consumption is often fed to animals [22,23]. Both fodder and concentrates have been reported to be contaminated with mycotoxins, with the latter being the major source of contamination [11,24–26]. This makes all dairy farming practiced in SSA at risk of contamination with mycotoxins. Table 2 is a summary of reported cases of dairy feed contamination in SSA, most of which are above the maximum regulatory levels set by national institutions, regional bodies, and the European Union (EU). Overall, aflatoxins (AFs) are the most commonly tested and detected mycotoxins in both finished feed and raw material with a maximum level of 9661 μg/kg [27]. Fumonisins (FUMs) were the second most common mycotoxins with sorted bad maize used for animal feed from Tanzania having the highest mean level of 14 mg/kg [28]. OTA was reported in South Africa, Kenya, Nigeria, and Sudan with the highest level of 19 μg/kg in Sudan. However, the sample sizes in these countries were too small for a proper comparison between countries. The means of the positive samples were between 2 and 15 μg/kg [29] and below the 50 μg/kg EU guidance limit for OTA. The occurrence of OTA seems to follow a similar pattern to AF in areas where AF commonly occur and this may be because they are both produced by *Aspergillus* species. HT-2 toxin was only reported in South Africa [11].

AF and FUM have been widely studied in SSA due to their frequent occurrence in food and feed [29] and high toxicity to animals and humans. This also makes them be regulated in most countries [10]. However, recently, other mycotoxins have been reported in dairy feed in South Africa [4,10,11], Kenya [6,21,30], Rwanda [13], Tanzania, Sudan, Ghana, and Nigeria [21]. Furthermore, raw materials used for compounded feed preparation can simultaneously be contaminated by different mycotoxins since some mycotoxigenic fungi grow and produce mycotoxins under similar conditions [31]. Synergistic, additive, and antagonistic effects due to co-occurrence of mycotoxins occur with, for example, FUM, reported to increase the uptake of AF and subsequently the carry-over to milk [32].

Table 2. Mycotoxins in dairy feed in Sub-Saharan Africa.

Aflatoxins									
Country	**Mycotoxin**	**Test**	**Feed**	***n***	**Positive (%)**	**Above EU Limit (%)**	**Max (μg/kg)**	**Mean (μg/kg)**	**Reference**
Ethiopia	AFB1	ELISA	Dairy feed (compounded feed, brewer yeast, silage, maize, and pea hull)	156	100%	100%	419	97	[33]
Ghana	AF	HPLC-FLD	Animal feed and raw materials	18	72%		199	26	[21]
Kenya	AF	ELISA	Animal feed (moldy maize)	207	56%		7.13	3.8	[17]
	AFB1	ELISA	Concentrates and forages	74	57%	56%	147.9	28.3	[6]
	AFB1	ELISA	Compound dairy feed (Manufacturer)	102		62%	4682	9.8	[27]
	AFB1		Dairy feed (Retailers)	31		90%	1198	25.6	
	AFB1		Dairy feed (Farmer)	114		73%	9661	13.7	
	AF	TLC	Dairy feed	72	100%	95%	1123		[34]
Nigeria	AF	HPLC-FLD	Animal feed and raw materials	27	78%		556	52	[21]
	AFB1	HPLC-FLD	Dairy feed	144	87%	66%	24.8	10.5	[35]
	AF	HPLC-FLD	Animal feeds and raw materials	50	94%		435.9	115	[21]
Sudan	AF	HPLC-FLD	Animal feed and raw materials	13	54%		75	90	[21]
South Africa	AFB1	UHPLC-QTOF-MS/MS	Dairy feed	40	48%	62%	3.3	0.7	[11]
	AFB2		Dairy feed	40	93%		23.9	3.1	
	AFG1		Dairy feed	40	55%		19.9	2.6	
	AFG2		Dairy feed	40	100%		116.0	41.3	
	AF	LC MS/MS	Compounded dairy feeds	25	52%	16%	71.8	14.7	[10]
	AF	HPLC-FLD	Animal feed and raw materials	77	6%		7	0.2	[21]
Tanzania	AFB1	ELISA	Spoilt maize	41	29%			3.5	[28]
	AFB1		Maize bran	20	60%			3.3	
	AFB1	HPLC	Sunflower based dairy feed	20	65%	62%	20.5		[36]

Table 2. *Cont.*

Type B Trichothecenes								
Country	**Mycotoxin**	**Test**	**Feed**	***n***	**Positive (%)**	**Max (μg/kg)**	**Mean (μg/kg)**	**Reference**
Ghana	Type B trichothecenes (DON, 3/15-Ac-DON, and NIV)	HPLC-UV	Animal feed and raw materials	18	50%	1550	955	[21]
Kenya	Type B trichothecenes (DON, 3/15-Ac-DON, and NIV)	HPLC-UV	Animal feed and raw materials	25	48	3859	422	
	Type B trichothecenes (DON, 3/15-Ac-DON, and NIV)	ELISA	Concentrates and forages	74	63%	180	49	[6]
Nigeria	Type B trichothecenes (DON, 3/15-Ac-DON, and NIV)	HPLC-UV	Animal feeds and raw materials	45	58%	463	316	[21]
Sudan	Type B trichothecenes (DON, 3/15-Ac-DON, and NIV)	HPLC-UV	Animal feed and raw materials	9	33%	353	100	[21]
South Africa	Type B trichothecenes (DON, 3/15-Ac-DON, and NIV)	HPLC-UV	Animal feed and raw materials	77	87%	11,022	1469	[21]
	DON	UHPLC-QTOF-MS/MS	Dairy feed	40	60%	82	20	[11]
	DON	LC-MS/MS	Compounded dairy feeds	25	96%	2280	891	[10]

Table 2. *Cont.*

Country	Mycotoxin	Test	Feed	*n*	Positive (%)	Max (µg/kg)	Mean (µg/kg)	Reference
Fumonisins								
Ghana	FUM	LC-MS	Animal feed and raw materials	18	89%	929	500	[21]
Kenya	FUM	LC-MS	Animal feed and raw materials	25	76	10,485	956	[21]
Nigeria	FUM	LC-MS	Animal feeds and raw materials	45	78%	2860	919	[21]
Sudan	FUM	LC-MS	Animal feed and raw materials	9	11%	23	23	[21]
South Africa	FUM	LC-MS/MS	Compounded dairy feeds	25	100%	2497	975	[10]
	FB1	UHPLC-QTOF-MS/MS	Dairy feed	40	85%	1390	373	[11]
	FUM	LC-MS	Animal feed and raw materials	77	57%	4398	454	[21]
Tanzania	FUM	ELISA	Spoilt maize	41	51%		14,450	[28]
	FUM	ELISA	Maize bran	20	60%		1630	

Country	Mycotoxin	Test	Feed	*n*	Positive (%)	Max (µg/kg)	Mean (µg/kg)	Reference
HT-2 Toxin								
South Africa	HT-2	UHPLC-QTOF-MS/MS	Dairy feed	40	88%	313	35	[11]

Country	Mycotoxin	Test	Feed	*n*	Positive (%)	Max (µg/Kg)	Mean (µg/Kg)	Reference
Ochratoxin a								
Kenya	OTA	HPLC-FLD	Animal feed and raw materials	2	50%	2	2	[21]
Nigeria	OTA	HPLC-FLD	Animal feeds and raw materials	5	100%	12	12	[21]
Sudan	OTA	HPLC-FLD	Animal feed and raw materials	6	67%	19	15	[21]
South Africa	OTA	LC-MS/MS	Compounded dairy feeds	25	16%	17	10	[10]

Table 2. *Cont.*

Zearalenone								
Country	Mycotoxin	Test	Feed	*n*	Positive (%)	Max (µg/kg)	Mean (µg/kg)	Reference
Ghana	ZEN	HPLC-FLD	Animal feed and raw materials	18	11%	310	178	[21]
Kenya	ZEN	HPLC-FLD	Animal feed and raw materials	25	56%	167	67	[21]
Nigeria	ZEN	HPLC-FLD	Animal feeds and raw materials	45	51%	80	46	[21]
South Africa	ZEN	UHPLC-QTOF-MS/MS	Dairy feed	40	60%	28	3	[11]
	ZEN	LC-MS/MS	Compounded dairy feeds	25	96%	123	72	[10]
	ZEN	HPLC-FLD	Animal feed and raw materials	77	29%	195	86	[21]

AF—Aflatoxins, AFB1—Aflatoxin B1, AFB2—Aflatoxin B2, AFG1—Aflatoxin G1, AFG2—Aflatoxin G2, Ac-DON—3/15-Acetyl-deoxynivalenol, DON—Deoxynivalenol, ELISA—Enzyme-linked Immunosorbent Assay, FB1—Fumonisin B1, FUM—Total Fumonisins, HPLC-UV—High-Performance Liquid Chromatography with Ultraviolet Detection, HPLC-FLD—High-Performance Liquid Chromatography with Fluorescent Detection MEAN-Mean of positives, LC-MS/MS—Liquid Chromatography Tandem Mass Spectrometry, NIV—Nivalenol, UHPLC-QTOF-MS/MS—Ultra High-performance Liquid Chromatography coupled with Quadrupole Time of Flight tandem Mass Spectrometry, ZEN—Zearalenone.

3.1. East Africa

Mycotoxins were reported in compounded dairy feeds, forages, and raw materials used for making compounded dairy feeds. Countries from East Africa that are near the equator have a higher occurrence of AF than other mycotoxins due to the warm and humid climate, which promotes the growth of *Aspergillus* species.

In Ethiopia, Gizachew et al. [33] reported a 100% incidence of AFB1 in compounded dairy feed, brewer yeast, silage, maize, and pea hull with all of the samples above the EU and EAC regulatory limit of 5 μg/kg. Similarly, in Kenya, Okoth and Kola [34] reported a 100% incidence of AF in compounded dairy feed, cottonseed cake, and sunflower seed cake at levels ranging from 5.13 to 1123 μg/kg. In the study, 95% of the samples were above the regulatory limit of 10 μg/kg with cottonseed cake and compounded dairy feed having a higher number of samples above the regulatory limit at 51.2% and 41.9%, respectively; 7% of the sunflower seed cake samples exceeded the regulatory limit. However, in Tanzania, Mohammed et al. [36] reported 61.5% of sunflower-based dairy feed having AFB1 above the EAC and EU regulatory limit and an overall incidence of 65%. Senerwa et al. [27] in Kenya reported a higher level of samples having AFB1 above the EAC and EU regulatory limit in compounded dairy feed at 90.3% but noted differences between different agro-ecological zones. Moldy maize used as animal feed in Kenya [23] and Tanzania [28] had AF and AFB1, respectively, at 56% and 29%; however, the means (3.84 μg/kg and 3.49 μg/kg respectively) were below the EAC and EU regulatory limit. In Sudan, Rodrigues et al. [21] reported an incidence of 54% of AF in compounded dairy feed and raw material used for making compounded dairy feed.

FUM are the second most common mycotoxins reported. Rodrigues et al. [21] reported 76% incidence of FUM in compounded animal feed, grains, and other feed commodities in Kenya with a mean of 956 μg/kg, which is below the EU guidance limit for FUM of 50,000 μg/kg. In Tanzania, Nyangi et al. [28] reported incidence of 60% and 51% of FUM in spoilt maize unfit for human consumption sorted from good maize after harvesting and meant for animal feed and maize bran, with the spoilt maize having a higher mean of 14,450 μg/kg, which is still below the EU guidance limit. Type B trichothecenes (DON, 3/15-acetyl-deoxynivalenol and nivalenol) were also reported in Kenya [6,21] and Sudan [21] at incidences of between 33% and 63% in compounded dairy feed, raw materials for animal feed, and forages. ZEN and OTA were also reported in dairy feed in Kenya and Sudan; however, few numbers of samples were analyzed for OTA for comparison [21].

3.2. Central Africa

Little information concerning mycotoxins contamination in feed is available for Central Africa. This is attributed to the lack of knowledge on the mycotoxin issue, poverty, and lack of research facilities, manpower, and skills in these countries [37]. However, in Burundi and the Democratic Republic of Congo, AFB1 has been reported in food samples and AFM1 in cow milk [38], but little data is available on mycotoxins contamination in dairy feed. Raphaël et al. [39] reported the occurrence of AF in poultry feed, maize, and peanut meal at levels as high as 950 μg/kg in Cameroon, but little data is available on dairy feed.

3.3. West Africa

AF and FUM were the most prevalent mycotoxins in Ghana (72% and 94%, respectively) and Nigeria (100% and 89%, respectively) [21,35]. Rodrigues et al. [21] reported the highest level of AF in compounded dairy feed and raw materials for making compounded feed in Nigeria at 435.9 μg/kg and a mean of positive of 115 μg/kg [21]. However, another study by Omeiza et al. [35] on compounded dairy feed and pasture showed a lower mean of 10.5 μg/kg, with 66.4% of the samples exceeding the EU regulatory limit. In the study, concentrates had the highest incidence of AFB1 at 93% with dry pasture having 60% and pasture mixed with concentrates having an incidence of 86.9%, indicating concentrates as the major source of AF. In Ghana, the highest level was 199 μg/kg with a mean of

26 μg/kg in compounded dairy feed and raw materials [21]. The highest level and mean of FUM reported in the region were in Nigeria (2860 μg/kg and 919 μg/kg, respectively) [21]. Other mycotoxins reported are ZEN, OTA, and type B trichothecenes (DON, 3/15-acetyl-deoxynivalenol, and nivalenol). This occurrence pattern of mycotoxins is similar to that of East Africa and this may be because countries in West Africa and East Africa are both near the equator, hence similar environmental conditions such as the higher levels of humidity.

3.4. *South Africa*

South Africa exhibited higher contamination of *Fusarium* mycotoxins FUM, DON, and ZEN [21]. This is in line with the findings by Gruber-Dorninger et al. [29] who reported DON, FUM, and ZEN as the most common mycotoxins in South Africa. FUM had an incidence of between 57% and 100%. The highest level of FUM was reported by Rodrigues et al. [21] at 4.4 mg/kg in commercial dairy feed, grain, and other feed commodities, with the highest mean level reported by Njobeh et al. [10] in compounded dairy feed being 0.98 mg/kg and below the EU guidance limit and South African limit of 50 mg/kg. Changwa et al. [11] and Njobeh et al. [10] reported similar occurrence patterns and incidences of DON and ZEN in dairy feed. This is also in line with the findings of Gruber-Dorninger et al. [29] where DON and ZEN are commonly reported to occur together since they are produced by the same species of fungi, but DON occurring at higher levels. Njobeh et al. [10] reported an incidence of 96% of both DON and ZEN in compounded feed and Changwa et al. [11] an incidence of 60% of both DON and ZEN in assorted dairy feeds. In both studies, compounded dairy feed had the highest level of DON. Rodrigues et al. [21] reported an incidence of 87% for type B trichothecenes (DON, 3/15-acetyl-deoxynivalenol, and nivalenol) and a lower incidence of 29% for ZEN. The highest reported level of DON (2280 μg/kg) [10] and ZEN (195 μg/kg) [21] were below the EU guidance limit for DON (5000 μg/kg) and ZEN (500 μg/kg), respectively, and the South African limit 3000 μg/kg and 500 μg/kg, respectively.

AF occurred at 6%–100%, with aflatoxin G2 (AFG2) being the most prevalent in compounded dairy feed and raw materials for animal feed. Changwa et al. [11] reported AFG2 and AFB2 as the most frequent AFs, with AFG2 having the highest mean concentration of 41 μg/kg in maize silage, grass, total mixed ratio, brewer yeast, molasses, and maize bran. AFB1 was the least frequent AF, with the highest level reported in lucerne based feeds (mean 2.1 μg/kg). Incidence and levels of AFG2 in the study were high, 62% of the positive samples exceeded the Food and Drug Agency (FDA) regulatory limit for AF of 20 μg/kg. However, Njobeh et al. [10] and Rodrigues et al. [21] reported lower levels of samples exceeding the regulatory limit (mean 14.7 μg/kg) in compounded dairy feed and 0.2 μg/kg in compounded feed and raw materials, respectively). Other mycotoxins reported were OTA and HT-2 toxin. Njobeh et al. [10] reported OTA at 16%, but Rodrigues et al. [21] and Changwa et al. [11] did not detect OTA. Changwa et al. [11] reported 87.5% incidence of HT-2 toxin with Njobeh et al. [10] and Rodrigues et al. [21] not detecting T-2 and HT-2 in compounded dairy feed and raw materials.

4. Impact of Mycotoxins in Dairy

Mycotoxins in Africa have an impact on food security and safety, animal and human health, international trade, and national budgets, leading to reduced self-sustainability and increased reliance on foreign aid [37]. In the dairy sector, contamination of feed with mycotoxins causes serious economic and food security and safety issues. Economic impact occurs through the direct market costs associated with lost trade or reduced revenues due to the rejection of contaminated animal products and reduced productivity, death of the animal especially calves which are more sensitive, and increased cost of treatment and mycotoxin mitigation [37]. Some mycotoxins, including AF and T-2, are immunosuppressive in cattle, leading to vaccination failure and increased susceptibility to infectious diseases [40] with hidden cost affecting animal health and productivity. The impact is borne by all participants in the dairy sector including feed manufacturers, dairy farmers, milk processors, and consumers [21]. Little has been done to financially quantify the cost of mycotoxins exposure in the dairy enterprise in Africa. In Kenya, Senerwa et al. [27] reported 61.4% of feed contaminated with AFB1 above the FAO/WHO/Kenya limit of 5 μg/kg. This translates to a possible economic cost per year for dairy feed manufacturers of 22.2 billion US $ and, additionally, a further 37.4 million US $ is incurred in losses by farmers annually due to reduced milk yield as a result of feeding cattle with feed contaminated with AFB1 [27]. In the same study, 10.3% of the milk samples had AF levels above the WHO/FAO limit of 0.5 μg/kg, which would cost dairy farmers 113.4 million US $ per year if legislation was enforced.

Animal health issues due to mycotoxicoses affect both the health and productivity of animals with the clinical signs manifested depending on the individual mycotoxin as shown in Table 3. There are two forms of mycotoxicoses, acute mycotoxicoses that occur due to consumption of a high single dose of mycotoxins and chronic mycotoxicoses due to chronic consumption of low levels of mycotoxins over time. Recorded toxic levels of mycotoxins that cause acute disease in dairy cattle are 100 μg/kg for AF, 400 μg/kg for ZEN, and above 100 μg/kg for T-2 [41]; however, chronic aflatoxicosis caused by low-level exposure of mycotoxins over time poses a more common health problem to the animals and also food safety concern to humans. Generally, mycotoxins cause reduced feed intake, alter ruminal fermentation and reduce feed utilization, suppressing immunity, alter reproduction, and cause hepatotoxicity and nephrotoxicity [42]. In comparison, ruminants can be less severely affected by certain mycotoxins compared to monogastrics, which is attributed to the microbial activity in the rumen that can modify the mycotoxin chemical structure into less toxic compounds. Upandaya et al. [43] conducted an in vitro study on the degradation of AFB1 by ruminal fluid from Holstein cattle using 80 μg/kg AFB1 and reported a degradation starting after 3 h incubation with an eventual 14% reduction of AFB1 by 12 h. In agreement, Jiang et al. [44] also performed an in vitro study with rumen fluid collected from Holsteins fed with two substrate alfalfa hay (HA) and ryegrass hay (HR) and, after 72 h incubation, there was a decrease of 83% for HA and 84% for HR of included AFB1 (960 ng/mL). However, in both studies, the metabolites formed due to the degradation were not reported.

Table 3. Effect of mycotoxins in dairy cattle.

Effect	AF	DON	FUM	OTA	T-2	ZEN
Reduced feed intake	√	√	√	√	√	√
Reduced milk yield	√	√	√	√	√	√
Reproductive effects	√		√		√	√
Immunosuppression	√				√	
Hepatotoxicity	√		√			
Nephrotoxicity	√		√			
Gastroenteritis				√	√	

AF—Aflatoxin, DON—Deoxynivalenol, FUM—Fumonisin, OTA—Ochratoxin A, T-2—T-2 toxin, ZEN—Zearalenone. √ Effect present.

Fumonisins are minimally absorbed by ruminants with most of it being excreted in the unmetabolized form in feces. Gurung et al. [45] in an in vitro study using 100 mg/kg FB1/kg reported minimal degradation of FB1 (10%) to hydroxylized FB1 (HFB1) by ruminal microbiota after incubation for 72 h. Similarly, goats fed diets containing 95 mg FB1/kg for 112 days excreted 50% of the FB1 in the unmetabolized form in feces [46].

The intact ruminal epithelium is an effective barrier against DON and ZEN [47]. DON is degraded by the ruminal microbiota to the less toxic metabolite de-epoxy DON (DOM-1) [48,49]. A study by Seeling et al. [48] reported 94%–99% biotransformation of DON to DOM-1. Keese et al. [47] reported no significant amount of unmetabolized DON passing through the ruminal epithelium in cattle fed 50% concentrate proportion and 5.3 mg DON/kg DM, nor if a ration with 60% concentrate and 4.6 mg/kg DM fed for a total period of 29 weeks; however, DOM-1 was present in serum, which indicates systemic uptake. Valgaeren et al. [49] studied the role of roughage provision on the absorption and disposition of DON and its acetylated derivatives in calves and observed an absolute DON oral bioavailability of

4.1% in ruminating calves compared to 50.7% in non-ruminating calves, indicating the ability of the rumen to degrade DON. Recently, Debevere et al. [50] showed the degradation of DON was hampered when ruminal pH was low (pH 5.8), as is the case in cattle suffering from sub-acute rumen acidosis. Interestingly, when rumen inoculum of non-lactating cows was used, a similarly reduced degradation of DON was seen.

An in vitro study by Kiessling et al. [51] on the effect of ruminal microbiota on OTA, ZEN, and T-2 after incubation between 30 min to 3 h showed degradation of OTA to ochratoxin α and phenylalanine, which are non-toxic metabolites mainly produced by the action of protozoa. This ruminal fauna of protozoa can also be affected by diet and the variation in the ruminal protozoa population will affect mycotoxin degradation. Kiessling et al. [51] reported a decrease of 20% of OTA degradation in sheep fed with a high hay concentrate ratio (3:7 weight/weight) compared to a low hay concentrate ratio (5:4 weight/weight). An in vivo study in sheep with 5 mg/kg OTA showed no OTA in blood [51]. For ZEN, the conversion was to α-zearalenol and to a lesser extent to β-zearalenol. α-zearalenol is more toxic and more easily absorbed from the intestines into the bloodstream as compared to the parent compound due to increased polarity, indicating the action of the ruminal microbiota can also increase the toxicity of ZEN. Kiessling et al. [51] also reported in vitro conversion of T-2 to HT-2 toxin after 30 min of incubation with ruminal fluid collected from sheep.

4.1. *Aflatoxins*

Field and experimental aflatoxicosis have been previously described in dairy cattle. Van Halderen et al. [52] reported a field outbreak mortality of 7 out of 25 calves in South Africa fed rations containing locally produced maize with 11,790 μg/kg AF. Clinical signs included loss in body mass, rough hair coat, diarrhea, and rectal prolapse. Outside SSA, Mckenzie et al. [53] reported acute aflatoxicosis that was believed to have caused mortality of 12 to 90 drought-stricken calves in Australia. The calves were fed peanut hay that was later analyzed and determined to contain 2230 μg/kg AF. More recently, Umar et al. [54] reported 45 field cases of aflatoxicosis on a local farm in Okara (Punjab, Pakistan). The cows were fed corn-rich forage with 33,500 μg/kg AF. The clinical signs were anorexia, depression, photosensitization, and diarrhea, with 15 animals dying.

Experimental studies have described aflatoxicosis, with clinical signs being reduced feed intake and feed conversion, reduced milk production, reduced reproduction capacity, lameness, immunosuppression, hepatotoxicity, and nephrotoxicity [41,42]. Applebaum et al. [55] reported a significant decrease in milk production in cattle fed with 13 mg AFB1 per day for 7 days. Likewise, Ogunade et al. [56] and Jiang et al. [57] also reported a numerical drop in milk yield in cows fed 75 μg/kg Dry Matter Intake (DMI) (1725 μg/head per day) for 5 days. Sulzberger et al. [58] reported depression in milk yield and feed conversion at 100 μg AFB1/kg of DMI. In contrary, studies by Queiroz et al. [59] on cattle fed 75 μg/kg DMI (1725 μg/head per day) for 4 days, Kutz et al. [60] feeding 112 μg of AFB1/kg of Total Mixed Ration (TMR) DMI to dairy cows in early to mid-lactation for 7 days, Sumantri et al. [61] on cattle fed 350 μg AFB1/cow/day for 10 days, and Mosoero et al. [62] using lower levels of AF of 0.16 μg/kg BW and 3.41 μg AFB1/cow/day for 3 days showed no health effect [62]. Exposure to AF also affects rumen fermentation, reducing the utilization of nutrients, and eventually may affect animal productivity. Mesgaran et al. [63] reported reduced gas production, dry matter digestibility, and ammonia-N concentrations caused by AFB1 in vitro. Jiang et al. [44] also reported similar results with AFB1 affecting in vitro fermentation characteristics in terms of reduced ammonia N and volatile fatty acids (VFA) concentrations but without reducing dry matter digestibility or affecting VFA pattern. Based on the levels of AF reported in dairy feed in SSA, chronic aflatoxicosis is a risk in the dairy sector.

4.2. *Deoxynivalenol*

DON, also called vomitoxin, induces anorexia and emesis in humans and animals. This is usually achieved by affecting the chemoreceptor trigger centers and causing gastrointestinal lesions. Pigs are the most sensitive species while cattle are less susceptible. DON affects ruminal fermentation and causes reduced milk yield [41]. Contradictory results have been shown on the effect of DON on feed intake. Trenholm et al. [64] studied the effect of a diet contaminated with DON in non-lactating Holstein dairy cows fed at a dose of 1.5 mg/kg and 6.4 mg/kg of feed for 6 weeks, there was no adverse effect observed, however, there was a slight decline in feed consumption following the change from the low DON dose (1.5 mg/kg) to the high DON dose (6.4 mg/kg). In contrary, Seeling et al. [41] reported no effect on feed intake on feeding approximately 3.4 mg DON per kg at a reference DM of 88% complete ration, similar to Winkler et al. using 5 mg DON/kg feed showed no effect on performance parameters on dairy cattle [65]. In calves, Valgaeren et al. [49] reported severe liver failure in 2-3 months old calves with no functioning rumen induced by 1.13 mg DON/kg feed, indicating the significance of rumen microbiota in DON degradation.

4.3. Fumonisins

Ruminants are more resistant to FUM toxicity than horses and pigs [66]. Field outbreaks have been reported in horses and pigs but not in dairy cattle. Fumonisins act by altering sphingolipid biosynthesis hence leading to the accumulation of sphinganine and causing toxicity. Oral administration to calves with a diet containing FB1 at 2.36 mg/kg/day increased to 3.54 mg/kg/day for 239 to 253 days showed elevated sphinganine/sphingosine ratios with mild hepatocellular morphology changes accompanied by mild bile duct epithelial changes [67]. Feeding trials with 75 mg/kg, 94 mg/kg, and 105 mg/kg FB1 for 14 days, 253 days, and 31 days, respectively, have also been reported to cause reduced milk yield, reduced feed intake, hepatotoxicity, nephrotoxicity, and reproduction problems and, hence, it can be concluded that oral administration with levels above 75 mg/kg is toxic to cattle [41,42,68]. Experimental administration of 1 mg/kg of FB1 intravenously to calves for 7 days caused lethargy, loss of appetite, hepatotoxicity, and nephrotoxicity [66].

4.4. Ochratoxin A

Ochratoxicosis is rarely reported in cattle. This is attributed to the ability of the rumen microbiota to easily degrade OTA to non-toxic forms as demonstrated by Kiessling et al. [51]. Ribelin et al. [69] reported anorexia, diarrhea, difficulty in rising and cessation of milk production with recovery on the 4th day in cattle fed a high single dose of OTA of 13.3 mg/kg; this dose can be rarely achieved in the field and low doses of 0.2 mg/kg, 0.75 mg/kg, and 1.66 mg/kg for 5 days produced no clinical disease. With the highest level of OTA reported in SSA being 19 μg/kg, which is way lower than those used in the experiment, it can, therefore, be concluded that OTA is rarely a problem in dairy cattle in SSA.

4.5. T-2 toxin

In dairy cattle, T-2 has been associated with hemorrhagic gastroenteritis [41], feed refusal, and gastrointestinal lesions [70]. Weaver et al. [70] reported severe depression, hindquarter ataxia, knuckling of the rear feet, listlessness, and anorexia in a calf fed 0.6 mg T-2/kg body weight for seven consecutive days. Reduction in milk yield and the absence of estrus have also been associated with T-2 [41].

4.6. Zearalenone

Zearalenone has an estrogenic response in cattle causing abortion and changes in the reproductive organs. A case report by Kallela and Ettala [71] reported early abortion in cattle feeding on hay containing 10 mg/kg ZEN. Abnormal estrus cycle, vaginitis, behavioral estrus in pregnant animals, mammary development in pre-puberty heifers, and sterility have also been reported in cattle fed with feed containing 1.5 mg ZEN/kg feed [72]. Experimental studies using 500 mg and 250 mg of 99% purified ZEN in a gelatin capsule orally in lactating dairy cattle and virgin heifers, respectively, showed no effects except for depression in the conception rate in the virgin heifers [73,74].

5. Food Safety and Hazards of Mycotoxins

Besides the effects on animal health, some mycotoxins can pass to milk causing food safety issues and posing a hazard to human health. Of all the studied mycotoxins, only AF has been described to be transferred to the milk of lactating cattle in significant levels of concern. This is of great importance to public health worldwide as the toxin is classified as a carcinogen and infants, the primary consumers of milk, are more susceptible. Negligible transfer of FUM, ZEN, OTA, and DON has been reported, but the health impacts of this are unknown. Carry-over studies and surveys of mycotoxins in milk in SSA have not been extensively carried out, except for surveys of AFM1.

Once ingested by ruminants, part of the ingested AFB1 is degraded in the rumen. Kiessling et al. [51] suggested that the type of microbiota in the rumen will determine the level of degradation and is dependent on species, age, sex, and breed. Upadhaya et al. [43] further reported AFB1 degradation of 14% in cattle compared to 25% in goats with type of feed also determining the level of degradation. Similarly, Jiang et al. [44] reported a higher degradation of AFB1 that was dependent on the type of feed. The type of feed has an effect on the rumen microbial ecosystem with a higher degradation in feed with cellulose such as roughages than those without. In both studies, AFB1 degradability rate was calculated as the difference between initially included AFB1 and residual AFB1 in the culture fluids with no formed metabolites tested for. The remaining AF is absorbed in the small intestines and hydroxylated in the liver to form AFM1, the major metabolite among other metabolites [75], with AFM1 being excreted in milk and urine, and is classified as class 1 carcinogen to humans [22]. The carry-over of AFB1 to milk varies from less than 1% to 6.2% [61,72,76]. The level of carry-over is usually determined by several factors such as the animal species, individual animal variability [72], feeding regimens and type of diet [77], presence of other mycotoxins [32], stage of lactation [76], and actual milk production [76]. AFM2 is also another metabolite of hydroxylation of AFB2 but is of little concern as compared to AFM1. Hernandez–Camarillo et al. [78] reported an occurrence of AFM2 in 20% of cheese in Mexico (mean 0.2 µg/kg) in comparison with 53% reported for AFM1 (mean 3.0 µg/kg) in the same samples [78]. With the higher occurrence of AFB1 in dairy feed in comparison with AFB2 in SSA, AFM1 is therefore of major concern as compared to AFM2. Studies on milk in SSA countries have shown a high incidence of AFM1 (Table 4).

Table 4. Aflatoxin M1 in milk in Sub-Saharan Africa.

Country	Test	Sample	*n*	Positive (%)	Above Eu Limit (%)	Max (µg/Kg)	Mean (µg/Kg)	Reference
Burundi	ELISA	Milk (fresh and yoghurt)	16	100%		0.08	0.03	[38]
D.R. Congo	ELISA	Milk (fresh and yogurt) and cheese	10	100%		0.26	0.03	[38]
Ethiopia	ELISA	Milk	110	100%	91.8%	4.98	0.4	[33]
Kenya	ELISA	Milk	96	100%	66.4%	4.63	0.29	[79]
	ELISA	Milk	291		51.9%	1.1	0.08	[80]
	ELISA	Milk	512	39.7%	10.4%	6.9	0.003	[27]
	ELISA	Milk	200		55%	1.67	0.128	[81]
Nigeria	HPLC	Milk powder	125	53.6%		0.46		[14]
	HPLC	Raw milk	100	75%	64%	0.46	0.11	[82]
Sudan	Fluorometry	Raw milk	35	100%	100%	2.52	0.92	[83]
		Imported powder milk	12			0.85	0.29	
South Africa	ELISA	Milk	30	100%	90.6%	0.15	0.09	[7]
		Milk	37	100%	62.1%	0.11	0.07	
Tanzania	HPLC	Milk	37	83.8%	100	2.01		[36]

ELISA—Enzyme-Linked Immunosorbent Assay. HPLC—High-Performance Liquid Chromatography, EU Limit—0.05 µg/kg.

Due to the rumen's capability to degrade DON, the carry-over of both unmetabolized DON and DOM-1 to milk is negligible. Keese et al. [47] detected no unmetabolized DON in milk using an

HPLC–UV method with prior β-glucuronidase incubation in cows fed either 5.3 mg DON/kg dry matter (DM) over 11 weeks, or a ration with 60% concentrate and 4.6 mg/kg DM for 29 weeks. Negligible amounts of DOM-1 (0.21 μg/kg) in two out of 24 samples in the study were detected with LC-MS/MS. In agreement, Seeling et al. [48] using an HPLC–UV method with β-glucuronidase incubation (limit of detection or LOD of 0.5 μg/kg) also reported no unmetabolized DON in milk and DOM-1 at 1.6 and 2.7 μg/kg in cows with 34 mg to 76 mg daily DON intake. Using a more sensitive GC-MS method (LOD of 0.1 μg/kg), DON was detected at levels between 0.1 and 0.3 μg/kg with DOM-1 at levels between 1.5 and 3.1 μg/kg.

Carry-over of FUM to milk is not significant and does not pose a hazard to human health. The occurrence of milk naturally contaminated with FUM does, however, occur, with Maragos and Richard [84] reporting only one sample out of 150 having detectable levels of FB1 using an LC with fluorescence detection method (LOD of 5 ng/mL). Experimentally, Richard et al. [68] did not detect any FUM in milk in two jersey cattle fed 3 mg/kg DMI daily (total 75 mg) for 14 days using two analytical methods (LOD of 5 ng/mL). Similarly, Scott et al. [85] detected no residues of FB1 in the milk of cows dosed with pure FB1 either orally (1.0 and 5.0 mg FB1/kg Body Weight (BW) or by i.v. injection (0.05 and 0.20 mg FB1/kg BW).

Both OTA and its metabolite ochratoxin α can be transferred to milk. Ribelin et al. [69] reported OTA in milk the next day in cows fed on 13.3 mg/kg OTA as a single dose, trace amount of OTA from day 3 to 5 in cows fed 1.66 mg/kg daily for four days, and no OTA in cows dosed with less than 1.66 mg/kg OTA. However, milk from all cows had traces of ochratoxin α. Other experimental studies have reported no carry-over of OTA. Zhang et al. [86] did not detect OTA and ochratoxin α using LC-MS/MS (LOD of 0.1–0.2 ng/mL) in cows administered a single dose of OTA at levels 30 μg/kg OTA BW) in feed. Similarly, Hashimoto et al. [87] detected no OTA in milk of cows fed 100 μg/kg DM for 28 days. Thus, with this negligible rate of carry-over and the low levels of OTA in feed in SSA, OTA poses no health hazard to humans through dairy products.

Little information is available on the levels of these other mycotoxins and their residues in milk in SSA.

5.1. East Africa

In East Africa, there is a high prevalence of AFM1 in milk that corresponds to the high levels of AFB1 reported in feeds. Ethiopia, Kenya, and Sudan have reported a 100% incidence of AFM1, with 91.8%, 66.4%, and 100% of the positive samples being above the EU regulatory limit of 0.05 μg/kg [33,80,83]. Imported milk powder in Sudan also had AFM1 at levels between 0.01 and 0.85 μg/kg [83], with 50% exceeding the EU regulatory limit and 33% above the CODEX and EAC regulatory limit of 0.5 μg/kg. Other studies in Kenya have reported AFM1 occurrence in milk between 39.7% and 99% and between 10.4% and 64% exceeded the EU regulatory limit of 0.05 μg/kg [80,81,88,89], with the highest level of 6.9 μg/kg that is way higher than the EU and EAC limit. Similar high AFM1 found in milk samples from Tanzania, with 83.8% of all positive samples exceeding the EU regulatory limit [36].

5.2. Central Africa

Despite little data being available of AFB1 in dairy feeds, AFM1 contaminates raw milk and milk products such as cheese and yogurt. A study on the levels of AFM1 in milk and milk products in markets in Burundi and DR Congo showed 100% positive samples with maximum levels of 0.082 and 0.261 μg/kg, respectively. These maximum levels are above the EU regulatory limit [38].

5.3. South Africa

High levels of AFM1 occur in milk in South Africa. Dutton et al. [90] reported a 100% incidence of AFM1 in milk from dairy farms, ranging from 0.02 μg/L to 1.5 μg/L. Retail milk was also contaminated with AFM1, at levels of 0.01–3.1 μg/L. Similarly, Mulunda et al. [7] in a study carried out in selected rural areas of Limpopo Province in South Africa reported 100% AFM1 occurrence with 90.6% and

62.1% of the positive samples above South Africa and EU regulatory limit of 0.05 μg/kg in Mapete and Nwanedi area, respectively [7].

5.4. West Africa

A high prevalence of AFM1 was found in raw milk and imported milk powder in Nigeria. Oyeyipo et al. [14] reported AF in repacked milk powder in five states in the South West region, Nigeria. Of the milk samples, 53.6% was contaminated with AFM1 but none exceeded the Nigerian regulatory limit of 0.5 μg/kg. However, the maximum level of 0.46 μg/kg was above the EU regulatory limit of 0.05 μg/kg. Interestingly, very high levels of AFB1 above the Nigerian and EU regulatory limit of 5 μg/kg were reported in milk (29.7–79.4 μg/kg) and this can be explained by the frequent presence of *Aspergillus* species that were found contaminating the milk due to the open-air repackaging of the milk powder. In another study on raw milk from free-grazing cows in Abeokuta, Nigeria, Oluwafemi et al. [82] reported a 75% occurrence of AFM1 with 64% exceeding the EU limit.

The high level of AFM1 in SSA is a major food safety concern. A risk assessment by Ahlberg et al. [91] on AFM1 exposure in low- and mid-income dairy consumers in Kenya reported 2.7% of children could hypothetically be stunted due to AFM1 exposure from milk, although stunting has not been proven to occur after exposure to AFM1. Exposure to AFM1 from milk in Kenya has been associated with a reduction in growth, although this is no evidence of causation [92]. However, the hepatocellular cancer risk was low at 0.004 cases per 100,000. In agreement, Sirma et al. [93] reported the annual incidence rates of cancer attributed to the consumption of AFM1 in milk in Kenya between 0.0014 to 0.0039 per 100,000 people. The potential risks to human health have led to several countries setting up legislation for mycotoxins in dairy feeds and milk.

6. Mycotoxin Mitigation Strategies

Due to the negative health and economic impact of mycotoxins on the dairy industry and the relative stability of mycotoxins to manufacturing processes, strategies have been developed to mitigate the effects of mycotoxins. Most of these strategies have been developed for control of AF but some are applicable for control of other mycotoxins. There is low awareness on mycotoxins by dairy farmers in SSA with little done in dissemination of information on appropriate control strategies [94]. Kangethe et al. [95] in a study in urban areas in Kenya reported the highest level of awareness on AF at 42%. Similarly, Kirino et al. [81] reported 58% of milk traders being aware of AF but very few had knowledge of AF carry-over to milk [81], and farmers also report feeding moldy maize to animals [23]. In Rwanda, 92.4% of livestock farmers and animal feed vendors were unaware of AF and FUM and their effects [13]. Similarly, Changwa et al. [11] reported a general awareness of mycotoxins between 17% and 92% in South Africa. James et al. [96] reported AF awareness levels of 20.8% among farmers, 26.7% among traders, 60% among poultry farmers, and 25.2% among consumers in Benin, Ghana, and Togo. A low level of awareness has also been reported in Tanzania and Ethiopia [94]. This low level of awareness may hinder the implementation of various mitigation strategies. These strategies are divided into two; pre-harvest strategies that are aimed at preventing the fungal contamination in the field and the post-harvest strategies that are applied to the harvested products during harvesting, processing, or storage to prevent contamination and reduce or eliminate the mycotoxin contamination [97,98]. Prevention of contamination is the preferable method, and post-harvest mitigation strategies are, therefore, very important. However, since this is not always sufficient in SSA, post-contamination options are also needed.

Post-harvest strategies are applied following harvesting. Rapid drying after harvesting reduces the moisture content that is essential in stopping the growth of fungi and mycotoxins production. Moisture content of 10%–13% is considered safe for cereals. However, proper drying and storage is often an issue in most SSA countries due to the high temperatures and humidity [98].

Storage of feed in dry condition with low humidity, proper aeration, and free from rodents and pests is essential for minimizing fungal contamination and mycotoxin production [98].

Most of the dairy feeds are usually bought or harvested and stored, the storage conditions are sometimes unfavorable and hence have a negative effect on feed quality [3].

The tropical climate in SSA is favorable for mycotoxin production by fungi as well as causing issues with food insecurity leading to practices such as diverting moldy grains to be used as animal feeds. This, therefore, makes decontamination the best strategy for the prevention of mycotoxins in the dairy chain [23,99]. Decontamination is applied to the already contaminated feed to eliminate the mycotoxin or to reduce the bioavailability of the toxin [42]. Chemical, physical, and biological methods have been widely applied to decontaminate feed from mycotoxins [5,42,100]. Substances used for decontamination are called detoxifiers and can be grouped into binders that prevent the absorption of mycotoxins and modifiers that break down the mycotoxins in the intestines into less toxic metabolites. Binders usually include clay minerals or yeast products, while modifiers include microorganisms and enzymes [101].

6.1. Chemical Decontamination

Chemical methods include the use of acids, bases, aldehydes, bisulphites, oxidizing agents, chlorinating agents, and various gases. These chemicals have been applied and found to be effective against some mycotoxins. Ammoniation has been shown to reduce the levels of OTA in cereals to undetectable levels with these grains being suitable for use in making animal feed without changing the nutritional value [102]. Bailey et al. [103] and Fremy et al. [104] demonstrated the effectiveness of ammonia treatment on the reduction of carry-over of AFM1 to milk in cattle; however, ammoniation is not effective against FUM with contaminated grains still retaining toxicity against rats [105,106]. Sodium bisulphite, hydrogen peroxide, and ozone are also effective in reducing AFB1 contamination in human food [107]. However, these methods are expensive and not easily acceptable by dairy farmers and may affect animal health in vivo due to the accumulation of chemical residues.

6.2. Physical Decontamination

Physical methods use adsorbents such as activated charcoal and aluminosilicate clay minerals such as smectite, bentonite, and montmorillonite. These adsorbents act by binding the mycotoxin to prevent its absorption and are effective and safe in ruminants [108–110].

Several adsorbents are effective on AF in terms of reducing the carry-over of AFM1 and the effect on animal health. Pietri et al. [109] reported a 41% and 31% decrease in AFM1 in milk in cows fed diet contaminated with 97.3 μg AFB1/kg DMI using 50 g/cow/day and 20 g/cow/day of a commercial detoxifier that contains bentonite, *Eubacterium* and yeast. Kissel et al. [111] in agreement found a 60.4% decrease in AFM1 carry-over in milk in cows fed a diet with 227 g/cow per day of bentonite, and Kutz et al. [60] reported a 45% and 48% decrease in excretion of AFM1 in the milk of cows fed 112 μg AFB1/kg DMI supplemented with two commercial sodium calcium aluminosilicate adsorbents at 0.56% of the diet. In another study, Xiong et al. [112], using sodium montmorillonite with live yeast, yeast culture, mannan oligosaccharide, and vitamin E at a dose of 0.25% of the diet, reported a decrease in the transfer of AFM1 from feed with 20 μg AFB1/kg Dry Matter Intake (DMI) compared with the AFB1 alone control (0.46 vs. 0.56%, respectively). Jiang et al. [57] also reported a reduced transfer of dietary AFB1 as AFM1 in milk and prevention in a decrease in milk yield caused by AFB1. The cows were fed a diet with 75 μg AFB1/kg DMI supplemented with bentonite clay (200 g/cow/day) with or without *Saccharomyces cerevisiae* fermentation product (35 g/cow/day). Sulzberger et al. [58] also further reported a 25% reduction in AF transfer from the rumen to milk using 0.5% clay, 18% (1% clay), and 41% (2% clay) in cows administered 100 μg AFB1/kg DMI. The cows received oral supplementation of 0.5%, 1%, and 2% clay containing vermiculite, nontronite, and montmorillonite. In other ruminants, Mugerwa et al. [110] reported a reduction of AFM1 carry-over in milk in lactating goats fed 100 μg AFB1/kg and 1% DMI activated charcoal and calcium bentonite. None of the adsorbents had any impact on feed intake and milk composition. The efficacy of these adsorbents to bind AF can be affected by the ratio of adsorbent to mycotoxin, the pH, and the temperature. Sumantri et al. [61] reported no effect of inclusion of bentonite at 0.25% and 2% in the diet of cows fed 350 μg AFB1/cow per day for 5 days. Ogunade et al. [56], despite also reporting no reduction in AFM1 concentration after feeding a diet with 75 μg AFB1/kg DMI and 20 g/head/day of a sequestering agent containing sodium bentonite and *S. cerevisiae* fermentation product, reported a reduction of the time required to reduce AFM1 in milk to safe levels after AFB1 contaminated feed withdrawal. However, there is a possibility of some of the adsorbents to bind other nutrients reducing feed nutritive value and feed palatability with EFSA recommending a maximum level of bentonite of 20,000 mg/kg for complete feed [108,113].

Adsorbents have low effectiveness against other mycotoxins. Bhatti et al. [114] in a study on the protective role of bentonite against OTA-induced immunotoxicity in broilers reported partial or no improvement on the negative effects induced by OTA. Binders containing humic acids and mixed-layered smectite-containing binders have been shown to have the capacity to bind to ZEN. De Mil et al. [101] in an in vitro study based on the low level of free ZEN concentration after 4 h incubation with a ZEN: binder ratio of 1:20,000 reported the binding of ZEN.

6.3. Biological Decontamination

Biological methods use enzymes or microorganisms to biotransform mycotoxins into less toxic metabolites. BBSH® 797 is a bacterial strain of the family *Coriobacteriaceae* that produces specific enzymes de-epoxidases that open the toxic epoxide ring of trichothecenes, such as DON and T-2 toxin, thus detoxifying them. The yeast strain *Trichosporon mycotoxinivorans* (Trichosporon MTV) is capable of degrading OTA and ZEN [115].

Enzymes used for biotransformation usually cleave the mycotoxin at the site responsible for toxicity in the gastrointestinal tract and produce a metabolite(s) with less toxicity than the parent mycotoxin. AF detoxifizyme (AFDF) from *Armillariella tabascens* [116] and laccase enzyme from *Peniophora* and *Pleurotus ostreatus* fungus [117] have been shown to be AF degrading enzymes. Carboxylesterase and amino-transferase have been shown to degrade FUM [118], with fumonisin esterase being commercially used for decontamination of feed contaminated with FUM [119]. Fumonisin esterase is an FB1 hydrolyzing enzyme that catalyzes the cleavage of FB1's tricarballylic acid side chains to form partially hydrolyzed FB1a and b (pHFB1a, pHFB1b) and eventually hydrolyzed FB1 (HFB1). This HFB1 is less toxic compared to FB1 [120]. Despite no information being available for their use in dairy cattle, the enzymes are effective in pigs and poultry [119,121].

In Kenya, mycotoxins detoxifiers are used in animal feeds for decontamination. Lack of regulation has led to a lack of information on the efficacy of these detoxifiers and this may expose the farmers to products that are not effective. Mutua et al. [99] in a study on the use of mycotoxin detoxifiers in Kenya showed that usage of binders is uncontrolled, and all nine types of products sold on the Kenyan market were imported as feed additives and not specified as mycotoxin binders. These binders are bought by feed processors and farmers formulating their feed. Information is still lacking on the use and regulation of these products. More information and regulation are, therefore, needed on the use of these detoxifiers in the dairy enterprise in SSA as an alternative strategy of controlling mycotoxins.

6.4. Vaccination against AFs

Vaccination as a recently investigated way of reducing AF toxicity, and AFM1 carry-over is a valuable option for AF mitigation in dairy. Polonelli et al. [122] reported up to 46% lower AFM1 levels in milk in vaccinated cows compared to control cows exposed to AFB1. Anaflatoxin B1(AnAFB1) conjugated to keyhole limpet hemocyanin (KLH) together with Freund's adjuvant was used as a vaccine (AnAFB1-KLH) and produced long-lasting titers of anti-aflatoxin B1 (AFB1) antibodies that also cross-reacted with other AFs. Giovati et al. [123] in an attempt to improve the vaccine used AnAFB1 conjugated to KLH and mixed with complete (priming) and incomplete Freund's adjuvant (boosters) and reported the average AFM1 concentration in milk collected from vaccinated cows being 74% lower than in milk of control animals. Similarly, the carry-over rate calculated in vaccinated cows (0.77%) was lower than control animals (3.40%). The results show that vaccination can be a feasible option for controlling hazards caused by AFs on animals and humans. However, the production of these vaccines is costly and hence not used as a mitigation strategy.

7. Conclusions

It is evident that mycotoxin contamination of dairy feeds is widespread in SSA. Due to low levels of awareness of mycotoxins and food insecurity in SSA, dairy animals will continue to be fed with mycotoxin-contaminated feed. This negatively impacts the dairy industry in SSA due to significant economic losses as a result of the impact on animal health and productivity and food safety due to the contamination of milk. Vital information concerning mycotoxin contamination of dairy feed and products in most African countries, especially Central Africa, is lacking and this may hinder the development of the dairy industry. Furthermore, this information may help in designing effective mycotoxin control strategies in SSA. Mycotoxin detoxifiers may play a significant role in controlling the effects of mycotoxins due to the high levels of contamination of dairy feeds shown in SSA; however, information is lacking on their use and regulation of these detoxifiers, exposing dairy farmers to the risk of using ineffective products. More research and regulation on the use of these detoxifiers may be an effective means of ensuring animal health as well as food safety and security.

Author Contributions: Conceptualization, G.A., J.F.L., and J.K.G.; writing—original draft preparation, D.C.K. and P.E.O.; writing—review and editing, everyone; supervision, G.A., J.K.G., J.F.L., S.O., S.C., J.F., and E.K.K.; project coordination, G.A. and S.C.; funding acquisition, G.A., S.C., and J.K.G. All authors have read and agreed to the published version of the manuscript.

Funding: This research was conducted within the ERA-NET LEAP-Agri MycoSafe-South project funded by the Belgian Federal Science Policy Office (BELSPO), Belgian National Fund for Scientific Research (NFSR), Research Council of Norway (RCN), Kenyan Ministry of Education, Science and Technology (MoEST), South Africa's National Research Foundation (NRF), and BIOMIN Holding GmbH and Harbro Ltd.

Acknowledgments: This research was conducted within the ERA-NET LEAP-Agri MycoSafe-South project. The time of JFL was supported by the CGIAR research program on Agriculture for Nutrition and Health.

Conflicts of Interest: The authors declare no conflict of interest.

References

1. Food and Agriculture Organization of the United Nations (FAOSTAT). *Food and Agriculture Organization of the United Nations (FAOSTAT 2017)*; FAOSTAT: Rome, Italy, 2017.
2. Ndambi, O.; Hemme, T.; Latacz-Lohmann, U. Dairying in Africa-Status and Recent Developments. *LRRD* **2007**, *19*, 25.
3. Odero-Waitituh, J.A. Smallholder Dairy Production in Kenya; a Review. *LRRD* **2017**, *29*, 139.
4. Mngadi, P.; Govinden, R.; Odhav, B. Co-Occuring Mycotoxins in Animal Feeds. *Afr. J. Biotechnol.* **2008**, *7*, 2239–2243.
5. Bhat, R.; Rai, R.V.; Karim, A.A. Mycotoxins in Food and Feed: Present Status and Future Concerns. *Compr. Rev. Food Sci. Saf.* **2010**, *9*, 57–81. [CrossRef]
6. Makau, C.M.; Matofari, J.W.; Muliro, P.S.; Bebe, B.O. Aflatoxin B1 and Deoxynivalenol Contamination of Dairy Feeds and Presence of Aflatoxin M1 Contamination in Milk from Smallholder Dairy Systems in Nakuru, Kenya. *Int. J. Food Contam.* **2016**, *3*, 6. [CrossRef]
7. Mulunda, F. Determination and Quantification of Aflatoxin M 1 in Fresh Milk Samples Obtained in Goats and Cattle in Selected Rural Areas of the Limpopo Province, South Africa. *J. Hum. Ecol.* **2016**, *56*, 183–187. [CrossRef]
8. Miller, J.D. Mycotoxins. In *Encylopedia of food science and technology*; Francis, F.J., Ed.; John Wiley: New York, NY, USA, 1999; pp. 1698–1706.
9. Cigic, I.K.; Prosen, H. An Overview of Conventional and Emerging Analytical Methods for the Determination of Mycotoxins. *Int. J. Mol. Sci.* **2009**, *10*, 62–115. [CrossRef]
10. Njobeh, P.B.; Dutton, M.F.; Aberg, A.T.; Haggblom, P. Estimation of Multi-Mycotoxin Contamination in South African Compound Feeds. *Toxins (Basel)* **2012**, *4*, 836–848. [CrossRef]

11. Changwa, R.; Abia, W.; Msagati, T.; Nyoni, H.; Ndleve, K.; Njobeh, P. Multi-Mycotoxin Occurrence in Dairy Cattle Feeds from the Gauteng Province of South Africa: A Pilot Study Using UHPLC-QTOF-MS/MS. *Toxins* **2018**, *10*, 294. [CrossRef] [PubMed]
12. Food and Agriculture Organization of the United Nations. *Worldwide Regulations for Mycotoxins in Food and Feed in 2003*; FAOSTAT: Rome, Italy, 2004.
13. Nishimwe, K.; Bowers, E.; Ayabagabo, J. de D.; Habimana, R.; Mutiga, S.; Maier, D. Assessment of Aflatoxin and Fumonisin Contamination and Associated Risk Factors in Feed and Feed Ingredients in Rwanda. *Toxins (Basel)* **2019**, *11*, 270. [CrossRef]
14. Oyeyipo, O.O.; Oyeyipo, F.M.; Ayah, I.R. Aflatoxin (M_1 and B_1) Contamination of Locally Repacked Milk Powder in South-Western Nigeria. *Sokoto J. Med. Lab. Sci.* **2017**, *2*, 175–187.
15. Food and Agriculture Organization of the United Nations/World Health Organization. *Codex Alimentarius Commission. Joint FAO/WHO Food Standards Programme- Codex Alimentarius Commission*; FAOSTAT: Rome, Italy; World Health Organization: Geneva, Switzerland, 2006.
16. Sirma, A.J.; Lindahl, J.F.; Makita, K.; Senerwa, D.; Mtimet, N.; Kang'ethe, E.K.; Grace, D. The Impacts of Aflatoxin Standards on Health and Nutrition in Sub-Saharan Africa: The Case of Kenya. *Glob. Food Sec.* **2018**, *18*, 57–61. [CrossRef]
17. Kang'ethe, E.K.; Gatwiri, M.; Sirma, A.J.; Ouko, E.O.; Mburugu-Musoti, C.K.; Kitala, P.M.; Nduhiu, G.J.; Nderitu, J.G.; Mungatu, J.K.; Hietaniemi, V.; et al. Exposure of Kenyan Population to Aflatoxins in Foods with Special Reference to Nandi and Makueni Counties. *Food Qual. Saf.* **2017**, *1*, 131–137. [CrossRef]
18. European Commission (EC). *Setting Maximum Levels for Certain Contaminants in Food stuffs*; Commission Regulation (EC) No. 1881/2006; EC: Brussels, Belgium, 2006.
19. USFDA. *Guidance for Industry: Fumonisin Levels in Human Foods and Animal Feeds*; Final Guidance; USFDA: Washington, DC, USA, 2001.
20. Blount, W.P. Turkey 'X' Disease. *J. Br. Turk. Fed.* **1961**, *9*, 52–54.
21. Rodrigues, I.; Handl, J.; Binder, E.M. Mycotoxin occurrence in commodities, feeds and feed ingredients sourced in the Middle East and Africa. *Food Addit. Contam. Part B Surveill.* **2011**, *4*, 168–179. [CrossRef] [PubMed]
22. World Health Organization; International Agency for Research on Cancer. Aflatoxins. In *IARC Monographs on the Evaluation of Carcinogenic Risks to Humans*; Some Traditional Herbal Medicines, Some Mycotoxins, Naphthalene and Styrene; IARC press: Lyon, France, 2002; Volume 82, Available online: https://monographs.iarc.fr/wp-content/uploads/2018/06/mono82.pdf (accessed on 24 February 2020).
23. Kang'ethe, E.K.; Korhonen, H.; Marimba, K.A.; Nduhiu, G.; Mungatu, J.K.; Okoth, S.A.; Joutsjoki, V.; Wamae, L.W.; Shalo, P. Management and Mitigation of Health Risks Associated with the Occurrence of Mycotoxins along the Maize Value Chain in Two Counties in Kenya. *Food Qual. Saf.* **2017**, *1*, 268–274. [CrossRef]
24. Nagaraj, D. Fumonisins: A Review on Its Global Occurrence, Epidemiology, Toxicity and Detection. *J. Vet. Med. Res.* **2017**, *4*, 1093.
25. Reddy, L.; Bhoola, K. Ochratoxins-Food Contaminants: Impact on Human Health. *Toxins* **2010**, *2*, 771–779. [CrossRef]
26. Hueza, I.M.; Raspantini, P.C.F.; Raspantini, L.E.R.; Latorre, A.O.; Górniak, S.L. Zearalenone, an Estrogenic Mycotoxin, Is an Immunotoxic Compound. *Toxins* **2014**, *6*, 1080–1095. [CrossRef]
27. Senerwa, D.M.; Mtimet, N.; Sirma, A.J.; Nzuma, J.; Kang'ethe, E.K.; Lindahl, J.F.; Grace, D. Direct market costs of aflatoxins in Kenyan dairy value chain. In Proceedings of the the Agriculture, Nutrition and Health (ANH) Academy Week, Addis Ababa, Ethiopia, 20–24 June 2016; University of Nairobi: Nairobi, Kenya.
28. Nyangi, C.; Mugula, J.K.; Beed, F.; Boni, S.; Koyano, E.; Sulyok, M. Aflatoxins and fumonisin contamination of marketed maize, maize bran and maize used as animal feed in northern Tanzania. *Afr. J. Food Agric. Nutr. Dev.* **2016**, *2016 16*, 11054–11065. [CrossRef]
29. Gruber-Dorninger, C.; Jenkins, T.; Schatzmayr, G. Global Mycotoxin Occurrence in Feed: A Ten-Year Survey. *Toxins (Basel)* **2019**, *11*, 375. [CrossRef] [PubMed]
30. Tanaka, T.; Hasegawa, A.; Yamamoto, S.; Lee, U.S.; Sugiura, Y.; Ueno, Y. Worldwide Contamination of Cereals by the *Fusarium* Mycotoxins Nivalenol, Deoxynivalenol, and Zearalenone. 1. Survey of 19 Countries. *J. Agric. Food Chem.* **1988**, *36*, 979–983. [CrossRef]

31. Atherstone, C.; Grace, D.; Waliyar, F.; Lindahl, J.; Osiru, M. *Aflatoxin Literature Synthesis and Risk Mapping: Special Emphasis on Sub-Saharan Africa*; ILRI project report; ILRI: Nairobi, Kenya, 2014.
32. Miazzo, R.F.P.; Magnoli, C.; Salvano, M.; Ferrero, S.; Chiacchiera, S.M.; Carvalho, E.; Rocha ROSA, C.; Dalcero, A. Efficacy of Sodium Bentonite as a Detoxifier of Broiler Feed Contaminated with Aflatoxin and Fumonisin. *Poult. Sci.* **2005**, *84*, 1–8. [CrossRef] [PubMed]
33. Gizachew, D.; Szonyi, B.; Tegegne, A.; Hanson, J.; Grace, D. Aflatoxin Contamination of Milk and Dairy Feeds in the Greater Addis Ababa Milk Shed, Ethiopia. *Food Control* **2016**, *59*, 773–779. [CrossRef]
34. Okoth, S.A.; Kola, M.A. Market Samples as a Source of Chronic Aflatoxin Exposure in Kenya. *Afr. J. Health Sci.* **2012**, *20*, 56–61.
35. Omeiza, G.K.; Kabir, J.; Kwaga, J.K.P.; Kwanashie, C.N.; Mwanza, M.; Ngoma, L. A Risk Assessment Study of the Occurrence and Distribution of Aflatoxigenic *Aspergillus Flavus* and Aflatoxin B1 in Dairy Cattle Feeds in a Central Northern State, Nigeria. *Toxicol. Rep.* **2018**, *5*, 846–856. [CrossRef]
36. Mohammed, S.; Munissi, J.J.E.; Nyandoro, S.S. Aflatoxin M1 in Raw Milk and Aflatoxin B1 in Feed from Household Cows in Singida, Tanzania. *Food Addit. Contam. Part B Surveill.* **2016**, *9*, 85–90. [CrossRef]
37. Gbashi, S.; Madala, N.E.; De Saeger, S.; De Boevre, M.; Adekoya, I.; Adebo, O.A.; Njobeh, P.B. The Socio-Economic Impact of Mycotoxin Contamination in Africa. In *Mycotoxins-Impact and Management Strategies*; Njobeh, P.B., Stepman, F., Eds.; IntechOpen: London, UK, 2018. Available online: https://www.intechopen.com/online-first/the-socio-economic-impact-of-mycotoxin-contamination-in-africa (accessed on 27 March 2019). [CrossRef]
38. Udomkun, P.; Mutegi, C.; Wossen, T.; Atehnkeng, J.; Nabahungu, N.L.; Njukwe, E.; Vanlauwe, B.; Bandyopadhyay, R. Occurrence of Aflatoxin in Agricultural Produce from Local Markets in Burundi and Eastern Democratic Republic of Congo. *Food Sci. Nutr.* **2018**, *6*, 2227–2238. [CrossRef]
39. Kana, J.R.; Gnonlonfin, B.G.J.; Harvey, J.; Wainaina, J.; Wanjuki, I.; Skilton, R.A.; Teguia, A. Assessment of Aflatoxin Contamination of Maize, Peanut Meal and Poultry Feed Mixtures from Different Agroecological Zones in Cameroon. *Toxins* **2013**, *5*, 884–894. [CrossRef]
40. Diekman, M.A.; Green, M.L. Mycotoxins and Reproduction in Domestic Livestock. *J. Anim. Sci.* **1992**, *70*, 1615–1627. [CrossRef]
41. Whitlow, L.; Hagler, W. Mold and Mycotoxin Issues in Dairy Cattle: Effects, Prevention and Treatment. *Adv. Dairy Technol.* **2008**, *20*, 195–209.
42. Gonçalves, B.; Corassin, C.; Oliveira, C. Mycotoxicoses in Dairy Cattle: A Review. *Asian J. Anim. Vet. Adv.* **2015**, *10*, 752–760. [CrossRef]
43. Upadhaya, S.D.; Sung, H.G.; Lee, C.H.; Lee, S.Y.; Kim, S.W.; Cho, K.J.; Ha, J.K. Comparative Study on the Aflatoxin B1 Degradation Ability of Rumen Fluid from Holstein Steers and Korean Native Goats. *J. Vet. Sci.* **2009**, *10*, 29–34. [CrossRef] [PubMed]
44. Jiang, Y.H.; Yang, H.J.; Lund, P. Effect of Aflatoxin B1 on in Vitro Ruminal Fermentation of Rations High in Alfalfa Hay or Ryegrass Hay. *Anim. Feed Sci. Tech.* **2012**, *175*, 85–89. [CrossRef]
45. Gurung, N.; Rankins, D.; Shelby, R. In Vitro Ruminal Disappearance of Fumonisin B1 and Its Effects on in Vitro Dry Matter Disappearance. *Vet. Hum. Toxicol.* **1999**, *41*, 196–199.
46. Gurung, N.K.; Rankins, D.L.J.; Shelby, R.A.; Goel, S. Effects of Fumonisin B1-Contaminated Feeds on Weanling Angora Goats. *J. Anim. Sci.* **1998**, *76*, 2863–2870. [CrossRef]
47. Keese, C.; Meyer, U.; Valenta, H.; Schollenberger, M.; Starke, A.; Weber, I.-A.; Rehage, J.; Breves, G.; Danicke, S. No Carry-over of Unmetabolised Deoxynivalenol in Milk of Dairy Cows Fed High Concentrate Proportions. *Mol. Nutr. Food Res.* **2008**, *52*, 1514–1529. [CrossRef]
48. Seeling, K.; Danicke, S.; Valenta, H.; Van Egmond, H.P.; Schothorst, R.C.; Jekel, A.A.; Lebzien, P.; Schollenberger, M.; Razzazi-Fazeli, E.; Flachowsky, G.; et al. Effects of *Fusarium* Toxin-Contaminated Wheat and Feed Intake Level on the Biotransformation and Carry-over of Deoxynivalenol in Dairy Cows. *Food Addit. Contam.* **2006**, *23*, 1008–1020. [CrossRef]
49. Valgaeren, B.; Theron, L.; Croubels, S.; Devreese, M.; De Baere, S.; Van Pamel, E.; Daeseleire, E.; De Boevre, M.; De Saeger, S.; Vidal, A.; et al. The Role of Roughage Provision on the Absorption and Disposition of the Mycotoxin Deoxynivalenol and Its Acetylated Derivatives in Calves: From Field Observations to Toxicokinetics. *Arch. Toxicol.* **2019**, *93*, 293–310. [CrossRef]

50. Debevere, S.; Cools, A.; Baere, S.D.; Haesaert, G.; Rychlik, M.; Croubels, S.; Fievez, V. In Vitro Rumen Simulations Show a Reduced Disappearance of Deoxynivalenol, Nivalenol and Enniatin B at Conditions of Rumen Acidosis and Lower Microbial Activity. *Toxins* **2020**, *12*, 101. [CrossRef]
51. Kiessling, K.H.; Pettersson, H.; Sandholm, K.; Olsen, M. Metabolism of Aflatoxin, Ochratoxin, Zearalenone, and Three Trichothecenes by Intact Rumen Fluid, Rumen Protozoa, and Rumen Bacteria. *Appl. Environ. Microbiol.* **1984**, *47 5*, 1070–1073. [CrossRef]
52. van Halderen, A.; Green, J.R.; Marasas, W.F.; Thiel, P.G.; Stockenstrom, S. A Field Outbreak of Chronic Aflatoxicosis in Dairy Calves in the Western Cape Province. *J. S. Afr. Vet. Assoc.* **1989**, *60*, 210–211. [PubMed]
53. McKenzie, R.A.; Blaney, B.J.; Connole, M.D.; Fitzpatrick, L.A. Acute Aflatoxicosis in Calves Fed Peanut Hay. *Aust. Vet. J.* **1981**, *57*, 284–286. [CrossRef] [PubMed]
54. Umar, S.; Munir, M.T.; Shah, M.A.A.; Shahzad, M.; Khan, R.A.; Khan, A.U.; Ameen, K.; Rafia-Munir, A.; Saleem, F. Outbreak of Aflatoxicosis on a Local Cattle Farm in Pakistan. *Veterinaria* **2015**, *3*, 13–17.
55. Applebaum, R.S.; Brackett, R.E.; Wiseman, D.W.; Marth, E.H. Responses of Dairy Cows to Dietary Aflatoxin: Feed Intake and Yield, Toxin Content, and Quality of Milk of Cows Treated with Pure and Impure Aflatoxin. *J. Dairy Sci.* **1982**, *65*, 1503–1508. [CrossRef]
56. Ogunade, I.M.; Arriola, K.G.; Jiang, Y.; Driver, J.P.; Staples, C.R.; Adesogan, A.T. Effects of 3 Sequestering Agents on Milk Aflatoxin M1 Concentration and the Performance and Immune Status of Dairy Cows Fed Diets Artificially Contaminated with Aflatoxin B1. *J. Dairy Sci.* **2016**, *99*, 6263–6273. [CrossRef]
57. Jiang, Y.; Ogunade, I.M.; Kim, D.H.; Li, X.; Pech-Cervantes, A.A.; Arriola, K.G.; Oliveira, A.S.; Driver, J.P.; Ferraretto, L.F.; Staples, C.R.; et al. Effect of Adding Clay with or without a Saccharomyces Cerevisiae Fermentation Product on the Health and Performance of Lactating Dairy Cows Challenged with Dietary Aflatoxin B1. *J. Dairy Sci.* **2018**, *101*, 3008–3020. [CrossRef]
58. Sulzberger, S.A.; Melnichenko, S.; Cardoso, F.C. Effects of Clay after an Aflatoxin Challenge on Aflatoxin Clearance, Milk Production, and Metabolism of Holstein Cows. *J. Dairy Sci.* **2017**, *100*, 1856–1869. [CrossRef]
59. Queiroz, O.C.M.; Han, J.H.; Staples, C.R.; Adesogan, A.T. Effect of Adding a Mycotoxin-Sequestering Agent on Milk Aflatoxin M(1) Concentration and the Performance and Immune Response of Dairy Cattle Fed an Aflatoxin B(1)-Contaminated Diet. *J. Dairy Sci.* **2012**, *95*, 5901–5908. [CrossRef]
60. Kutz, R.E.; Sampson, J.D.; Pompeu, L.B.; Ledoux, D.R.; Spain, J.N.; Vazquez-Anon, M.; Rottinghaus, G.E. Efficacy of Solis, NovasilPlus, and MTB-100 to Reduce Aflatoxin M1 Levels in Milk of Early to Mid-Lactation Dairy Cows Fed Aflatoxin B1. *J. Dairy Sci.* **2009**, *92*, 3959–3963. [CrossRef]
61. Sumantri, I.; Murti, T.W.; van der Poel, A.F.B.; Boehm, J.; Agus, A. Carry-Over of Aflatoxin B1-Feed into Aflatoxin M1-Milk In Dairy Cows Treated With Natural Sources of Aflatoxin and Bentonite. *J. Indones. Trop. Anim. Agric.* **2012**, *37*, 271–277. [CrossRef]
62. Masoero, F.; Gallo, A.; Moschini, M.; Piva, G.; Diaz, D. Carryover of Aflatoxin from Feed to Milk in Dairy Cows with Low or High Somatic Cell Counts. *Animal* **2007**, *1*, 1344–1350. [CrossRef] [PubMed]
63. Mesgaran, M.D.; Mojtahedi, M.; Vakili, S.A.; Hayati-Ashtiani, M. Effect of Aflatoxin B1 on in Vitro Rumen Microbial Fermentation Responses Using Batch Culture. *Res. Annu.* **2013**, *3*, s686–s693. Available online: http://www.journalarrb.com/index.php/ARRB/article/view/24848 (accessed on 2 November 2019).
64. Trenholm, H.L.; Thompson, B.K.; Hartin, K.E.; Greenhalgh, R.; McAllister, A.J. Ingestion of Vomitoxin (Deoxynivalenol)-Contaminated Wheat by Nonlactating Dairy Cows. *J. Dairy Sci.* **1985**, *68*, 1000–1005. [CrossRef]
65. Winkler, J.; Kersten, S.; Valenta, H.; Meyer, U.; Engelhardt, U.H.; Danicke, S. Development of a Multi-Toxin Method for Investigating the Carryover of Zearalenone, Deoxynivalenol and Their Metabolites into Milk of Dairy Cows. *Food Addit. Contam. Part A Chem. Anal. Control Expo. Risk Assess.* **2015**, *32*, 371–380. [CrossRef] [PubMed]
66. Mathur, S.; Constable, P.D.; Eppley, R.M.; Waggoner, A.L.; Tumbleson, M.E.; Haschek, W.M. Fumonisin B(1) Is Hepatotoxic and Nephrotoxic in Milk-Fed Calves. *Toxicol. Sci.* **2001**, *60*, 385–396. [CrossRef]
67. Baker, D.C.; Rottinghaus, G.E. Chronic Experimental Fumonisin Intoxication of Calves. *J. Vet. Diagn. Investig.* **1999**, *11*, 289–292. [CrossRef]
68. Richard, J.L.; Meerdink, G.; Maragos, C.M.; Tumbleson, M.; Bordson, G.; Rice, L.G.; Ross, P.F. Absence of Detectable Fumonisins in the Milk of Cows Fed *Fusarium* Proliferatum (Matsushima) Nirenberg Culture Material. *Mycopathologia* **1996**, *133*, 123–126. [CrossRef]

69. Ribelin, W.E.; Fukushima, K.; Still, P.E. The Toxicity of Ochratoxin to Ruminants. *Can. J. Comp. Med.* **1978**, *42*, 172–176.
70. Weaver, G.A.; Kurtz, H.J.; Mirocha, C.J.; Bates, F.Y.; Behrens, J.C.; Robison, T.S.; Swanson, S.P. The Failure of Purified T-2 Mycotoxin to Produce Hemorrhaging in Dairy Cattle. *Can. Vet. J.* **1980**, *21*, 210–213.
71. Kallela, K.; Ettala, E. The Oestrogenic *Fusarium* Toxin (Zearalenone) in Hay as a Cause of Early Abortions in the Cow. *Nord Vet. Med.* **1984**, *36*, 305–309.
72. Coppock, R.W.; Christian, R.G.; Jacobsen, B.J. Aflatoxins. In *Veterinary Toxicology: Basic and Clinical Principles*; Gupta, R.C., Ed.; Elsevier Inc.: San Diego, CA, USA, 2012.
73. Weaver, G.A.; Kurtz, H.J.; Behrens, J.C.; Robison, T.S.; Seguin, B.E.; Bates, F.Y.; Mirocha, C.J. Effect of Zearalenone on the Fertility of Virgin Dairy Heifers. *Am. J. Vet. Res.* **1986**, *47*, 1395–1397.
74. Weaver, G.A.; Kurtz, H.J.; Behrens, J.C.; Robison, T.S.; Seguin, B.E.; Bates, F.Y.; Mirocha, C.J. Effect of Zearalenone on Dairy Cows. *Am. J. Vet. Res.* **1986**, *47*, 1826–1828. [PubMed]
75. Kuilman, M.E.M. Bovine Hepatic Metabolism of Aflatoxin B1. *J. Agric. Food Chem.* **1998**, *46*, 2707–2713. [CrossRef]
76. Britzi, M.; Friedman, S.; Miron, J.; Solomon, R.; Cuneah, O.; Shimshoni, J.A.; Soback, S.; Ashkenazi, R.; Armer, S.; Shlosberg, A. Carry-over of Aflatoxin B1 to Aflatoxin M1 in High Yielding Israeli Cows in Mid- and Late-Lactation. *Toxins (Basel)* **2013**, *5*, 173–183. [CrossRef] [PubMed]
77. van der Fels-Klerx, H.J.; Camenzuli, L. Effects of Milk Yield, Feed Composition, and Feed Contamination with Aflatoxin B1 on the Aflatoxin M1 Concentration in Dairy Cows' Milk Investigated Using Monte Carlo Simulation Modelling. *Toxins (Basel)* **2016**, *8*, 290. [CrossRef] [PubMed]
78. Hernández Camarillo, E.; Carvajal-Moreno, M.; Robles-Olvera, V.; Vargas-Ortiz, M.; Salgado-Cervantes, M.; Roudot, A.; Rodriguez-Jimenes, G. Quantifying the Levels of the Mutagenic, Carcinogenic Hydroxylated Aflatoxins (AFM 1 and AFM 2) in Artisanal Oaxaca-Type Cheeses from the City of Veracruz, Mexico. *J. Microb. Biochem. Technol.* **2016**, *8*, 491–497. [CrossRef]
79. Kuboka, M.M.; Imungi, J.K.; Njue, L.; Mutua, F.; Grace, D.; Lindahl, J.F. Occurrence of Aflatoxin M1 in Raw Milk Traded in Peri-Urban Nairobi, and the Effect of Boiling and Fermentation. *Infect. Ecol. Epidemiol.* **2019**, *9*, 1625703. [CrossRef]
80. Lindahl, J.F.; Kagera, I.N.; Grace, D. Aflatoxin M1 Levels in Different Marketed Milk Products in Nairobi, Kenya. *Mycotoxin Res.* **2018**, *34*, 289–295. [CrossRef]
81. Kirino, Y.; Makita, K.; Grace, D.; Lindahl, J. Survey of Informal Milk Retailers in Nairobi, Kenya and Prevalence of Aflatoxin M1 in Marketed Milk. *African J. Food Agric. Nutr. Dev.* **2016**, *16*, 11022–11038. [CrossRef]
82. Oluwafemi, F.; Badmos, A.; Kareem, S.; Ademuyiwa, O.; Kolapo, A. Survey of Aflatoxin M1 in Cows' Milk from Free-Grazing Cows in Abeokuta, Nigeria. *Mycotoxin Res.* **2014**, *30*. [CrossRef]
83. Ali, M.A.I.; El-Zubeir, I.; Fadel Elseed, A.M. Aflatoxin M 1 in Raw and Imported Powdered Milk Sold in Khartoum State, Sudan. *Food Addit. Contam. Part. B Surveill.* **2014**, *7*, 208–212. [CrossRef]
84. Maragos, C.M.; Richard, J.L. Quantitation and Stability of Fumonisins B1 and B2 in Milk. *J. AOAC Int.* **1994**, *77*, 1162–1167. [CrossRef] [PubMed]
85. Scott, P.M.; Delgado, T.; Prelusky, D.B.; Trenholm, H.L.; Miller, J.D. Determination of Fumonisins in Milk. *J. Environ. Sci. Health B* **1994**, *29*, 989–998. [CrossRef] [PubMed]
86. Zhang, Z.; Fan, Z.; Nie, D.; Zhao, Z.; Han, Z. Analysis of the Carry-Over of Ochratoxin A from Feed to Milk, Blood, Urine, and Different Tissues of Dairy Cows Based on the Establishment of a Reliable LC-MS/MS Method. *Molecules* **2019**, *24*, 2823. [CrossRef] [PubMed]
87. Hashimoto, Y.; Katsunuma, Y.; Nunokawa, M.; Minato, H.; Yonemochi, C. Influence of Repeated Ochratoxin A Ingestion on Milk Production and Its Carry-over into the Milk, Blood and Tissues of Lactating Cows. *Anim. Sci. J.* **2016**, *87*, 541–546. [CrossRef]
88. Kagera, I.; Kahenya, P.; Mutua, F.; Anyango, G.; Kyallo, F.; Grace, D.; Lindahl, J. Status of Aflatoxin Contamination in Cow Milk Produced in Smallholder Dairy Farms in Urban and Peri-Urban Areas of Nairobi County: A Case Study of Kasarani Sub County, Kenya. *Infect. Ecol. Epidemiol.* **2018**, *9*, 1547095. [CrossRef]
89. Sirma, A.; Senerwa, D.; Lindahl, J.; Makita, K.; Kang'ethe, E.; Grace, D. Aflatoxin M1 Survey in Dairy Households in Kenya. In Proceedings of the the FoodAfrica Midterm Seminar, Helsinki, Finland, 16 June 2014.
90. Dutton, M.F.; Mwanza, M.; de Kock, S.; Khilosia, L.D. Mycotoxins in South African Foods: A Case Study on Aflatoxin M1 in Milk. *Mycotoxin Res.* **2012**, *28*, 17–23. [CrossRef]

91. Ahlberg, S.; Grace, D.; Kiarie, G.; Kirino, Y.; Lindahl, J. A Risk Assessment of Aflatoxin M1 Exposure in Low and Mid-Income Dairy Consumers in Kenya. *Toxins (Basel)* **2018**, *10*, 348. [CrossRef]
92. Kiarie, G.; Dominguez-Salas, P.; Kang'Ethe, S.; Grace, D.; Lindahl, J. Aflatoxin Exposure among Young Children in Urban Low-Income Areas of Nairobi and Association with Child Growth. *Afr. J. Food Agric. Nutr. Dev.* **2016**, *16*, 10967–10990. [CrossRef]
93. Sirma, A.J.; Makita, K.; Grace, D.; Senerwa, D.; Lindahl, J.F. Aflatoxin Exposure from Milk in Rural Kenya and the Contribution to the Risk of Liver Cancer. *Toxins (Basel)* **2019**, *11*, 469. [CrossRef]
94. Stepman, F. Scaling-Up the Impact of Aflatoxin Research in Africa. The Role of Social Sciences. *Toxins (Basel)* **2018**, *10*, 136. [CrossRef] [PubMed]
95. Kang'ethe, E.K.; Lang'a, K.A. Aflatoxin B1 and M1 Contamination of Animal Feeds and Milk from Urban Centers in Kenya. *Afr. Health Sci.* **2009**, *9*, 218–226. [PubMed]
96. James, B.; Adda, C.; Cardwell, K.; Annang, D.; Hell, K.; Korie, S.; Edorh, M.; Gbeassor, F.; Nagatey, K.; Houenou, G.; et al. Public Information Campaign on Aflatoxin Contamination of Maize Grains in Market Stores in Benin, Ghana and Togo. *Food Addit. Contam. A* **2007**, *24*, 1283–1291. [CrossRef] [PubMed]
97. Falade, T. Aflatoxin Management Strategies in Sub-Saharan Africa [Online First]. In *Fungi and Mycotoxins—Their Occurrence, Impact on Health and the Economy as well as Pre- and Postharvest Management Strategies*; Njobeh, P.B., Stepman, F., Eds.; IntechOpen: London, UK, 2018. [CrossRef]
98. Hell, K.; Mutegi, C.; Fandohan, P. Aflatoxin Control and Prevention Strategies in Maize for Sub-Saharan Africa. *Julius-Kühn-Archiv* **2010**, *425*, 534.
99. Mutua, F.; Lindahl, J.; Grace, D. Availability and Use of Mycotoxin Binders in Selected Urban and Peri-Urban Areas of Kenya. *Food Secur.* **2019**. [CrossRef]
100. Xiong, J.L.; Wang, Y.M.; Ma, M.R.; Liu, J. Seasonal Variation of Aflatoxin M1 in Raw Milk from the Yangtze River Delta Region of China. *Food Control* **2013**, *34*, 703–706. [CrossRef]
101. De Mil, T.; Devreese, M.; De Baere, S.; Van Ranst, E.; Eeckhout, M.; De Backer, P.; Croubels, S. Characterization of 27 Mycotoxin Binders and the Relation with in Vitro Zearalenone Adsorption at a Single Concentration. *Toxins* **2015**, *7*, 21. [CrossRef]
102. Chelkowski, J.; Szebiotko, K.; Golinski, P.; Buchowski, M.; Godlewska, B.; Radomyska, W.; Wiewiorowska, M. Mycotoxins in Cereal Grain. Part 5. Changes of Cereal Grain Biological Value after Ammoniation and Mycotoxins (Ochratoxins) Inactivation. *Food Nahrung.* **1982**, *26*, 1–7. [CrossRef]
103. Bailey, G.S.; Price, R.L.; Park, D.L.; Hendricks, J.D. Effect of Ammoniation of Aflatoxin B1-Contaminated Cottonseed Feedstock on the Aflatoxin M1 Content of Cows' Milk and Hepatocarcinogenicity in the Trout Bioassay. *Food Chem. Toxicol.* **1994**, *32*, 707–715. [CrossRef]
104. Fremy, J.M.; Gautier, J.P.; Herry, M.P.; Terrier, C.; Calet, C. Effects of Ammoniation on the "carry-over" of Aflatoxins into Bovine Milk. *Food Addit. Contam.* **1988**, *5*, 39–44. [CrossRef]
105. Norred, W.P.; Voss, K.A.; Bacon, C.W.; Riley, R.T. Effectiveness of Ammonia Treatment in Detoxification of Fumonisin-Contaminated Corn. *Food Chem. Toxicol.* **1991**, *29*, 815–819. [CrossRef]
106. Chourasia, H.K. Efficacy of Ammonia in Detoxification of Fumonisin Contaminated Corn. *Indian J. Exp. Biol.* **2001**, *39*, 493–495. [PubMed]
107. Karlovsky, P.; Suman, M.; Berthiller, F.; De Meester, J.; Eisenbrand, G.; Perrin, I.; Oswald, I.P.; Speijers, G.; Chiodini, A.; Recker, T.; et al. Impact of Food Processing and Detoxification Treatments on Mycotoxin Contamination. *Mycotoxin Res.* **2016**, *32*, 179–205. [CrossRef] [PubMed]
108. EFSA Panel on Additives and Products or Substances used in Animal Feed (FEEDAP); Rychen, G.; Aquilina, G.; Azimonti, G.; Bampidis, V.; de Lourdes Bastos, M.; Bories, G.; Chesson, A.; Cocconcelli, P.S.; Flachowsky, G.; et al. Safety and Efficacy of Bentonite as a Feed Additive for All Animal Species. *EFSA J.* **2017**, *15*, e05096. [CrossRef]
109. Pietri, A.; Bertuzzi, T.; Piva, G.; Binder, E.M.; Schatzmayr, D.A.; Rodrigues, I. Aflatoxin Transfer from Naturally Contaminated Feed to Milk of Dairy Cows and the Efficacy of a Mycotoxin Deactivating Product. *Int. J. Dairy Sci.* **2009**, *4*, 34–42. [CrossRef]
110. Mugerwa, S.; Kabirizi, J.; Zziwa, E. Effect of Supplementing Lactating Goats Fed on Aflatoxin Contaminated Feed with Calcium Bentonite and Activated Charcoal on Aflatoxin M1 Concentration, Excretion and Carryover in Milk. *Uganda J. Agric. Sci.* **2015**, *16*, 83–92. [CrossRef]

111. Kissell, L.; Davidson, S.; Hopkins, B.A.; Smith, G.W.; Whitlow, L.W. Effect of Experimental Feed Additives on Aflatoxin in Milk of Dairy Cows Fed Aflatoxin-Contaminated Diets. *J. Anim. Physiol. Anim. Nutr. (Berl.)* **2013**, *97*, 694–700. [CrossRef] [PubMed]
112. Xiong, J.L.; Wang, Y.M.; Nennich, T.D.; Li, Y.; Liu, J.X. Transfer of Dietary Aflatoxin B1 to Milk Aflatoxin M1 and Effect of Inclusion of Adsorbent in the Diet of Dairy Cows. *J. Dairy Sci.* **2015**, *98*, 2545–2554. [CrossRef]
113. Upadhaya, S.D.; Park, M.A.; Ha, J.K. Mycotoxins and Their Biotransformation in the Rumen: A Review. *Asian-Australas J. Anim. Sci.* **2010**, *23*, 1250–1260. [CrossRef]
114. Bhatti, S.A.; Khan, M.Z.; Saleemi, M.K.; Saqib, M.; Khan, A.; ul-Hassan, Z. Protective Role of Bentonite against Aflatoxin B1- and Ochratoxin A-Induced Immunotoxicity in Broilers. *J. Immunotoxicol.* **2017**, *14*, 66–76. [CrossRef]
115. Schatzmayr, G.; Zehner, F.; Täubel, M.; Schatzmayr, D.; Klimitsch, A.; Loibner, A.P.; Binder, E.M. Microbiologicals for Deactivating Mycotoxins. *Mol. Nutr. Food Res.* **2006**, *50*, 543–551. [CrossRef]
116. Liu, D.L.; Yao, D.S.; Liang, R.; Ma, L.; Cheng, W.Q.; Gu, L.Q. Detoxification of Aflatoxin B1 by Enzymes Isolated from Armillariella Tabescens. *Food Chem. Toxicol.* **1998**, *36*, 563–574. [CrossRef]
117. Alberts, J.F.; Engelbrecht, Y.; Steyn, P.S.; Holzapfel, W.H.; van Zyl, W.H. Biological Degradation of Aflatoxin B1 by Rhodococcus Erythropolis Cultures. *Int. J. Food Microbiol.* **2006**, *109*, 121–126. [CrossRef] [PubMed]
118. Heinl, S.; Hartinger, D.; Thamhesl, M.; Vekiru, E.; Krska, R.; Schatzmayr, G.; Moll, W.-D.; Grabherr, R. Degradation of Fumonisin B1 by the Consecutive Action of Two Bacterial Enzymes. *J. Biotechnol.* **2010**, *145*, 120–129. [CrossRef] [PubMed]
119. EFSA Panel on Additives and Products or Substances used in Animal Feed (FEEDAP); Rychen, G.; Aquilina, G.; Azimonti, G.; Bampidis, V.; de Lourdes Bastos, M.; Bories, G.; Chesson, A.; Cocconcelli, P.S.; Flachowsky, G.; et al. Safety and Efficacy of Fumonisin Esterase (FUMzyme®) as a Technological Feed Additive for All Avian Species. *EFSA J.* **2016**, *14*, e04617. [CrossRef]
120. Schwartz, H.; Hartinger, D.; Doupovec, B.; Gruber-Dorninger, C.; Aleschko, M.; Schaumberger, S.; Nagl, V.; Hahn, I.; Berthiller, F.; Schatzmayr, D.; et al. Application of Biomarker Methods to Investigate FUM Zyme Mediated Gastrointestinal Hydrolysis of Fumonisins in Pigs. *World Mycotoxin J.* **2018**, *11*, 1–14. [CrossRef]
121. EFSA Panel on Additives and Products or Substances used in Animal Feed (EFSA FEEDAP Panel); Rychen, G.; Aquilina, G.; Azimonti, G.; Bampidis, V.; de Lourdes Bastos, M.; Bories, G.; Chesson, A.; Cocconcelli, P.S.; Flachowsky, G.; et al. Safety and Efficacy of Fumonisin Esterase from Komagataella Phaffii DSM 32159 as a Technological Feed Additive for Pigs and Poultry. *EFSA J.* **2018**, *16*, e05269. [CrossRef]
122. Polonelli, L.; Giovati, L.; Magliani, W.; Conti, S.; Sforza, S.; Calabretta, A.; Casoli, C.; Ronzi, P.; Grilli, E.; Gallo, A.; et al. Vaccination of Lactating Dairy Cows for the Prevention of Aflatoxin B1 Carry over in Milk. *PLoS ONE* **2011**, *6*, e26777. [CrossRef]
123. Giovati, L.; Gallo, A.; Masoero, F.; Cerioli, C.; Ciociola, T.; Conti, S.; Magliani, W.; Polonelli, L. Vaccination of Heifers with Anaflatoxin Improves the Reduction of Aflatoxin B1 Carry over in Milk of Lactating Dairy Cows. *PLoS ONE* **2014**, *9*, e94440. [CrossRef]

© 2020 by the authors. Licensee MDPI, Basel, Switzerland. This article is an open access article distributed under the terms and conditions of the Creative Commons Attribution (CC BY) license (http://creativecommons.org/licenses/by/4.0/).

Article

Mycotoxin Occurrence in Maize Silage—A Neglected Risk for Bovine Gut Health?

Nicole Reisinger [1], Sonja Schürer-Waldheim [1], Elisabeth Mayer [1], Sandra Debevere [2,3], Gunther Antonissen [2,4], Michael Sulyok [5] and Veronika Nagl [1,*]

1 BIOMIN Research Center, Technopark 1, 3430 Tulln, Austria; nicole.reisinger@biomin.net (N.R.); sonja.schuerer@outlook.com (S.S.-W.); e.mayer@biomin.net (E.M.)
2 Department of Pharmacology, Toxicology and Biochemistry, Faculty of Veterinary Medicine, Ghent University, Salisburylaan 133, 9820 Merelbeke, Belgium; sandra.debevere@UGent.be (S.D.); gunther.antonissen@UGent.be (G.A.)
3 Department of Animal Sciences and Aquatic Ecology, Faculty of Bioscience Engineering, Ghent University, Coupure Links 653, 9000 Ghent, Belgium
4 Department of Pathology, Bacteriology and Avian Diseases, Faculty of Veterinary Medicine, Ghent University, Salisburylaan 133, 9820 Merelbeke, Belgium
5 Institute for Bioanalytics and Agro-Metabolomics, University of Natural Resources and Life Sciences, Vienna (BOKU), Konrad Lorenz-Straße 20, 3430 Tulln, Austria; michael.sulyok@boku.ac.at
* Correspondence: veronika.nagl@biomin.net; Tel.: +43-2272-81166-0

Received: 12 September 2019; Accepted: 1 October 2019; Published: 4 October 2019

Abstract: Forages are important components of dairy cattle rations but might harbor a plethora of mycotoxins. Ruminants are considered to be less susceptible to the adverse health effects of mycotoxins, mainly because the ruminal microflora degrades certain mycotoxins. Yet, impairment of the ruminal degradation capacity or high ruminal stability of toxins can entail that the intestinal epithelium is exposed to significant mycotoxin amounts. The aims of our study were to assess (i) the mycotoxin occurrence in maize silage and (ii) the cytotoxicity of relevant mycotoxins on bovine intestinal cells. In total, 158 maize silage samples were collected from European dairy cattle farms. LC-MS/MS-based analysis of 61 mycotoxins revealed the presence of emerging mycotoxins (e.g., emodin, culmorin, enniatin B1, enniatin B, and beauvericin) in more than 70% of samples. Among the regulated mycotoxins, deoxynivalenol and zearalenone were most frequently detected (67.7%). Overall, 87% of maize silages contained more than five mycotoxins. Using an in vitro model with calf small intestinal epithelial cells B, the cytotoxicity of deoxynivalenol, nivalenol, fumonisin B1 and enniatin B was evaluated (0–200 μM). Absolute IC50 values varied in dependence of employed assay and were 1.2–3.6 μM, 0.8–1.0 μM, 8.6–18.3 μM, and 4.0–6.7 μM for deoxynivalenol, nivalenol, fumonisin B1, and enniatin B, respectively. Results highlight the potential relevance of mycotoxins for bovine gut health, a previously neglected target in ruminants.

Keywords: modified mycotoxin; co-occurrence; corn silage; CIEB; WST-1; NR; SRB; sphingolipid metabolism; Sa/So

Key Contribution: By analyzing 158 samples from 10 different countries, we provide a comprehensive overview on mycotoxin contamination patterns in European maize silages. For the first time, the cytotoxicity of frequently occurring Fusarium toxins was determined on bovine intestinal cells, resulting in a toxicity ranking of NIV > DON > ENNB > FB1.

---

1. Introduction

Mycotoxins are toxic secondary metabolites of different molds, such as *Aspergillus* spp., *Fusarium* spp., *Penicillium* spp. or *Alternaria* spp., and often found in animal feeds. They impair animal health by manifold modes of action, causing hepatotoxic, nephrotoxic, immunomodulatory, genotoxic, and neurotoxic effects as well as reproductive and developmental disorders [1]. During the last decade, the intestine has moved into the spotlight of mycotoxin research. It represents the first barrier to these feed contaminants and is often exposed to higher mycotoxin concentrations than other body tissues. Here, mycotoxins do not only affect digestion and nutrient uptake, but also intestinal histomorphology, gut barrier integrity, mucin production, microbiota composition, and the local immune system [2,3].

Due to their frequent occurrence and negative impact on animal health, many countries have established regulations for mycotoxins in feed. In the European Union (EU), maximum limits are in place for aflatoxin B1 (AFB1) and ergot alkaloids [4], while guidance levels have been set for deoxynivalenol (DON), zearalenone (ZEN), ochratoxin A (OTA) and the sum of fumonisin B1 (FB1) and fumonsin B2 (FB2) [5]. These regulations neither take the presence of multiple mycotoxins into account, nor the occurrence of so-called emerging mycotoxins. This heterogenous group of fungal metabolites is not clearly defined, but commonly referred to as "mycotoxins, which are neither routinely determined, nor legislatively regulated; however, the evidence of their incidence is rapidly increasing" [6]. Proper risk assessment of emerging mycotoxins, e.g., enniatins, culmorins, beauvericin, moniliformin, roquefortine C or fusaric acid, is challenging, as data on toxicity and occurrence are still scarce [7].

Forages are especially prone to contamination by emerging mycotoxins [8,9]. Fresh, dried and ensiled forages are important components of ruminant diets, representing 50–75% of the total diet [10]. Ensiling describes the preservation of green forage by lactic fermentation under anaerobic conditions and shows geographic variations concerning the quantity and type of silage produced [11]. In the EU-28 alone, approximately 2.4 million tons of green maize, which is mainly grown for silage, were harvested in 2018 [12]. Silages can contain a wide range of mycotoxins, that originate either from pre-harvest contamination, or from spoilage with (acid-tolerant and micro-aerobe) toxigenic fungi during storage [8]. Hence, ruminants might be exposed to a plethora of mycotoxins, in particular compared to chicken or swine, which have less diverse diets [9]. However, this risk has been poorly addressed so far, and the need for thorough mycotoxin monitoring in ruminant forages has been highlighted only recently [9].

In general, ruminants are considered to be less susceptible to mycotoxins than other livestock species, mainly because their ruminal microflora is capable of degrading certain mycotoxins to less toxic metabolites [8]. Most prominently, DON is extensively metabolized to de-epoxy-deoxynivalenol (DOM-1), reaching conversion rates of up to 81–99% in dairy cattle [13,14]. The close connection between a functional ruminal microflora and DON toxicity was impressively depicted by Valgaeren et al. [15]. Driven by clinical cases of DON toxicosis in 2- to 3-month-old beef calves, the authors showed that the oral bioavailability of DON is markedly increased in non-ruminating calves (50.7%) compared to ruminating calves (4.1%). Moreover, it was recently demonstrated that a low ruminal pH-value can impair the degradation of DON, NIV, ZEN, and enniatin B (ENNB) in vitro [16]. Especially in the light of subacute rumen acidosis, one of the most important nutritional diseases in dairy cattle [17], these findings are of significant practical relevance. In addition, certain mycotoxins, e.g., ENNB [18], exert antimicrobial activity. It has therefore been suggested that such mycotoxins might alter the ruminal microflora and its degradation capacity [8]. Finally, some mycotoxins hardly undergo ruminal metabolism [8]. For example, limited degradation of 10–18% was reported for FB1 [19,20]. Hence, major amounts of mycotoxins might reach the small intestine and affect gut health both in non-ruminating calves and dairy cattle.

The aims of our study were twofold. First, we investigated the mycotoxin exposure of dairy cattle. To this end, a total of 158 maize silage samples were collected in Europe and analyzed for 61 mycotoxins, including regulated as well as emerging mycotoxins. Second, an in vitro model using calf small intestinal epithelial cells B (CIEB) was established to assess the cytotoxicity of DON, NIV, FB1 and ENNB. Thus, our study does not only deliver comprehensive mycotoxin occurrence data, but also new toxicological information regarding the relevance of mycotoxins for bovine gut health, a previously neglected target in ruminants.

2. Results

2.1. Mycotoxin Occurrence in Maize Silage

Maize silage samples (n = 158) were collected during a 5-year period (2014–2018) at European dairy cattle farms and were analyzed for mycotoxin occurrence with a liquid chromatography–tandem mass spectrometry (LC-MS/MS)-based multi-mycotoxin method. Table 1 gives an overview on proportions of positive samples and detected mycotoxin concentrations in fresh silages.

Table 1. Occurrence of tested mycotoxins in 158 dairy maize silage samples collected in Europe from 2014 to 2018. Mycotoxin concentrations are expressed as µg/kg fresh silage. Numbers in bold indicate the top five values per category (e.g., highest number of positive samples, highest median, etc.).

Mycotoxin	Positive Samples [1] (n)	Positive Samples [1] (%)	Median Concentration [2] (µg/kg)	75th Percentile [2] (µg/kg)	95th Percentile[2] (µg/kg)	Maximum Concentration (µg/kg)
Regulated mycotoxins (except ergot alkaloids) [3]						
Aflatoxin B1 [4]	0	0.0	-	-	-	-
Deoxynivalenol	107	67.7	303	556	1490	3060
Fumonisin B1	55	34.8	60.0	147	262	553
Fumonisin B2	46	29.1	20.4	34.4	101	133
Ochratoxin A	4	2.5	2.38	2.51	2.62	2.65
Zearalenone	107	67.7	15.2	61	1110	1670
Ergot alkaloids [3]						
Ergine	0	0.0	-	-	-	-
Ergocornine	0	0.0	-	-	-	-
Ergocorninin	0	0.0	-	-	-	-
Ergocristine	0	0.0	-	-	-	-
Ergocristinine	0	0.0	-	-	-	-
Ergocryptine	2	1.3	5.41	7.77	9.65	10.12
Ergocryptinine	0	0.0	-	-	-	-
Ergometrine	1	0.6	49.6	49.6	49.6	49.6
Ergometrinine	1	0.6	3.20	3.20	3.20	3.20
Ergosin	2	1.3	1.89	1.90	1.91	1.91
Ergosinin	0	0.0	-	-	-	-
Ergotamine	1	0.6	1.54	1.54	1.54	1.54
Ergotaminine	0	0.0	-	-	-	-
Ergovalin	0	0.0	-	-	-	-
Type-A trichothecenes						
Diacetoxyscirpenol	0	0.0	-	-	-	-
HT-2 toxin	34	21.5	14.7	21.4	51.9	90.2
Monoacetoxyscirpenol	4	2.5	9.91	15.9	29.5	32.9
Neosolaniol	0	0.0	-	-	-	-
T-2 toxin	6	3.8	2.55	2.89	3.79	4.08
Type-B trichothecenes						
3-Acetyldeoxynivalenol	0	0.0	-	-	-	-
15-Acetyldeoxynivalenol	8	5.1	274	480	624	687
Nivalenol	94	59.5	113	237	623	5770
Modified mycotoxins						
Deoxynivalenol-3-glucoside	40	25.3	17.1	49.2	121	129
HT-2-toxin-3-glucoside [5]	1	0.6	6.28	6.28	6.28	6.28
Nivalenol-3-glucoside [5]	5	3.2	6.01	6.06	9.68	10.6
α-zearalenol	12	7.6	4.84	6.93	18.1	22.2
β-zearalenol	8	5.1	4.90	6.66	12.4	12.6
Emerging mycotoxins						
Alternariol	45	28.5	3.11	4.45	12.1	48.1
Alternariol methylether	37	23.4	1.95	3.46	5.79	30.8
Apicidin	79	50.0	9.49	25.0	102	175
Aurofusarin	108	68.4	97.8	307	3840	4710
Beauvericin	120	76.0	9.16	19.0	75.9	214
Bikaverin	42	26.6	20.3	58.8	248	415
Butenolid	30	19.0	28.9	70.9	249	583
Culmorin	125	79.1	190	719	2930	6680
5-Hydroxyculmorin	19	12.0	571	989	1400	1480
15-Hydroxyculmorin	84	53.2	229	504	1520	1670
15-Hydroxyculmoron	22	13.9	204	396	441	484
Emodin	131	82.9	4.38	14.1	211	1640
Enniatin A	30	19.0	2.45	5.23	32.5	50.1
Enniatin A1	98	62.0	2.70	8.73	25.2	173.9
Enniatin B	121	76.6	7.07	13.8	47.4	429
Enniatin B1	124	78.5	5.68	15.5	46.7	555
Enniatin B2	8	5.1	3.40	5.49	16.0	20.7
Enniatin B3	0	0.0	-	-	-	-
Equisetin	86	54.4	4.75	8.42	17.4	45.4
Fusaproliferin	4	2.5	170	286	316	322.3
Fusaric acid	35	22.2	229	998	1800	4120
Kojic acid	67	42.4	96.3	185	876	25,930
Moniliformin	71	44.9	7.84	18.5	61.6	113
Mycophenolic Acid	9	5.7	14.8	80	262	352
Roquefortine C	7	4.4	11.7	21.3	326	454
Sterigmatocystin	3	1.9	2.38	5.89	8.65	9.35
Tenuazonic acid	42	26.6	60.6	182	574	727

[1] Samples with values > limit of detection (LOD); [2] Excluding data < LOD. In case values were between LOD and limit of quantification (LOQ), LOQ/2 was used for calculation; [3] According to regulations/recommendations set by the European Commission for dairy feeds [4,5]; [4] All samples below < LOD for aflatoxin B2, aflatoxin G1 and aflatoxin G2; [5] Included in analysis from 2016 onwards.

Only two out of 158 samples (1.2%) showed no mycotoxin contamination (all mycotoxins < limit of detection). The presence of regulated mycotoxins was absent or marginal in the case of aflatoxin B1 and ochratoxin A. Similarly, ergot alkaloids were only found in 2.5% of samples, with no dominant pattern on co-occurrence of individual alkaloids. However, since concentrations of ergot alkaloids were rather low, and epimerization is promoted using acidic extraction solvents, those results should not be over interpreted. In comparison, the Fusarium toxins ZEN and DON showed a high prevalence of 67.7% each. Among the regulated mycotoxins, the highest median and maximum values were obtained for DON with 303 µg/kg and 3060 µg/kg, respectively. None of the samples exceeded the EU maximum/guidance levels set for aflatoxins, DON, FB1+FB2, ochratoxin A, or ergot alkaloids [4,5]. For the latter, it should be noted that the respective EU directive refers to the content of rye ergot (*Claviceps purpurea*; 1000 mg/kg), whereas concentrations of individual ergot alkaloids were determined in our study. Values of the 14 ergot alkaloids were adjusted to a dry matter content of 88% and summed up, yielding a maximum of 103 µg/kg total ergot alkaloids found in a silage sample from Germany. In contrast, eight samples (5.1%) contained ZEN levels ≥ 2000 µg/kg, which represents the EU guidance value for cereals and cereal products, including forages. Those samples originated from two different countries (Austria, the Netherlands) in two consecutive years (2014, 2015).

Besides DON, the highest prevalence among trichothecenes was found for NIV (59.5%) and HT-2 toxin (21.5%). Despite moderate median values, maximum NIV levels reached 5770 µg/kg in a maize silage sample from Denmark (collected in 2015). Notably, another sample from the Netherlands contained 2260 µg/kg NIV (2018), implying that prominent NIV levels were not limited to one country or season. Indicative levels for T-2+HT-2 toxin in feed were not exceeded [21]. Interestingly, 3-acetyldeoxynivalenol was not found in any of the samples, whereas the median value of 15-acetyldeoxynivaleol (274 µg/kg) was similar to the one obtained for DON.

Furthermore, maize silages were analyzed for several modified mycotoxins, including deoxynivalenol-3-glucoside (DON-3-Glc), HT-2-toxin-3-glucoside (HT2–3-Glc) and nivalenol-3-glucoside (NIV-3-Glc). Concentrations of the modified mycotoxins did not exceed the levels of the respective parent toxins. DON-3-Glc was found in 25.3 % of samples, albeit at low levels and with an average molar percentage of D3G/DON of 2.7% (0.3–9.3%). Molar percentage for HT2–3-Glc/HT-2 toxin and NIV-3-Glc/NIV were 10.9% and 1.3% (0.9–1.6%), respectively.

Overall, the five most frequently detected mycotoxins all belonged to the group of emerging mycotoxins: emodin (EMO) was found in 82.9% of samples, followed by culmorin (CUL; 79.1%), enniatin B1 (ENNB1; 78.5%), enniatin B (ENNB; 76.6%), and beauvericin (BEA; 76.0%). In addition, members of the emerging mycotoxins showed the highest median (5-hydroxyculmorin, 571 µg/kg) and maximum values (kojic acid 25,930 µg/kg) observed in our survey. Only six of the analyzed emerging mycotoxins were present in less than 10% of the samples, namely mycophenolic acid, enniatin B2, roquefortine C, fusaproliferin, sterigmatocystin, and enniatin B3.

Finally, we evaluated the co-occurrence of mycotoxins in maize silages. On average, 13 mycotoxins per sample were found (range: 0–32), and 87% of samples contained more than five mycotoxins (Figure 1, left). For assessment of the most frequently co-occurring mycotoxin combinations, toxins with an individual prevalence of ≥ 20% were considered. The most prevalent combinations were ENNB & ENNB1 (in 74.1% of samples), CUL & ENNB (67.7%), CUL & ENNB1 (67.7%), CUL & DON (66.5%), and CUL & BEA (65.8%). Figure 1 (right) illustrates all mycotoxin combinations analyzed.

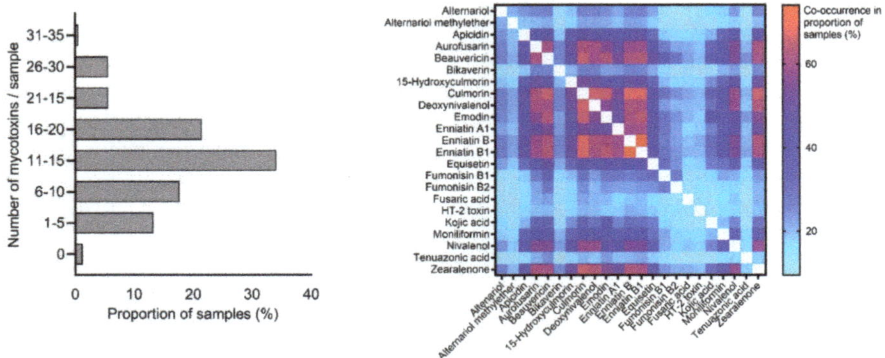

Figure 1. Mycotoxin co-occurrence in maize silage samples collected at European dairy cattle farms. Left: Number of mycotoxins detected per sample. Right: Prevalence of different mycotoxin combinations (only mycotoxins with individual prevalence of ≥ 20% were used for calculations).

2.2. Cytotoxicity of Mycotoxins on Calf Small Intestinal Epithelial Cells

First, the species origin of used CIEB was verified via DNA Barcoding. In addition, the absence of mycoplasma contamination was confirmed prior to and throughout the experimental period. CIEB formed a cell monolayer and showed typical epithelial, cobblestone morphology (Figure 2A). For further characterization, immunohistochemistry was employed. Cytokeratins were expressed as a network radiating from the nucleus to the plasma membrane (Figure 2B), whereas villin was uniformly distributed in the cytoplasma of CIEB (Figure 2C). Vimentin was strongly expressed, forming a filamentous network throughout the cytoplasm with increased density around the nucleus (Figure 2D). As expected, isotype control antibody (Mouse IgG1,) did not show a positive reaction (Supplementary Figure S1).

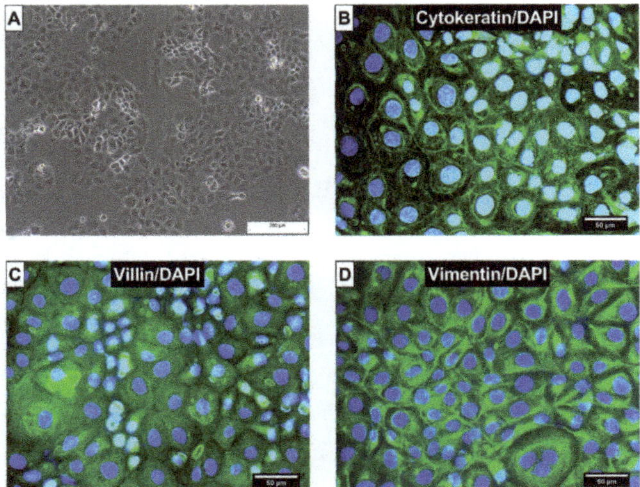

Figure 2. (**A**) Morphology of calf small intestinal epithelial cells B (CIEB) visualized with inverse light microscopy (passage 10, 100 × magnification). Immunostaining of CIEB in chamber slides with (**B**) cytokeratin as an epithelial cell marker, (**C**) villin as marker for intestinal cells, and (**D**) vimentin as mesenchymal marker. 4′,6-Diamidin-2-phenylindol (DAPI) was used as cell nuclei counterstain (400× magnification).

The cytotoxicity of DON, NIV, FB1, and ENNB on CIEB was evaluated based on metabolic activity (WST-1 assay), lysosomal activity (NR assay), and total protein content (SRB assay). To this end, cells were treated for 48 h with increasing toxin concentrations (0–200 µM). While DON, NIV and FB1 were dissolved in culture medium, DMSO had to be used in case of ENNB due to its lower solubility. Since the DMSO proportions in the three highest ENNB concentrations (0.33–1.33% DMSO for 50–200 µM ENNB) affected the lysosomal activity of CIEB (Supplementary Table S1), respective data were excluded for calculations of IC50 values.

All mycotoxins tested had a dose-dependent effect on metabolic and lysosomal activity as well as on total protein content of CIEB. Obtained absolute IC50 values varied depending on the mycotoxin and assay (Figure 3). Still, some general patterns were observed. First, in all assays employed, NIV showed the highest cytotoxicity with IC50 values ranging between 0.8 and 1.0 µM, followed by DON (IC50 values 1.2–3.6 µM) and ENNB (IC50 values 4.0–6.7 µM). In comparison, FB1 showed less pronounced cytotoxic effects (IC50 values 8.6–18.3 µM). Second, the WST-1 assay showed the highest sensitivity for all tested mycotoxins except for ENNB. Here, the lowest IC50 value was obtained with the NR assay. Calculation of the absolute IC50 value for the SRB assay was not possible because the protein content never deceeded 50% in ENNB-treated cells.

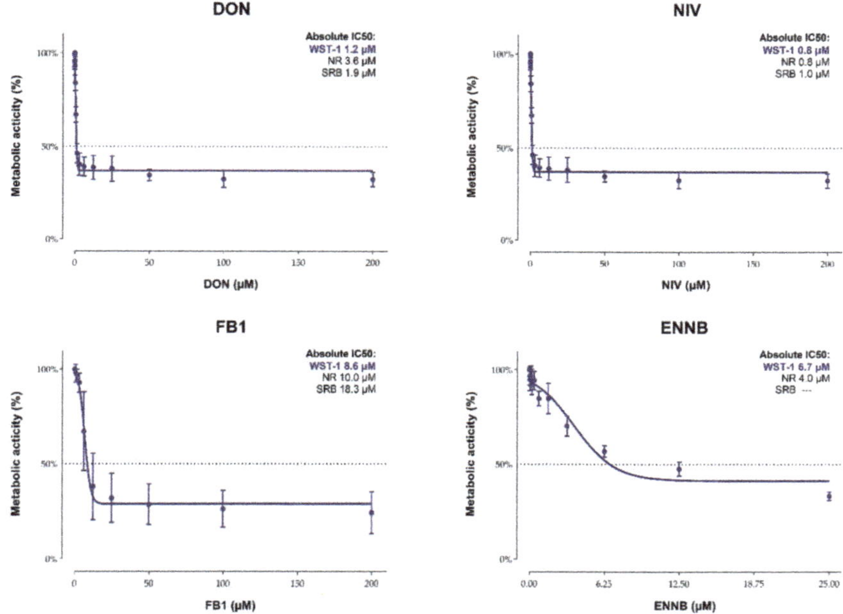

Figure 3. Impact of deoxynivalenol (DON), nivalenol (NIV), fumonisin B1 (FB1), and enniatin B (ENNB) on metabolic activity (%) of calf small intestinal epithelial cells B assessed via the WST-1 assay (48 h incubation, six independent experiments, three replicates per experiment). For comparison, absolute IC50 values for all three assays (WST-1, NR, SRB) are listed.

To monitor alterations of the sphingolipid metabolism in FB1-treated cells, sphinganine (Sa) and sphingosine (So) were determined in cell supernatants via LC-MS/MS. Sa was significantly increased from 25 µM FB1 onwards, whereas no influence on So concentrations was observed (Table 2). Compared to the control, a significant elevation of the Sa/So ratio was evident at 6.25–200 µM FB1. From 12.5 µM FB1 onwards, the numerical increase of the Sa/So ratio was less distinct, indicating a plateau in the response.

Table 2. Sphinganine (Sa) and sphingosine (So) concentrations as well as sphinganine to sphingosine ratio (Sa/So) in supernatants of calf small intestinal epithelial cells B treated with increasing concentrations of FB1 (0–200 μM; n = 4 independent experiments). [a,b] Superscripts indicate significant differences to cells incubated without FB1 (0 μM).

FB1 (μM)	Sa (ng/mL)	So (ng/mL)	Sa/So
0	0.21 ± 0.81 [a]	1.40 ± 0.40	0.15 ± 0.02 [a]
0.781	0.28 ± 0.10 [a]	1.42 ± 0.33	0.20 ± 0.03 [a]
1.563	0.53 ± 0.25 [a]	1.70 ± 0.47	0.31 ± 0.06 [a]
3.125	3.89 ± 1.51 [a]	1.61 ± 0.25	2.36 ± 0.55 [a]
6.25	15.60 ± 5.91 [a]	1.53 ± 0.25	9.98 ± 2.20 [b]
12.5	33.61 ± 13.29 [a]	1.96 ± 0.64	16.96 ± 2.11 [b]
25	44.57 ± 17.19 [b]	2.36 ± 1.00	19.11 ± 1.71 [b]
50	50.64 ± 25.01 [b]	2.49 ± 1.11	20.08 ± 0.86 [b]
100	53.57 ± 23.27 [b]	2.29 ± 0.90	23.25 ± 1.15 [b]
200	56.13 ± 26.26 [b]	2.16 ± 0.83	25.58 ± 2.11 [b]
p-value	<0.0001	0.0462	<0.0001

3. Discussion

Mycotoxin occurrence is influenced by multiple factors, including plant species and variety, region, temperature, humidity, insect damage, storage conditions, and other agricultural practices [22]. Our survey focused on the presence of mycotoxins in maize silage, because this feed component can be the main source for dietary mycotoxin intake in dairy cattle [23]. Since sample numbers per country and/or year were limited in our survey, definite conclusions on regional or yearly trends of mycotoxin occurrence were not justified and therefore omitted. Respective information can be retrieved from other excellent feed surveys [24] and reviews [9,11]. Similar to the approach of Storm et al. [25], mycotoxin concentrations were expressed as μg/kg fresh weight except for the comparison with EU maximum/levels, for which levels were normalized to a dry matter content of 88%. Because literature reports do not uniformly express mycotoxin concentrations in silage (using either fresh or dried weight), the suffix "fresh weight" is used in the following whenever clearly indicated in the respective study, or when samples were not dried prior to analysis.

In 98.8% of silage samples at least one mycotoxin was detected. The top five positions in terms of prevalence were all held by emerging mycotoxins, namely EMO, CUL, ENNB1, ENNB, and BEA. Although data on the presence of emerging mycotoxins in feed are scarce, high incidences of enniatins and BEA have been described previously. For example, ENNB1, ENNB, and BEA were found in 97%, 90%, and 100% of maize silages collected in Poland, respectively [26]. Reported median values (6.0–20.9 μg/kg fresh weight) were in a similar range in our study. In silage samples from Spain [27], ENNB showed yearly variations in prevalence (31–72%) with higher average concentrations (151–163 μg/kg). In contrast, moderate incidences of around 25% for ENNB [25] and BEA [28] were reported in Denmark. Differences between studies might stem from distinct fungal contamination patterns and/or variations in methodology (e.g., sampling procedure, limits of detection). Reports on the toxicity of these *Fusarium* toxins in ruminants are completely lacking so far [29]. In this respect, the described antimicrobial activity of enniatins and BEA, potentially affecting the composition and function of the rumen microbiota, might be of special interest. In addition, ENNB and BEA were demonstrated to impair the barrier function in intestinal porcine enterocytes (IPEC-J2; [30]). Since a certain proportion of ENNB might by-pass the rumen [16], negative effects on the bovine gut cannot be excluded.

To the best of our knowledge, the occurrence of EMO and CUL in European maize silages has not been addressed yet. In line with our data, a survey conducted in Israel showed high prevalence of EMO in maize silages (100%; [31]). For CUL, results deviate from our study, mostly in terms of incidence (6.6% versus 79.1%) but also concerning median values obtained (46 μg/kg fresh weight versus 190 μg/kg). This mycotoxin has recently caught scientific attention because of its potency to inhibit DON glucuronidation [32], and we confirmed the commonly observed co-occurrence of CUL and DON [7] for maize silages. Still, the relevance of CUL for dairy cattle remains debatable, as metabolization to DOM-1 is the primary detoxification pathway for DON in ruminants. Another emerging mycotoxin that has gained certain interest is fusaric acid. Shimshoni et al. [31], authors of the aforementioned survey in Israel, pointed out both its high prevalence and concentration in maize silage. Evaluated on a larger sample size and in a different region, our findings corroborate a certain relevance for fusaric acid (detected in nearly one quarter of silages, maximum concentration of 4,120 μg/kg fresh weight). Concerns for bovine health were related to the growth inhibition of important rumen microorganisms and the toxin's potential carry over to milk [31]. However, like for other emerging mycotoxins, toxicodynamic and toxicokinetic studies are warranted to verify these assumptions and to elucidate the role of fusaric acid for food safety.

FB1+FB2 were detected in approximately one third of the samples, albeit at low concentrations. As unveiled by Latorre et al. [33], the majority of fumonisins in maize silage are present in a modified form. These so-called "hidden fumonisins" escape routine analysis, but are expected to be released upon mammalian digestion [34]. For assessing the total fumonisin burden, alkaline hydrolysis of samples is required [33]. This was not performed in the present case and thus represents a limitation of our study. Similarly, total exposure to type-A or -B trichothecenes is underestimated in surveys that do not account for acetylated or modified forms. In our study, DON showed a high prevalence of 67.7% with moderate median concentrations of 303 μg/kg fresh weight. In the past, higher average DON values of 1,629 μg/kg fresh weight [25] or 854–1316 μg/kg [10,27] were monitored, and incidences varied substantially from 6.1–86% [25–27,35]. While average molar DON-3-Glc/DON percentages of 20% were proposed for cereals [36], we found markedly lower values of 2.72%. Further studies are necessary to assess whether this observation is related to the commodity maize silage as such or merely to our sample set. The same applies to our findings on NIV-3-Glc and HT2–3-Glc, both showing negligible prevalence. Although this indicates that NIV-3-Glc does not contribute significantly to the total NIV burden of dairy cattle, the prominent prevalence of the parent toxin (59.5%) must be underlined. Maximum NIV values exceeded previously reported data [25,26,37], revealing that the NIV exposure can be extremely high for individual dairy cattle herds.

ZEN was the only mycotoxin found at levels above the EU maximum/guidance limits [4,5], with 5.1% of samples exceeding $\geq$ 2,000 μg/kg ZEN. In most mycotoxin surveys, maize silages complied to the EU regulations [10,25,26,37], whereas Dangac et al. [27] reported 1.4% of samples exceeding the recommended maximum levels for ZEN. It should be noted that a different limit was employed in that study (500 μg/kg ZEN for complete feedstuffs), which hampers a direct comparison of results. Still, data emphasize the need to monitor ZEN in maize silage and to control its formation pre- and postharvest. This is especially important in the light of potential synergistic effects with other mycotoxins. Naturally, ZEN often co-occurred with DON (63.3%). While exposure to diets co-contaminated with ZEN and DON did not affect the performance of dairy cows [38], alterations of health-related blood parameters were observed by Dänicke et al. [39]. In addition, authors suggested an influence on ketogenesis at the cellular level. Clearly, more studies are needed to decipher the interactions of DON and ZEN in ruminants. The same is valid for other mycotoxin combinations.

Co-occurrence of mycotoxins might be of relevance for animal health even at comparably low concentrations. As summarized by Chehli et al. [40], the type and intensity of mycotoxin interactions can vary dose-dependently. Our study confirmed that mycotoxin co-occurrence in feed is rather the rule than the exception. Strikingly, silage samples contained 13 mycotoxins on average, and in 87% of samples more than five mycotoxins were found. These high values are partly attributed to the broad palette of mycotoxins tested in our study, and therefore, further expand existing knowledge on mycotoxin co-contamination in maize silage (e.g., Refs. [9,26,27]).

Next, we investigated the impact of silage mycotoxins on bovine gut health. Based on our survey results, we focused on *Fusarium* toxins and assessed the cytotoxic potential of DON, NIV, FB1, and ENNB on bovine intestinal cells. Toxins were selected due to their high prevalence (DON,

ENNB), maximum concentrations (NIV) or ruminal stability even under physiological conditions (FB1; [20]). The toxicity of mycotoxins in bovine intestinal cells is currently unknown, mainly because of two reasons. First, the rumen microbiota was long thought to neutralize the toxicity of mycotoxins. However, recent studies indicate that the ruminal degradation capacity might be impaired under specific conditions, such as altered ruminal development in calves [15] or rumen acidosis [16]. Second, the small number of commercially available bovine lines restrains research in this field.

Since CIEB are not widely used, we first confirmed the species identity and the absence of mycoplasma contamination. Although these aspects are of paramount importance for reliable and reproducible in vitro results, they are often neglected. Mycoplasma contamination can alter the properties of cell lines, and infected CIEB were described to exhibit low viability and poor growth [41]. In the same study, authors reported misidentification of three out of eight tested cell lines. The dimension of this issue is even more striking when retrieving information from the International Cell Line Authentication Committee, which has documented 451 false identified cell lines [42]. Resources wasted in the last 50 years due to misidentification of cell lines, stemming from cross-contamination, wrongly labelled samples or inadequate protocols, can only be estimated [43]. Consequently, increased attention should be paid to adequate quality controls for in vitro experiments, also in mycotoxin research.

Further characterization of CIEB was performed by immunofluorescence staining. CIEB showed a positive reaction for cytokeratins and villin. Cytokeratin proteins, which are characteristic components of the cytoskeleton, are commonly used for identification of epithelial cells [44]. Villin is an actin-binding protein in the microvilli of epithelial cell [45]. Expression of both proteins has been used to verify the intestinal epithelial nature of bovine cells before [46]. Besides, CIEB were immuno-positive for vimentin. This protein is a typical marker for non-epithelial cells, such as fibroblasts [47]. However, unequivocal identification of fibroblasts remains challenging. For example, the expression of vimentin was reported for the intestinal porcine epithelial cell line (IPEC-J2; [48]). Another study even excluded the presence of fibroblasts in cells isolated from calf intestine although they showed a positive reaction for vimentin [49]. It seems that the expression of vimentin is not a unique property of mesenchymal cells but can also be found in intestinal epithelial cells and should be evaluated in combination with the presence/absence of cytokeratin expression. Altogether, our immunohistochemistry results asserted the epithelial intestinal origin of CIEBs.

Absolute IC50 values of mycotoxins were calculated based on viability tests performed with three different assays (WST-1, SRB, NR). Independent of the assay, NIV was the most cytotoxic mycotoxin (IC50 0.8–1.0 μM), closely followed by DON (1.2–3.6 μM). The higher comparative cytotoxicity of NIV is in accordance to experiments performed in human (epithelial colorectal adenocarcinoma cells, Caco-2; [50]) and porcine intestinal cells (IPEC-1, IPEC-J2; [51,52]). Likewise, the absolute IC50 values obtained for NIV and DON in CIEB are in a similar range as reported previously. For example, IC50 values for NIV were 0.9–2.1 μM in Caco-2, IPEC-1 and IPEC-J2 cells [50–52], and 0.9–3.6 for DON [50–53]. Opposed to that, individual studies found higher IC50 values, e.g., 6.9 μM for NIV [54] or up to 44.8 μM for DON [55]. Differences between studies can derive from experimental conditions, such as cultivation medium, tested concentration range, exposure period, chosen endpoint, calculation of IC50 values, or differentiation status of cells [40]. Overall, data indicate that CIEB are at least as sensitive to NIV and DON as human or porcine intestinal cells.

For ENNB, a higher cytotoxicity compared to NIV [54] and DON [56] was observed in Caco-2, which could not be confirmed for CIEB. Interestingly, the most sensitive IC50 value for ENNB (4.0 μM) was generated by the NR assay, which measures lysosomal activity. Indeed, destabilization of lysosomes has been suggested as an upstream event of ENNB-induced cell death [57]. As IC50 values after incubation periods of up to 48 h varied strongly in Caco-2 (2.1 to > 30 μM; [18]), comparison of results is challenging. Still, in line with the present study, it was reported that the NR assay yields lower IC50 values for ENNB than assays measuring metabolic activity [57,58]. Since mitochondria are one of the major cellular targets of enniatins [18], we originally assumed a strong response in the WST-1 assay. However, among other effects on these cell organelles, enniatins induced swelling of rat liver mitochondria [59]. Interestingly, the same phenomenon was described for IPEC-J2 cells exposed to DON, and here the comparably weaker cytotoxic response assessed by the WST-1 assay was partly attributed to alterations of the mitochondrial morphology and metabolic activity [30]. Although further mechanistic studies are needed, cumulative data suggest that the metabolic activity does not represent the most sensitive endpoint for cytotoxicity assessment of ENNB. For the SRB assay, calculation of an IC50 value was not possible. To the best of our knowledge, no other study has employed this test to determine the cytotoxicity of ENNB so far. Hence, we cannot conclude whether the total protein content is the least sensitive endpoint for this mycotoxin or whether this finding is rather limited to our experimental conditions. Concordant with Springler et al. [55], our study underlines the importance of multi-parameter analysis in the cytotoxicity assessment of mycotoxins.

FB1 showed the lowest cytotoxicity among the toxins tested in our study. Considering previous studies demonstrating minor cytotoxicity of this mycotoxin in Caco-2 [60–63] and IPEC-J2 [52,64], these results are not surprising. Still, it should be noted that IC50 values obtained in CIEB are markedly lower than the ones reported previously for other intestinal cells (if computable at all). This might be partly related to the narrow concentration range tested in some of the studies [52,60,63]. FB1-induced effects on the intestine were proposed to originate from disruption of the sphingolipid metabolism which causes intracellular accumulation of the sphingoid bases sphinganine (Sa) and sphingosine (So) [65]. In line with reports addressing intestinal tissues/cells from monogastric livestock species [66,67], a dose-dependent increase of the Sa/So ratio was observed in CIEB. Compared to Loiseau et al. [66], who found a significant elevation of the Sa/So ratio after 48 h of exposure to 100 μM FB1 in IPEC-J2, CIEB reacted to lower toxin concentrations, reaching statistical significance at 6.25 μM FB1. Yet, absolute Sa/So values were smaller in our study, which might be explained by the type of matrix used for analysis (cell extract [66] or supernatant).

Although ZEN showed high prevalence in silage samples and was the only mycotoxin exceeding the EU guidance levels, we did not include this compound in our cytotoxicity experiments. This decision was mainly based on the primary mode of action of ZEN, which is the activation of estrogen receptors [68]. Compared to other mycotoxins, the effects of ZEN on the intestine are less detrimental [69]. For example, IC50 values for ZEN obtained by measuring metabolic activity in Caco-2 were 313 μM [70] and 25 μM [62] after 48 and 72 h of incubation, respectively, and thus even higher than those observed for FB1 in the same experiments. However, increased sensitivity of CIEB to ZEN cannot be ruled out at this stage and should be addressed in future studies. In vitro models represent an essential tool to unravel the toxicological relevance and mode of action of substances. Yet, direct extrapolation to in vivo conditions is often limited, mainly because in-vitro experiments cannot fully reflect the complexity of an intact organism [40]. In an attempt to compare concentrations used in our in vitro experiment to mycotoxin levels in dairy feed, we used the dataset provided by Seeling et al. [14]. In this study, 14 duodenal fistulated cows were exposed to DON-contaminated feed, which allowed the assessment of the toxin's duodenal flow. On average, 1.3% on ingested DON reached the duodenum in unmetabolized form. This low proportion partly stemmed from ruminal absorption of the toxin, but mostly from metabolization to DOM-1 (94–99%). Calculating with this percentage, the IC50 value for DON in CIEB (356 μg/L; WST-1) theoretically corresponds to a feed concentration of approximately 27,400 μg/kg. Although this value exceeds the maximum DON levels detected in fresh maize silage by a factor of ten, it should not be overlooked that minor changes in the ruminal degradation capacity would have a marked impact on the outcome of this estimation. As such, it highlights the practical relevance of our findings.

4. Conclusions

Our survey reports a high prevalence of emerging mycotoxins, namely EMO, CUL, enniatins and BEA, in European maize silages. In addition, the well-known *Fusarium* toxins ZEN, DON and NIV were frequently detected, often co-occurring with the listed emerging mycotoxins. Based on the comparison of obtained IC50 values, our data indicate that CIEB are at least as sensitive to NIV, DON,

ENNB and FB1 as human or porcine intestinal cells. Thus, our study stresses the potential negative health impact of mycotoxins on bovine gut health and highlights the need for further research in this field. In particular, effects of mycotoxin combinations on the composition and functionality of the rumen microbiota as well as on bovine gut health in vivo should be addressed

5. Materials and Methods

5.1. Mycotoxin Survey

In total, 158 maize silage samples were collected at European dairy cattle farms from January 2014 to December 2018. Per year, 19 (2014), 20 (2015), 51 (2016), 36 (2017), and 31 (2018) samples were taken. The 10 countries of origin are displayed in Figure 4.

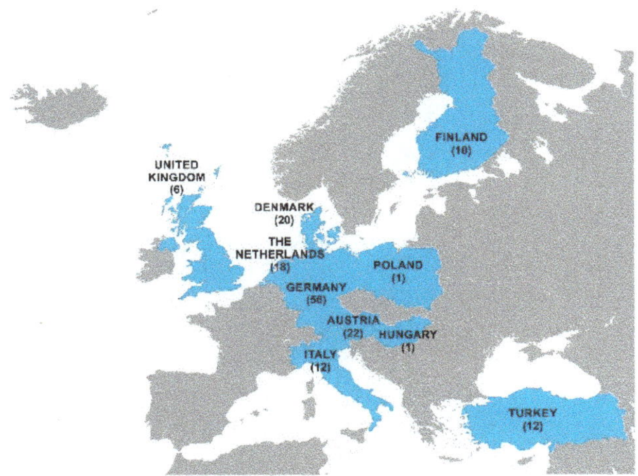

Figure 4. Number of maize silage samples per country of origin.

Samples were provided by the BIOMIN Mycotoxin Survey Program and collected as described previously [71]. Feed was sent for analysis in paper bags or bags with ventilation to avoid humidity building up. Prior to LC-MS/MS analysis, aliquots of 500 g were dried (60 °C, 48 h; drying cabinet FP 24, Binder GmbH, Tuttlingen, Germany), and the dry matter content was determined simultaneously. Thereafter, dried samples were homogenized, extracted and subjected to LC-MS/MS-based multi-mycotoxin analysis according to Malachovà et al. [72]. Details regarding the identification and quantification of mycotoxins as well as the method performance are reported in the aforementioned publication. The accuracy of the method is verified by regular participation in proficiency testing schemes including samples of complex animal feed [72,73].

Final mycotoxin concentrations were corrected for dry matter content and expressed as µg/kg fresh weight (average dry matter content 36.5 ± 8.6%). Samples with mycotoxin levels below the limit of detection (LOD) were considered negative. In case samples shown a mycotoxin concentration between the LOD and limit of quantification (LOQ), LOQ/2 was used to calculate median and percentile values. For evaluation of samples exceeding the EU legislation on maximum/guidance/indicative mycotoxin levels in feed [4,5,21], values were normalized to a dry matter content of 88%. Specifically, the guidance values for the category "Feed materials – Cereals and cereal products" were used for DON (8,000 µg/kg), ZEN (2,000 µg/kg) and ochratoxin A (250 µg/kg). In the case of FB1, the guidance level for "Feed materials – Maize and maize products" was selected (60,000 µg/kg), whereas for the sum of T-2 and HT-2 toxins, the indicative level for "Cereal products for feed and compound feed – Other cereal products" (500 µg/kg) was used.

5.2. *In vitro Experiments*

5.2.1. Cell Line

Calf small intestinal epithelial cells B (CIEB) are a spontaneously immortalized cell line from *bos taurus* (NCBI Taxonomy: 9913). CIEB clone 9 (RRID:CVCL_6A77) were originally purchased from Bionutritec (Iunel, France).

Prior to the conduction of experiments, cells were checked for the absence of mycoplasma contamination with the broth-agar microbiological culture method (Friis and PH broth and agar media; German Collection of Microorganisms and Cell Cultures, Braunschweig, Germany). In addition, CIEB were sent to DSMZ for confirmation of species identification. Cells were tested to be free of mitochondrial DNA sequences from human (detection limit: 10^{-3}) as well as from mouse, rat, and Syrian and Chinese hamster (detection limit:10^{-5}). DNA Barcoding by PCR amplification of the 5′-coding region of Cytochrome C Oxidase I and sequencing of the respective PCR product was used to confirm the species origin of CIEB.

5.2.2. Routine Cultivation of CIEB

Cells were maintained in high glucose (4.5 g/L) Gibco®D-MEM medium (Invitrogen, Thermo Scientific, Vienna, Austria). Media was supplemented with 10% fetal bovine serum (Life Technologies, Thermo Scientific, Vienna, Austria), 16 mM HEPES (Sigma Aldrich, Vienna, Austria), Gibco®2.5 mM GlutaMAX™, and penicillin/streptomycin (100 units/mL; 100 µg/mL; Sigma Aldrich)). CIEB were cultivated at 37 °C in a 5% CO2 atmosphere (Galaxy 48 S, New Brunswick, Eppendorf, Hamburg, Austria). At first passage, coated (Coating Matrix Kit, Life Technologies) 25-flasks (Star Lab, Hamburg, Germany) were used. Thereafter, cells were cultivated in uncoated 75-flasks (Star Lab). CIEB were subcultured two to three times a week upon reaching 80% confluence. Cells were used until passage 24 and regularly confirmed to be free of mycoplasma contamination via PCR (Venor®GeM Mycoplasma Detection Kit; Minerva Biolabs, Berlin, Germany).

5.2.3. Characterization of CIEB by Immunohistochemistry (Cytokeration, Vimentin and Villin)

For antibody staining, eight well cell imaging slides (Eppendorf, Hamburg, Germany) were coated with a Coating Matrix Kit (Life Technologies). CIEB were seeded at a density of 2×10^4 cells/well and incubated for 48 h at 37 °C and 5% CO2. Cells were washed twice with PBS before fixing them with 4% paraformaldehyde solution for 10 minutes at 4 °C. Subsequently, cells were washed again with PBS. For permeabilization, cells were incubated with PBS containing 0.1% Triton X-100 at 200 rpm followed by another washing step with PBS. Blocking was performed by incubating cells for 1 h at 200 rpm with a 2% BSA solution at room temperature, followed by a washing step with PBS. Thereafter, cells were incubated with primary antibodies for 1 h at room temperature: mouse anti-pan cytokeratin antibody [C11] (1:250 dilution, ab7753, Abcam), mouse anti-vimentin antibody [RV202] (1:150, dilution, ab8978, Abcam), mouse anti-villin antibody [3E5G11] - N-terminal (1:250 dilution, ab201989, Abcam). Mouse IgG1, Kappa Monoclonal [B11/6]–Isotype Control (1:250 dilution, ab91353, Abcam) was used as isotype control. After another washing step, cells were incubated for 2 h at room temperature in the dark with the secondary antibody: goat anti-mouse IgG H&L FITC (1:2,000, ab6785, Abcam). Thereafter, cells were washed again, and three drops/well of a fluoroshield mounting medium containing DAPI (Abcam) were added for 5 minutes at room temperature in the dark. Finally, the chambers of the imaging slides were removed, and cover glasses were added.

5.2.4. Cytotoxicity Tests

Solid mycotoxin standards were purchased from Romer Labs (Tulln, Austria; DON, NIV and FB1) or Sigma Aldrich (ENNB). For the preparation of stock solutions, 5 mg of DON, NIV and FB1 were dissolved in appropriate amounts of culture medium to yield a concentration of 1 mM, respectively.

In the case of ENNB, 5 mg of the standard were dissolved in DMSO to achieve a concentration of 15 mM. All stock solutions were stored at −20 °C.

CIEB were seeded in 96-well plates (Eppendorf, Vienna, Austria) with a density of 2 × 10^4 cells/well and 200 μL medium/well. After 24 h, cells were incubated with DON, NIV, FB1 (0–200 μM) or ENNB (0–200 μM; DMSO 0–1.33%) in triplicate for 48 h. Thereafter, three different assays were performed to assess the cytotoxicity of increasing mycotoxin concentrations. First, the water-soluble tetrazolium (WST-1; Roche, Rotkreuz, Switzerland) was performed to assess the metabolic activity of CIEB. While the sulforhodamine B assay (SRB; Xenometrix, Allschwil, Switzerland/Sigma Aldrich) was used to measure the total protein synthesis of cells, the neutral red assay (NR, Xenometrix) was used the measure the ability of viable cells to incorporate and bind neutral red within lysosomes. All tests were performed according to the instructions of the manufacturer. Six independent experiments were performed for each mycotoxin and test, respectively.

5.2.5. Sphinganine and Sphingosine Analysis

In supernatants of FB1-treated cells, the concentration of the sphingoid bases sphinganine (Sa) and sphingosine (So) were determined via LC-MS/MS as described in Reisinger et al. [74]. Supernatants from triplicates were pooled, and four independent experiments were performed.

5.2.6. Data Analysis

Statistical analysis was performed with GraphPad Prism (Prism version 8 for Windows, GraphPad Software, La Jolla, California, USA). Absolute IC50 values of data from the WST-1, NR and SRB assays were calculated with relative numbers. After data normalization, a four-parameter nonlinear regression curve [log (inhibition) versus response with variable slope (least squares ordinary fit, with the condition that the Hillslope is < 0)] was applied to calculate the IC50 values. Sa/So ratios were calculated by dividing the Sa concentration by the So concentration of each experiment. As data were not normally distributed, the Kruskal-Wallis test was performed as a non-parametric test. Dunnett's test was used as a post-hoc test to compare different FB1 concentrations against cell control (0 μM FB1).

For the extrapolation of DON concentrations used in vitro to DON levels in feed, data from Seeling et al. [14] were used. Average DON recovery in duodenum (% of ingested DON) was calculated based on the individual data provided for 14 cows in table 3 of the publication.

Supplementary Materials: The following are available online at http://www.mdpi.com/2072-6651/11/10/577/s1, Figure S1: Immunostaining of calf small intestinal cells B (CIEB) in chamber slides with isotype control antibody for cytokeratin, villin and vimentin, Table S1: Impact of the solvent DMSO (0–10%) on metabolic activity (%) of calf small intestinal epithelial cells B assessed via the WST-1, NR, and SRB assay (48 h incubation, four independent experiments, three replicates per experiment).

Author Contributions: Conceptualization, N.R. and V.N.; methodology, N.R., S.S.-W., M.S.; formal analysis, N.R., S.S.-W., E.M., M.S. and V.N.; resources, S.D, G.A.; writing—original draft preparation, N.R. and V.N.; writing—review and editing, N.R., S.S.-W., E.M., S.D., G.A., M.S. and V.N.; visualization, N.R., S.S.-W., E.M. and V.N.; supervision, N.R., E.M. and M.S.; project administration, N.R., V.N.; funding acquisition, V.N.

Funding: This research received funding from the Austrian Research Promotion Agency (Österreichische Forschungsförderungsgesellschaft FFG, Frontrunner program line, grant number 866384).

Acknowledgments: We appreciate the support of Angela Bruckner, Irene Leitgeb, Brigitte Galler, and Armin Humpel during sample preparation of maize silages. In addition, we thank Oliver Greitbauer for performing the LC-MS/MS based analysis of sphingoid basis and Ines Taschl for her helpful assistance during analysis of survey data.

Conflicts of Interest: N.R., S.S.-W., E.M. and V.N. are employed by BIOMIN Holding GmbH, that operates the BIOMIN Research Center and is a producer of animal feed additives. This, however, did not influence the design of the experimental studies or bias the presentation and interpretation of results.

References

1. Bennett, J.W.; Klich, M. Mycotoxins. *Clin. Microbiol. Rev.* **2003**, *16*, 497–516. [CrossRef] [PubMed]
2. Grenier, B.; Applegate, T. Modulation of intestinal functions following mycotoxin ingestion: Meta-analysis of published experiments in animals. *Toxins* **2013**, *5*, 396–430. [CrossRef] [PubMed]
3. Robert, H.; Payros, D.; Pinton, P.; Théodorou, V.; Mercier-Bonin, M.; Oswald, I.P. Impact of mycotoxins on the intestine: Are mucus and microbiota new targets? *J. Toxicol. Environ. Health B Crit. Rev.* **2017**, *20*, 249–275. [CrossRef]
4. European Commission. Directive 2002/32/EC of the European Parliament and of the Council of 7 May 2002 on undesirable substances in animal feed. *Off. J. Eur. Union* **2002**, *L140*, 10–22.
5. European Commission. Commission recommendation of of 17 August 2006 on the presence of deoxynivalenol, zearalenone, ochratoxin a, T-2 and Ht-2 and fumonisins in products intended for animal feeding. *Off. J. Eur. Union* **2006**, *L229*, 7–9.
6. Vaclavikova, M.; Malachova, A.; Veprikova, Z.; Dzuman, Z.; Zachariasova, M.; Hajslova, J. 'Emerging'mycotoxins in cereals processing chains: Changes of enniatins during beer and bread making. *Food Chem.* **2013**, *136*, 750–757. [CrossRef]
7. Gruber-Dorninger, C.; Novak, B.; Nagl, V.; Berthiller, F. Emerging mycotoxins: Beyond traditionally determined food contaminants. *J. Agric. Food Chem.* **2016**. [CrossRef]
8. Fink-Gremmels, J. Mycotoxins in cattle feeds and carry-over to dairy milk: A review. *Food Addit. Contam.* **2008**, *25*, 172–180. [CrossRef]
9. Gallo, A.; Giuberti, G.; Frisvad, J.; Bertuzzi, T.; Nielsen, K. Review on mycotoxin issues in ruminants: Occurrence in forages, effects of mycotoxin ingestion on health status and animal performance and practical strategies to counteract their negative effects. *Toxins* **2015**, *7*, 3057–3111. [CrossRef]
10. Driehuis, F.; Spanjer, M.; Scholten, J.; Te Giffel, M. Occurrence of mycotoxins in maize, grass and wheat silage for dairy cattle in the Netherlands. *Food Addit. Contam.* **2008**, *1*, 41–50. [CrossRef]
11. Alonso, V.A.; Pereyra, C.M.; Keller, L.A.M.; Dalcero, A.M.; Rosa, C.; Chiacchiera, S.M.; Cavaglieri, L.R. Fungi and mycotoxins in silage: An overview. *J. Appl. Microbiol.* **2013**, *115*, 637–643. [CrossRef] [PubMed]
12. Eurostat. Green Maize by Area, Production and Humidity. Available online: https://ec.europa.eu/eurostat/databrowser/view/tag00101/default/table?lang=en (accessed on 13 August 2019).
13. Hedman, R.; Pettersson, H. Transformation of nivalenol by gastrointestinal microbes. *Arch. Tierernahr.* **1997**, *50*, 321–329. [CrossRef] [PubMed]
14. Seeling, K.; Dänicke, S.; Valenta, H.; Van Egmond, H.P.; Schothorst, R.C.; Jekel, A.A.; Lebzien, P.; Schollenberger, M.; Razzazi-Fazeli, E.; Flachowsky, G. Effects of fusarium toxin-contaminated wheat and feed intake level on the biotransformation and carry-over of deoxynivalenol in dairy cows. *Food Addit. Contam.* **2006**, *23*, 1008–1020. [CrossRef] [PubMed]
15. Valgaeren, B.; Théron, L.; Croubels, S.; Devreese, M.; De Baere, S.; Van Pamel, E.; Daeseleire, E.; De Boevre, M.; De Saeger, S.; Vidal, A.; et al. The role of roughage provision on the absorption and disposition of the mycotoxin deoxynivalenol and its acetylated derivatives in calves: From field observations to toxicokinetics. *Arch. Toxicol.* **2018**, *93*, 1–18. [CrossRef] [PubMed]
16. Debevere, S.; De Baere, S.; Haesaert, G.; Rychlik, M.; Croubels, S.; Fievez, V. In vitro rumen simulations show a lower disappearance of deoxynivalenol, nivalenol, zearalenone and enniatin B at conditions of rumen acidosis and at dry conditions. Manuscript in preparation.
17. Enemark, J.M. The monitoring, prevention and treatment of sub-acute ruminal acidosis (SARA): A review. *Vet. J.* **2008**, *176*, 32–43. [CrossRef]
18. Prosperini, A.; Berrada, H.; Ruiz, M.J.; Caloni, F.; Coccini, T.; Spicer, L.J.; Perego, M.C.; Lafranconi, A. A review of the mycotoxin enniatin b. *Front. Public Health* **2017**, *5*, 304. [CrossRef] [PubMed]
19. Caloni, F.; Spotti, M.; Auerbach, H.; den Camp, H.O.; Gremmels, J.F.; Pompa, G. In vitro metabolism of fumonisin b1 by ruminal microflora. *Vet. Res. Commun.* **2000**, *24*, 379–387. [CrossRef] [PubMed]
20. Gurung, N.; Rankins, J.D.; Shelby, R. In vitro ruminal disappearance of fumonisin B1 and its effects on in vitro dry matter disappearance. *Vet. Hum. Toxicol.* **1999**, *41*, 196–199.
21. European Commission. 2013/165/eu: Commission recommendation of 27 March 2013 on the presence of t-2 and ht-2 toxin in cereals and cereal products. *Off. J. Eur. Union* **2013**, *L91*, 12–15.

22. Jouany, J.P. Methods for preventing, decontaminating and minimizing the toxicity of mycotoxins in feeds. *Anim. Feed Sci. Technol.* **2007**, *137*, 342–362. [CrossRef]
23. Driehuis, F.; Spanjer, M.; Scholten, J.; Te Giffel, M. Occurrence of mycotoxins in feedstuffs of dairy cows and estimation of total dietary intakes. *J. Dairy Sci.* **2008**, *91*, 4261–4271. [CrossRef] [PubMed]
24. Gruber-Dorninger, C.; Jenkins, T.; Schatzmayr, G. Global mycotoxin occurrence in feed: A ten-year survey. *Toxins* **2019**, *11*, 375. [CrossRef] [PubMed]
25. Storm, I.M.; Rasmussen, R.R.; Rasmussen, P.H. Occurrence of pre-and post-harvest mycotoxins and other secondary metabolites in Danish maize silage. *Toxins* **2014**, *6*, 2256–2269. [CrossRef] [PubMed]
26. Panasiuk, L.; Jedziniak, P.; Pietruszka, K.; Piatkowska, M.; Bocian, L. Frequency and levels of regulated and emerging mycotoxins in silage in Poland. *Mycotoxin Res.* **2018**, *35*, 1–9. [CrossRef] [PubMed]
27. Dagnac, T.; Latorre, A.; Fernández Lorenzo, B.; Llompart, M. Validation and application of a liquid chromatography-tandem mass spectrometry based method for the assessment of the co-occurrence of mycotoxins in maize silages from dairy farms in nw spain. *Food Addit. Contam Part A Chem. Anal. Control. Expos. Risk Asess.* **2016**, *33*, 1850–1863. [CrossRef] [PubMed]
28. Sørensen, J.L.; Nielsen, K.F.; Rasmussen, P.H.; Thrane, U. Development of a LC-MS/MS method for the analysis of enniatins and beauvericin in whole fresh and ensiled maize. *J. Agric. Food Chem.* **2008**, *56*, 10439–10443. [CrossRef] [PubMed]
29. Fraeyman, S.; Croubels, S.; Devreese, M.; Antonissen, G. Emerging fusarium and alternaria mycotoxins: Occurrence, toxicity and toxicokinetics. *Toxins* **2017**, *9*, 228. [CrossRef]
30. Springler, A.; Vrubel, G.-J.; Mayer, E.; Schatzmayr, G.; Novak, B. Effect of fusarium-derived metabolites on the barrier integrity of differentiated intestinal porcine epithelial cells (IPEC-J2). *Toxins* **2016**, *8*, 345. [CrossRef]
31. Shimshoni, J.; Cuneah, O.; Sulyok, M.; Krska, R.; Galon, N.; Sharir, B.; Shlosberg, A. Mycotoxins in corn and wheat silage in Israel. *Food Addit. Contam Part A Chem. Anal. Control. Expos. Risk Asess.* **2013**, *30*, 1614–1625. [CrossRef]
32. Woelflingseder, L.; Warth, B.; Vierheilig, I.; Schwartz-Zimmermann, H.; Hametner, C.; Nagl, V.; Novak, B.; Šarkanj, B.; Berthiller, F.; Adam, G. The fusarium metabolite culmorin suppresses the in vitro glucuronidation of deoxynivalenol. *Arch. Toxicol.* **2019**, *93*, 1–15. [CrossRef] [PubMed]
33. Latorre, A.; Dagnac, T.; Lorenzo, B.F.; Llompart, M. Occurrence and stability of masked fumonisins in corn silage samples. *Food Chem.* **2015**, *189*, 38–44. [CrossRef] [PubMed]
34. Dall'Asta, C.; Falavigna, C.; Galaverna, G.; Dossena, A.; Marchelli, R. In vitro digestion assay for determination of hidden fumonisins in maize. *J. Agric. Food Chem.* **2010**, *58*, 12042–12047. [CrossRef] [PubMed]
35. Kosicki, R.; Błajet-Kosicka, A.; Grajewski, J.; Twarużek, M. Multiannual mycotoxin survey in feed materials and feedingstuffs. *Anim. Feed Sci. Technol.* **2016**, *215*, 165–180. [CrossRef]
36. Berthiller, F.; Crews, C.; Dall'Asta, C.; Saeger, S.D.; Haesaert, G.; Karlovsky, P.; Oswald, I.P.; Seefelder, W.; Speijers, G.; Stroka, J. Masked mycotoxins: A review. *Mol. Nutr. Food Res.* **2013**, *57*, 165–186. [CrossRef] [PubMed]
37. Rasmussen, R.R.; Storm, I.M.L.D.; Rasmussen, P.H.; Smedsgaard, J.; Nielsen, K.F. Multi-mycotoxin analysis of maize silage by lc-ms/ms. *Anal. Bioanal. Chem.* **2010**, *397*, 765–776. [CrossRef]
38. Winkler, J.; Kersten, S.; Meyer, U.; Engelhardt, U.; Dänicke, S. Residues of zearalenone (ZEN), deoxynivalenol (DON) and their metabolites in plasma of dairy cows fed fusarium contaminated maize and their relationships to performance parameters. *Food Chem. Toxicol.* **2014**, *65*, 196–204. [CrossRef] [PubMed]
39. Dänicke, S.; Winkler, J.; Meyer, U.; Frahm, J.; Kersten, S. Haematological, clinical–chemical and immunological consequences of feeding fusarium toxin contaminated diets to early lactating dairy cows. *Mycotoxin Res.* **2017**, *33*, 1–13. [CrossRef]
40. Cheli, F.; Giromini, C.; Baldi, A. Mycotoxin mechanisms of action and health impact: 'In vitro'or 'in vivo' tests, that is the question. *World Mycotoxin J.* **2015**, *8*, 573–589. [CrossRef]
41. Steube, K.G.; Koelz, A.-L.; Uphoff, C.C.; Drexler, H.G.; Kluess, J.; Steinberg, P. The necessity of identity assessment of animal intestinal cell lines: A case report. *Cytotechnology* **2012**, *64*, 373–378. [CrossRef]
42. International Cell Line Authentication Committee. Register of Misidentified Cell Lines. Available online: https://iclac.org/databases/cross-contaminations/ (accessed on 13 August 2019).
43. Chatterjee, R. Cell biology: Cases of mistaken identity. *Science* **2007**, *16*, 928–931. [CrossRef] [PubMed]
44. Ordóñez, N.G. Broad-Spectrum immunohistochemical epithelial markers: A review. *Hum. Pathol.* **2013**, *44*, 1195–1215. [CrossRef] [PubMed]

45. George, S.P.; Wang, Y.; Mathew, S.; Srinivasan, K.; Khurana, S. Dimerization and actin-bundling properties of villin and its role in the assembly of epithelial cell brush borders. *J. Biol. Chem.* **2007**, *282*, 26528–26541. [CrossRef] [PubMed]
46. Miyazawa, K.; Hondo, T.; Kanaya, T.; Tanaka, S.; Takakura, I.; Itani, W.; Rose, M.T.; Kitazawa, H.; Yamaguchi, T.; Aso, H. Characterization of newly established bovine intestinal epithelial cell line. *Histochem. Cell Biol.* **2010**, *133*, 125–134. [CrossRef] [PubMed]
47. Chang, Y.; Li, H.; Guo, Z. Mesenchymal stem cell-like properties in fibroblasts. *Cell Physiol. Biochem.* **2014**, *34*, 703–714. [CrossRef]
48. Zakrzewski, S.S.; Richter, J.F.; Krug, S.M.; Jebautzke, B.; Lee, I.-F.M.; Rieger, J.; Sachtleben, M.; Bondzio, A.; Schulzke, J.D.; Fromm, M. Improved cell line IPEC-J2, characterized as a model for porcine jejunal epithelium. *PLoS ONE* **2013**, *8*, e79643. [CrossRef] [PubMed]
49. Loret, S.; Rusu, D.; El Moualij, B.; Taminiau, B.; Heinen, E.; Dandrifosse, G.; Mainil, J. Preliminary characterization of jejunocyte and colonocyte cell lines isolated by enzymatic digestion from adult and young cattle. *Res. Vet. Sci.* **2009**, *87*, 123–132. [CrossRef]
50. Nielsen, C.; Casteel, M.; Didier, A.; Dietrich, R.; Märtlbauer, E. Trichothecene-induced cytotoxicity on human cell lines. *Mycotoxin Res.* **2009**, *25*, 77–84. [CrossRef]
51. Alassane-Kpembi, I.; Puel, O.; Oswald, I.P. Toxicological interactions between the mycotoxins deoxynivalenol, nivalenol and their acetylated derivatives in intestinal epithelial cells. *Arch. Toxicol.* **2014**, *89*, 1–10. [CrossRef]
52. Wan, L.Y.M.; Turner, P.C.; El-Nezami, H. Individual and combined cytotoxic effects of fusarium toxins (deoxynivalenol, nivalenol, zearalenone and fumonisins B1) on swine jejunal epithelial cells. *Food Chem. Toxicol.* **2013**, *57*, 276–283. [CrossRef]
53. Dänicke, S.; Hegewald, A.-K.; Kahlert, S.; Kluess, J.; Rothkötter, H.-J.; Breves, G.; Döll, S. Studies on the toxicity of deoxynivalenol (DON), sodium metabisulfite, don-sulfonate (DONS) and de-epoxy-don for porcine peripheral blood mononuclear cells and the intestinal porcine epithelial cell lines IPEC-J1 and IPEC-J2, and on effects of don and dons on piglets. *Food Chem. Toxicol.* **2010**, *48*, 2154–2162. [PubMed]
54. Vejdovszky, K.; Warth, B.; Sulyok, M.; Marko, D. Non-synergistic cytotoxic effects of fusarium and alternaria toxin combinations in caco-2 cells. *Toxicol. Lett.* **2016**, *241*, 1–8. [CrossRef] [PubMed]
55. Springler, A.; Hessenberger, S.; Reisinger, N.; Kern, C.; Nagl, V.; Schatzmayr, G.; Mayer, E. Deoxynivalenol and its metabolite deepoxy-deoxynivalenol: Multi-parameter analysis for the evaluation of cytotoxicity and cellular effects. *Mycotoxin Res.* **2017**, *33*, 25–37. [CrossRef] [PubMed]
56. Fernández-Blanco, C.; Font, G.; Ruiz, M.-J. Interaction effects of enniatin B, deoxinivalenol and alternariol in Caco-2 cells. *Toxicol. Lett.* **2016**, *241*, 38–48. [CrossRef] [PubMed]
57. Ivanova, L.; Egge-Jacobsen, W.; Solhaug, A.; Thoen, E.; Fæste, C. Lysosomes as a possible target of enniatin b-induced toxicity in Caco-2 cells. *Chem. Res. Toxicol.* **2012**, *25*, 1662–1674. [CrossRef] [PubMed]
58. Prosperini, A.; Juan-García, A.; Font, G.; Ruiz, M. Reactive oxygen species involvement in apoptosis and mitochondrial damage in caco-2 cells induced by enniatins A, A1, B and B1. *Toxicol. Lett.* **2013**, *222*, 36–44. [CrossRef] [PubMed]
59. Tonshin, A.A.; Teplova, V.V.; Andersson, M.A.; Salkinoja-Salonen, M.S. The fusarium mycotoxins enniatins and beauvericin cause mitochondrial dysfunction by affecting the mitochondrial volume regulation, oxidative phosphorylation and ion homeostasis. *Toxicology* **2010**, *276*, 49–57. [CrossRef] [PubMed]
60. Wentzel, J.F.; Lombard, M.J.; Du Plessis, L.H.; Zandberg, L. Evaluation of the cytotoxic properties, gene expression profiles and secondary signalling responses of cultured cells exposed to fumonisin B1, deoxynivalenol and zearalenone mycotoxins. *Arch. Toxicol.* **2017**, *91*, 2265–2282. [CrossRef] [PubMed]
61. Creppy, E.E.; Chiarappa, P.; Baudrimont, I.; Borracci, P.; Moukha, S.; Carratù, M.R. Synergistic effects of fumonisin B1 and ochratoxin A: Are in vitro cytotoxicity data predictive of in vivo acute toxicity? *Toxicology* **2004**, *201*, 115–123. [CrossRef]
62. Kouadio, J.H.; Mobio, T.A.; Baudrimont, I.; Moukha, S.; Dano, S.D.; Creppy, E.E. Comparative study of cytotoxicity and oxidative stress induced by deoxynivalenol, zearalenone or fumonisin B1 in human intestinal cell line Caco-2. *Toxicology* **2005**, *213*, 56–65. [CrossRef] [PubMed]
63. Sobral, M.M.C.; Faria, M.A.; Cunha, S.C.; Ferreira, I.M. Toxicological interactions between mycotoxins from ubiquitous fungi: Impact on hepatic and intestinal human epithelial cells. *Chemosphere* **2018**, *202*, 538–548. [CrossRef] [PubMed]

64. Bouhet, S.; Hourcade, E.; Loiseau, N.; Fikry, A.; Martinez, S.; Roselli, M.; Galtier, P.; Mengheri, E.; Oswald, I.P. The mycotoxin fumonisin B1 alters the proliferation and the barrier function of porcine intestinal epithelial cells. *Toxicol. Sci.* **2004**, *77*, 165–171. [CrossRef] [PubMed]
65. Bouhet, S.; Oswald, I.P. The intestine as a possible target for fumonisin toxicity. *Mol. Nutr. Food Res.* **2007**, *51*, 925–931. [CrossRef] [PubMed]
66. Loiseau, N.; Debrauwer, L.; Sambou, T.; Bouhet, S.; Miller, J.D.; Martin, P.G.; Viadère, J.-L.; Pinton, P.; Puel, O.; Pineau, T. Fumonisin b1 exposure and its selective effect on porcine jejunal segment: Sphingolipids, glycolipids and trans-epithelial passage disturbance. *Biochem. Pharmacol.* **2007**, *74*, 144–152. [CrossRef]
67. Grenier, B.; Schwartz-Zimmermann, H.E.; Caha, S.; Moll, W.D.; Schatzmayr, G.; Applegate, T.J. Dose-dependent effects on sphingoid bases and cytokines in chickens fed diets prepared with fusarium verticillioides culture material containing fumonisins. *Toxins* **2015**, *7*, 1253–1272. [CrossRef]
68. Fink-Gremmels, J.; Malekinejad, H. Clinical effects and biochemical mechanisms associated with exposure to the mycoestrogen zearalenone. *Anim. Feed Sci. Technol.* **2007**, *137*, 326–341. [CrossRef]
69. Liew, W.-P.-P.; Mohd-Redzwan, S. Mycotoxin: Its impact on gut health and microbiota. *Front. Cell. Infect. Microbiol.* **2018**, *8*, 60. [CrossRef]
70. Cetin, Y.; Bullerman, L.B. Cytotoxicity of fusarium mycotoxins to mammalian cell cultures as determined by the mtt bioassay. *Food Chem. Toxicol.* **2005**, *43*, 755–764. [CrossRef]
71. Kovalsky Paris, M.P.; Schweiger, W.; Hametner, C.; Stückler, R.; Muehlbauer, G.J.; Varga, E.; Krska, R.; Berthiller, F.; Adam, G. Zearalenone-16-O-glucoside: A new masked mycotoxin. *J. Agric. Food Chem.* **2014**, *62*, 1181–1189. [CrossRef]
72. Malachová, A.; Sulyok, M.; Beltrán, E.; Berthiller, F.; Krska, R. Optimization and validation of a quantitative liquid chromatography—Tandem mass spectrometric method covering 295 bacterial and fungal metabolites including all regulated mycotoxins in four model food matrices. *J. Chromatogr.* **2014**, *1362*, 145–156. [CrossRef]
73. Malachová, A.; Sulyok, M.; Beltran, E.; Berthiller, F.; Krska, R. Multi-toxin determination in food-the power of "dilute and shoot" approaches in LC-MS/MS. *LC GC Eur.* **2015**, *28*, 542–555.
74. Reisinger, N.; Dohnal, I.; Nagl, V.; Schaumberger, S.; Schatzmayr, G.; Mayer, E. Fumonisin B1 (FB1) induces lamellar separation and alters sphingolipid metabolism of in vitro cultured hoof explants. *Toxins* **2016**, *8*, 89. [CrossRef] [PubMed]

© 2019 by the authors. Licensee MDPI, Basel, Switzerland. This article is an open access article distributed under the terms and conditions of the Creative Commons Attribution (CC BY) license (http://creativecommons.org/licenses/by/4.0/).

Article

Contamination of Pet Food with Mycobiota and *Fusarium* Mycotoxins—Focus on Dogs and Cats

Natalia Witaszak [1,*], Agnieszka Waśkiewicz [2], Jan Bocianowski [3] and Łukasz Stępień [1]

1 Institute of Plant Genetics, Polish Academy of Sciences, 60-479 Poznań, Poland; lste@igr.poznan.pl
2 Department of Chemistry, Poznań University of Life Scienses, 60-625 Poznań, Poland; agnieszka.waskiewicz@up.poznan.pl
3 Department of Mathematical and Statistical Methods, Poznań University of Life Sciences, 60-637 Poznań, Poland; jan.bocianowski@up.poznan.pl
* Correspondence: nwit@igr.poznan.pl; Tel.: +48-61-6550-237

Received: 28 January 2020; Accepted: 17 February 2020; Published: 19 February 2020

Abstract: A wide range of pet food types are available on the market; the dominant type is dry food formulated in croquets. One of the most common ingredients of dry food are cereals—vectors of harmful mycotoxins posing the risk to pet health. In this study, 38 cat and dog dry food samples available on the Polish market were investigated. Morphological and molecular methods were applied to identify fungal genera present in pet food. Quantification of ergosterol and *Fusarium* mycotoxins: Fumonisin B_1, deoxynivalenol, nivalenol, and zearalenone were performed using high performance liquid chromatography. Obtained results indicated five genera of mycotoxigenic fungi: *Alternaria* sp., *Aspergillus* sp., *Cladosporium* sp., *Penicillium* sp., and *Fusarium* sp., including *Fusarium verticillioides* and *Fusarium proliferatum.* Ergosterol and mycotoxins of interest were detected in both cat and dog food samples in the amounts ranging from 0.31 to 4.05 μg/g for ergosterol and 0.3–30.3, 1.2–618.4, 29.6–299.0, and 12.3–53.0 ng/g for zearalenone, deoxynivalenol, nivalenol, and fumonisin B_1, respectively. The conclusion is the presence of mycotoxins in levels much lower than recommended by EU regulations does not eliminate the risk and caution is advised concerning that long-term daily intake of even small doses of mycotoxins can slowly damage pet's health.

Keywords: pet food; *Fusarium*; ergosterol; mycotoxins; trichothecenes; fumonisin B_1; zearalenone; HPLC

Key Contribution: Due to cereal additions, pet food contains plant pathogens, like *Fusarium* sp. Tested samples were contaminated with mycotoxins: fumonisin B_1, deoxynivalenol, nivalenol and zearalenone. Therefore, mycotoxin analyses are needed in pet food together with microbiological tests.

1. Introduction

Companion animals play a significant role in people's lives since their domestication. In the past they were treated as workers or hunters and now people call them "family members" and "friends". People often assign human emotions, behavior, and personality traits to animals. This phenomenon is called anthropomorphism [1]. There are numerous research reports available concerning health and emotional benefits coming from the companionship of pets. For example, the presence of a dog or cat in a human's lifetimes decreases blood pressure and heart rate and reduces stress. Pets can also support treatment during depression and other psychic and emotional disorders [2]. It has been proven that taking care of animals by children teaches them responsibility, builds self-confidence, and ensures appropriate emotional development [3].

Increasing attention and care for pets' health and well-being is observed because of the relationship between human and companion animals. Therefore, the owners pay heed to diet for their pets more often than they did in the past and this became easier because of a large availability of pet food—a type of animal feed dedicated for companion animals. The pet food industry offers a wide range of food products for dogs and cats, especially wet and dry food. According to the statistical data on statista.com, the mean value of pet food sales in years 2010–2016 was 8.5 million tons per year. It is quite obvious why consumers choose commercial pet food—it is a simple, fast, and cheap way to obtain balanced and differential food for dog or cat [4]. Because dogs are omnivores and cats are carnivores, every food must have a suitable composition and nutrient value. This difference entails consequences in the varying nutrient requirements of these animals. For example, cats use proteins as a main source of energy so they food should contain more proteins (>26%) comparing to dog food (18–22%) according to recommendations of the Association of American Feed Control Officials. Moreover, cat food must include higher doses of taurine, vitamin A, arachidonic acid, and arginine [5]. On the other hand, a dog's genome contains genes responsible for starch digestion and glucose absorption [6], while cats do not have these genes and are generally not able to digest dietary fiber, which may eventually be fermented by gut bacteria [7,8]. The excess of carbohydrates derived by cereals in dry cat food might affect the animal health in opposition to cooked starch, which is efficiently digested by cats [8].

The main types of pet food are home-cooked and raw food as well as commercially available dry, canned, and semi-moist food. Animal derivatives or by-products are the most important ingredient in dry food followed by vegetables and cereals, which are the main source of starch and, therefore, they play a role of fillers improving croquet consistency. Maize, wheat, rice, and barley are most commonly used to the production of dry pet food. In grain-free pet food cereals are replaced by potatoes and beet pulp. Besides the impact on pets' gastrointestinal tract (changes in commensal microflora populations), cereals are among the most important sources of toxic contaminations of the pet food [7].

Mycotoxin-producing fungi from the *Fusarium* genus are the primary cereal pathogens, causing every year yield losses world-wide. The most common pathogens of maize are *Fusarium verticillioides* and *F. proliferatum*, while *F. culmorum*, *F. avenaceum*, and *F. graminearum* are common wheat pathogens. Three groups of mycotoxins produced by *Fusarium* fungi are of major importance: fumonisins, trichothecenes and zearalenone, causing sphingolipid metabolism disruption; replication, transcription and translation disturbances and estrogen receptors overexpression, respectively [9]. Only a few research studies are available dealing with the toxicity of mycotoxins to dogs' and cats' health. According to data contained in EFSA 2017 report, there are only eight reports about deoxynivalenol (DON) toxicity on pets' health [10]. Trichothecenes (especially deoxynivalenol) cause appetite and weight loss, diarrhea, vomiting and even gastrointestinal hemorrhage [11,12]. Zearalenone (ZEN) induces oviducts and uterus hyperplasia and edema in bitches and reduces spermatogenesis in dogs [13]. There are no publications about fumonisins affecting companion animals' health.

During pet food processing dry foods are subjected to extrusion under pressure of 34–37 bars and temperature of 100–200 °C. This process, lowering the moisture to about 8–10% and hermetic storage are common practices used to protect the pellet from contamination and mold development [14,15]. However, this issue does not solve the problem of mycotoxins, which are thermostable and still pose a health risk. Many research reports state food and feed quality problems and their influence on human and animal health but there is limited information about pet food quality and the effect of mycotoxins on their health [16–18]. There are enormous numbers of publications and websites about pet nutrition but only few papers about pet food quality, and dog food is more often concerned than cat food. Main health institutions in Europe, USA, and Canada are aware that *Fusarium* mycotoxins pose a health risk. Therefore, there are established regulations of their permissible content in pet food and feed (especially for farm animals) [19,20]. Unfortunately, they still do not apply to the food for dogs and cats. The aim of this study was to investigate dry food for dogs and cats in respect to their microbiological quality and contamination by mycotoxins to estimate its purity and possible risk for companion animals' vigor.

2. Results

Pet food producers commonly apply cereals to dry pet food as fillers for croquets consistency improvement. According to the cereal additions' information, five groups of pet food samples: "maize", "maize and wheat", "wheat", "no cereals", and "no data" were distinguished. Dog food samples represented all five groups while only three groups were present for cat food because of the lack of samples containing solely wheat and no cereals. Each sample was screened for the presence of filamentous fungi as well as ergosterol content and main *Fusarium* mycotoxins were analyzed (Table 1). Five genera of fungi were identified: *Alternaria* sp., *Aspergillus* sp., *Cladosporium* sp., *Fusarium* sp., and *Penicillium* sp. *Penicillium* sp. was the most common genus contaminating food samples (38%), followed by *Fusarium* sp. (33%) (Figure 1). *F. verticillioides* was commonly observed among *Fusarium* population but *F. proliferatum* and *F. oxysporum* were also identified. *F. verticillioides* and *F. proliferatum* produce fumonisin B_1, which was detected in eleven samples out of 33 dry pet foods for dogs and cats, regardless of cereal component in the pet food, primarily in samples containing wheat or maize and wheat.

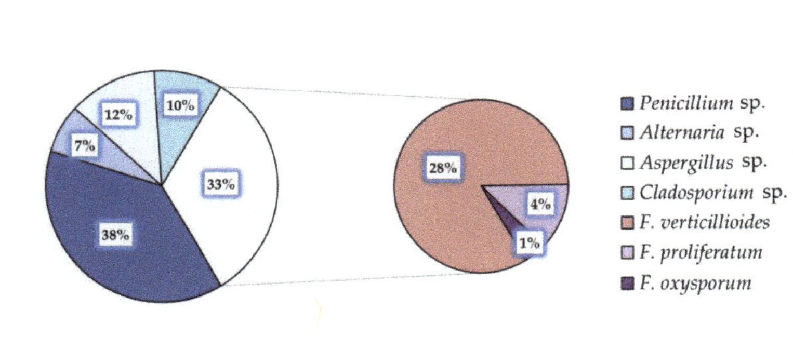

Figure 1. Frequencies of fungal genera and *Fusarium* species isolated and identified in dry pet food samples for cats and dogs.

Among all material collected, only two samples did not contain any fungal contaminant, however, the indicator of fungal biomass–ergosterol (ERG) was detected (Table 1). Both samples belonged to the dog food group, and one of them (dF59) still contained fungal mycotoxins. This indicates the need of supporting the mycotoxins investigation with microbiological screening of the samples.

The recommended maximum allowed concentration of ergosterol in grain is 7 µg/g [21]. In our study, ergosterol was detected in all investigated samples and the range of its concentration was 0.31–4.15 µg/g which means that the legal limit has not been exceeded.

Zearalenone, trichothecenes and fumonisin B_1 were also found but these mycotoxins were not present in all investigated samples. The minimum and maximum values of ZEN, DON, NIV, and FB_1 were 0.3–30.3, 22.2–618.4, 29.6–299.0, and 17.9–53.0 ng/g, respectively. Deoxynivalenol was the most abundant mycotoxin among all food samples (74%) and it also prevailed in dog food (77%), while zearalenone was dominating in cat food samples. The least frequent mycotoxin was FB_1, which is very interesting finding, considering the fact that the most frequent isolated species were *F. proliferatum* and *F. verticillioides*—main producers of fumonisin B_1. In addition, *Fusarium* representatives biosynthesizing trichothecenes and zearalenone, e.g., *F. culmorum*, *F. graminearum*, or *F. equiseti*, were not detected.

Table 1. Fungal species and genera isolated and identified in dog and cat food samples, along with ergosterol, zearalenone, deoxynivalenol, nivalenol and fumonisin B_1 concentrations measured in the respective samples. The samples were grouped based on their cereal component.

Pet	Sample No.	*Fusarium* sp.	Other Fungi	ERG [μg/g]	Mycotoxins			
					ZEN [ng/g]	DON [ng/g]	NIV [ng/g]	FB_1 [ng/g]
maize and wheat								
cat	cF1	Fv, Fp	P	0.73 ± 0.04	3.9 ± 0.3	55.3 ± 3.6	n.d.	17.9 ± 2.9
	cF2	Fv	P	1.04 ± 0.07	4.1 ± 0.4	104.5 ± 8.6	n.d.	n.d.
	cF4	Fv	P	4.14 ± 0.19	n.d.	22.2 ± 4.6	n.d.	n.d.
	cF15	Fv	P	1.75 ± 0.05	8.7 ± 1.5	53.0 ± 9.4	n.d.	n.d.
dog	dF36	–	P, C	0.88 ± 0.03	2.0 ± 0.5	n.d.	29.6 ± 4.0	n.d.
	dF39	Fv	P, Alt	1.36 ± 0.07	n.d.	n.d.	n.d.	28.9 ± 2.4
	dF40	Fo	P	0.43 ± 0.04	n.d.	n.d.	n.d.	n.d.
	dF42	–	P	0.85 ± 0.05	0.3 ± 0.1	97.3 ± 7.6	n.d.	n.d.
	dF45	Fv	P, C	1.32 ± 0.17	30.3 ± 2.6	303.4 ± 18.9	86.0 ± 13.6	53.0 ± 4.2
	dF46	Fv	P, A	1.19 ± 0.13	n.d.	279.6 ± 18.7	76.9 ± 3.7	46.9 ± 2.3
	dF47	–	P, A	0.50 ± 0.03	n.d.	76.3 ± 3.7	100.1 ± 14.1	n.d.
	dF58	–	P	0.41 ± 0.02	n.d.	129.9 ± 13.6	102.9 ± 9.4	n.d.
	dF59	–	–	0.83 ± 0.02	n.d.	24.6 ± 5.0	61.3 ± 3.2	n.d.
maize								
cat	cF12	Fv, Fp	P, C	0.85 ± 0.04	4.5 ± 0.8	n.d.	42.0 ± 3.1	n.d.
	cF13	–	P	2.04 ± 0.14	2.8 ± 0.2	217.5 ± 18.4	46.7 ± 4.1	n.d.
	cF14	Fv	–	4.05 ± 0.07	2.0 ± 0.2	n.d.	64.8 ± 4.2	12.5 ± 2.9
	cF29	Fv	P, C	1.09 ± 0.14	9.1 ± 2.1	n.d.	n.d.	n.d.
dog	dF33	Fv	P	2.95 ± 0.18	n.d.	554.3 ± 40.4	299.0 ± 25.9	31.9 ± 2.6
	dF51	Fv	P, A	1.18 ± 0.12	3.0 ± 0.3	99.4 ± 8.3	66.1 ± 4.8	33.5 ± 3.6
	dF52	Fv	P, A	0.83 ± 0.04	0.5 ± 0.0	114.7 ± 12.3	n.d.	n.d.
	dF53	Fv	P, A, Alt	1.18 ± 0.07	5.3 ± 0.3	n.d.	n.d.	52.4 ± 2.7
	dF54	–	C, P	0.31 ± 0.02	n.d.	203.1 ± 10.1	n.d.	n.d.
	dF56	Fv	P	0.74 ± 0.04	1.7 ± 0.48	300.1 ± 20.8	n.d.	n.d.
	dF57	–	A	1.58 ± 0.05	3.7 ± 0.3	195.1 ± 9.1	96.5 ± 8.2	n.d.
	dF60	–	P, C	0.78 ± 0.03	1.2 ± 0.3	307.8 ± 15.1	37.7 ± 3.6	n.d.
	dF80	–	A, Alt	1.05 ± 0.08	1.9 ± 0.2	n.d.	n.d.	n.d.
	dF81	–	P, Alt	1.02 ± 0.07	2.1 ± 0.2	36.8 ± 3.7	n.d.	n.d.
wheat								
dog	dF43	Fv	P	1.39 ± 0.12	n.d.	195.4 ± 13.5	n.d.	33.7 ± 3.8
	dF50	–	–	0.38 ± 0.02	n.d.	n.d.	n.d.	n.d.
	dF55	–	P	0.50 ± 0.03	n.d.	195.8 ± 8.8	n.d.	n.d.
no cereals								
dog	dF49	Fp	P, A	0.66 ± 0.04	n.d.	37.3 ± 5.4	n.d.	n.d.
	dF79	–	P, Alt	1.01 ± 0.04	n.d.	24.5 ± 3.4	n.d.	n.d.
no data								
cat	cF24	Fv	–	1.28 ± 0.16	27.2 ± 6.1	618.4 ± 35.1	87.2 ± 8.1	n.d.
	cF25	Fv	A	1.28 ± 0.13	17.2 ± 5.3	124.3 ± 8.2	265.8 ± 21.4	n.d.
	cF26	–	P, C	1.83 ± 0.19	1.6 ± 0.9	38.2 ± 6.4	31.6 ± 3.1	n.d.
	cF30	–	P	0.95 ± 0.06	23.4 ± 5.6	n.d.	n.d.	12.3 ± 2.1
dog	dF48	Fv	–	0.58 ± 0.02	1.2 ± 0.2	112.0 ± 17.3	67.3 ± 4.4	n.d.
	dF78	Fv	A	0.71 ± 0.02	2.2 ± 0.2	93.1 ± 7.7	n.d.	29.1 ± 3.4

cF–cat food, dF–dog food, Fv–F. verticillioides, Fp–F. proliferatum, Fo–F. oxysporum, P–Penicillium, C–Cladosporium, A–Aspergillus, Alt–Alternaria, n.d.–not detected.

Based on cereal composition, three groups were distinguished among cat food samples—food containing: maize and wheat, only maize and samples without composition data (Table 2). The highest mean concentration of ergosterol occurred in cat maize-containing food, and it reached 2.01 μg/g. In samples without the information about cereal composition, significantly different values of the three mycotoxins were observed compared to the samples with maize or wheat and maize. The following concentrations of zearalenone, deoxynivalenol and nivalenol were measured: 17.3, 260.3, and 128.2 ng/g, respectively. In samples containing maize and wheat, nivalenol was not detected. It is worth highlighting that the concentrations of fumonisin B_1 in cat food with maize were lower than in food containing maize and wheat. Similar trend was observed for dog food but probably relating to a larger set of samples rather than higher concentrations of FB_1 in food composed with maize and wheat comparing to food with pure maize.

Table 2. Content of ergosterol and *Fusarium* mycotoxins in cat food samples in relation to their cereal composition.

Cereal Composition	ERG [μg/g]		Mycotoxins [ng/g]							
			ZEN		DON		NIV		FB_1	
	Range	Mean	Range	Mean	Range	Mean	Range	Mean	Range	Mean
maize	0.81–4.12	2.01 ^{a}	1.9–11.3	4.6 ^{b}	196.4–230.2 *	217.5 ^{b}	43.0–69.0	51.2 ^{b}	10.0–15.6 *	12.5 ^{a}
maize and wheat	0.70–4.35	1.93 ^{a}	3.6–10.3	5.6 ^{b}	18.7–112.3	59.7 ^{b}	n.d.	n.d. ^{b}	15.0–20.8 *	17.9 ^{a}
no data	0.90–2.03	1.34 ^{a}	0.9–32.2	17.3 ^{a}	32.6–652.9	260.3 ^{a}	29.2–281.0	128.2 ^{a}	10.1–14.4 *	12.3 ^{a}
$LSD_{0.05}$		0.94		5.71		134.22		53.07		5.51
ANOVA *F*		0.301		<0.001		0.047		0.003		0.841

n.d–not detected; a, b–in columns, means followed by the same letters are not significantly different; * calculations based on one sample.

In contrast to cat food samples, in dog food the highest concentrations of ERG and mycotoxins were detected in the samples containing either maize or maize and wheat (Table 3). Examined material presented the highest mean values for trichothecenes (226.4 and 124.8 ng/g for DON and NIV, respectively) in samples containing maize. In turn, dog food with maize and wheat contained maximum concentrations of ZEN (10.9 ng/g) as well as FB_1 (42.9 ng/g). It is worth noting that dog food sample without cereals contained the lowest level of ergosterol (837.6 ng/g) and zearalenone, nivalenol and fumonisin B_1 were not detected. Only deoxynivalenol was present in a very low amount—30.9 ng/g.

Table 3. Content of ergosterol and *Fusarium* mycotoxins in dog food samples in relation to their cereal composition.

Cereal Composition	ERG [μg/g]		Mycotoxins [ng/g]							
			ZEN		DON		NIV		FB_1	
	Range	Mean	Range	Mean	Range	Mean	Range	Mean	Range	Mean
maize	0.29–3.13	1.16 ^{a}	0.5–5.6	2.4 ^{a}	32.8–593.1	226.4 ^{a}	33.6–325.5	124.8 ^{a}	29.5–55.5	39.3 ^{a}
maize and wheat	0.38–1.45	0.86 ^{ab}	0.2–32.6	10.9 ^{a}	19.3–324.6	151.9 ^{ab}	26.5–114.8	76.1 ^{a}	26.5–57.0	42.9 ^{a}
wheat	0.36–1.50	0.76 ^{ab}	n.d.	n.d.	184.7–210.5	195.6 ^{ab}	n.d.	n.d.	29.6–37.1 *	33.7 ^{a}
no cereals	0.62–1.06	0.84 ^{ab}	n.d.	n.d.	21.0–42.2	30.9 ^{b}	n.d.	n.d.	n.d.	n.d.
no data	0.56–0.72	0.65 ^{b}	1.0–2.4	1.7 ^{a}	86.2–127.1	102.5 ^{ab}	62.5–71.1 *	67.3 ^{a}	29.6–32.8	29.1 ^{a}
$LSD_{0.05}$		0.458		5.22		115.3		56.58		16.88
ANOVA *F*		0.048		0.441		0.043		0.123		0.572

n.d.–not detected; a, b–in columns, means followed by the same letters are not significantly different; * calculations based on one sample.

The comparison of food samples for dogs and cats distinguished two trends in mycotoxins distribution (Figure 2). In the first one, dog food samples with maize/maize and wheat revealed higher concentrations of all mycotoxins (except ZEN) than cat food samples with those cereals. The reverse trend was observed in the samples without cereal addition data—mycotoxin levels in cat food were higher comparing to dog food. The only exception was FB_1 which level was higher in dog food than in cat food just like in cereal-containing samples.

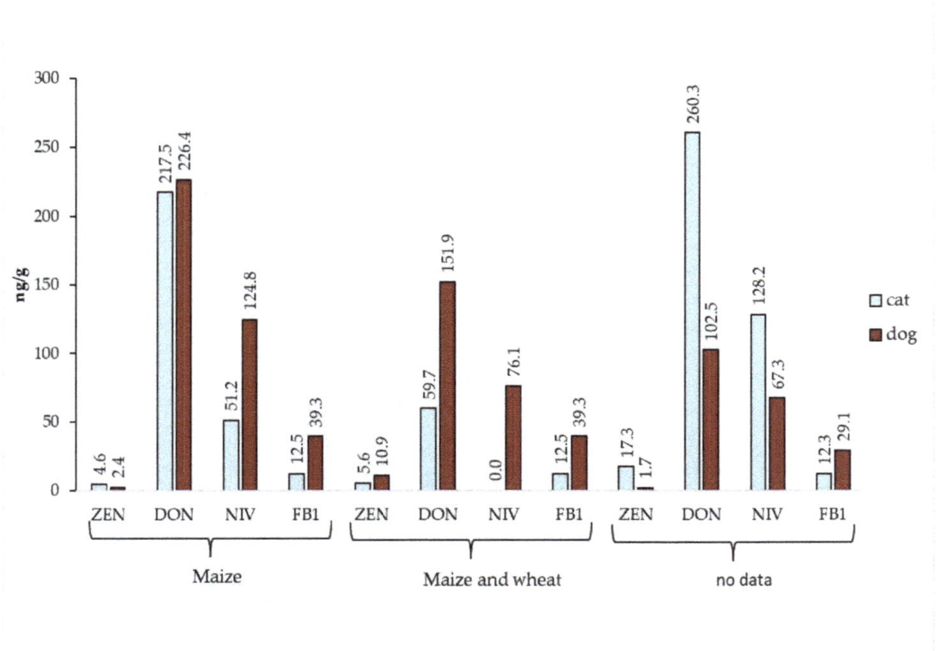

Figure 2. Comparison of *Fusarium* mycotoxins content in cat and dog food samples in relation to their cereal composition.

Only 58% of samples contained detectable *Fusarium* spp. among collected material, while *Fusarium* mycotoxins were detected in 94% of samples. We compared the samples in terms of *Fusarium* species content and their mycotoxins contribution (Figure 3). As expected, the concentrations of mycotoxins were higher in food samples containing *Fusarium* sp. than in food samples with other fungal contaminants. Only one mycotoxin (fumonisin B_1) was not detected in the sample with the absence of *Fusarium* sp.

So far, no mycotoxin limit values have been established for pet food. In 2006, the European Commission issued a recommendation regarding the permitted levels of mycotoxins in animal feed products [19,20]. The guidance values of DON, ZEN and FB_1 + FB_2 in feed with a moisture content of 12% were as follows: 0.9–12, 0.1–3 and 5–60 mg/kg, respectively. Based on these regulations we could conclude that none of the investigated samples exceed the recommended level of mycotoxins.

The analysis of correlation proved high significant correlations between ergosterol and mycotoxins content in dog food samples (with the exception of ZEN) (Table 4). Similarly, high correlations were detected in the case of ZEN/FB_1 as well as DON/NIV. NIV/FB_1 and DON/FB_1 showed significant and high significant correlations, respectively. In turn, in the case of cat food samples only one significant correlation was observed, i.e. between ERG and ZEN. Concentrations of DON and ZEN were strongly correlated.

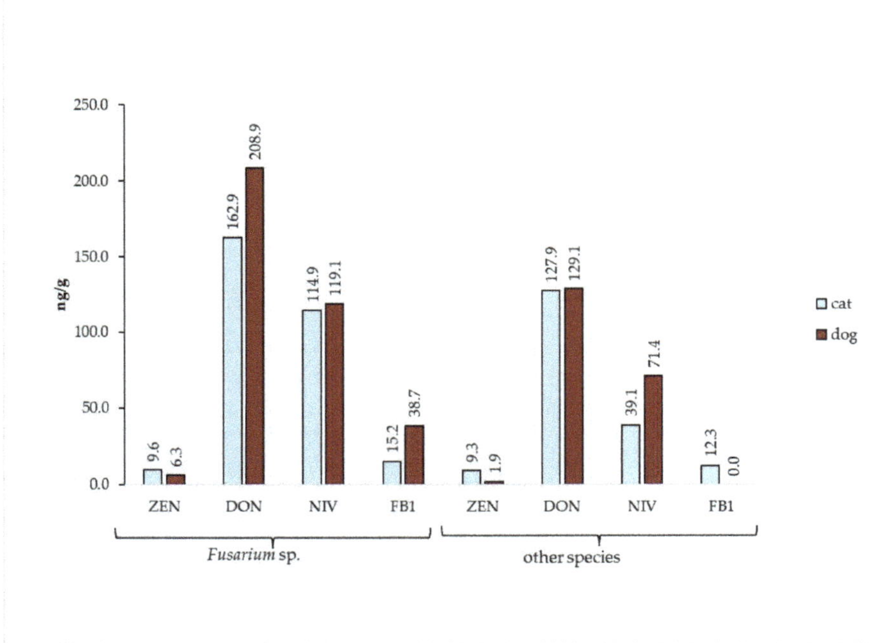

Figure 3. Comparison of *Fusarium* mycotoxins content in cat and dog food samples in relation to their *Fusarium* sp. content.

Table 4. Correlation coefficient factor between ergosterol and mycotoxins. Data highlighted in blue represent samples of cat food and in orange—dog food.

Trait	ERG	ZEN	DON	NIV	FB_1
ERG	1	−0.4239 *	−0.1611	−0.0423	0.0056
ZEN	0.1736	1	0.5604 ***	0.3533 *	0.0148
DON	0.5451 ***	0.2197	1	0.2924	−0.26
NIV	0.6633 ***	0.1228	0.6627 ***	1	−0.197
FB_1	0.5195 ***	0.4944 ***	0.2968 **	0.2458 *	1

* $p < 0.05$; ** $p < 0.01$; *** $p < 0.001$.

3. Discussion

Pet food industry is a developing area of world economy. According to the data reported by statista.com, there were about 8.5×10^4 of dogs and over 1×10^5 cats in the European Union in 2018 [4]. Different types of pet food are available on the market but dry and canned food are prevailing. In canned food the main ingredient is meat while dry food contains mainly cereal grains. Additives of cereals (mainly rice, wheat and maize) improve pulp consistency and facilitate formulation of croquettes. Unfortunately, most cereals used for pet food production contain pathogens belonging to the *Fusarium* genus, which are responsible for biosynthesis of strongly toxic secondary metabolites (mycotoxins).

Mycotoxins are present in pet food since these compounds have stable chemical structures and are insensitive to most manufacturing processes like cleaning, milling, and boiling [15]. During pet food production manufacturers apply mainly heating and extrusion processes and several studies were conducted concerning the influence of food processing on mycotoxins' concentrations. It has been shown that high temperature baking as well as extrusion are the most relevant in the case of mycotoxins content reduction, however, the effectiveness of these processes depends on conditions used, like temperature, duration of the process, or humidity. The decrease of mycotoxins concentrations during different types and conditions of food processing ranged between 20–65%, 23–92%, 24–70% for FB_1, ZEN, and DON, respectively [22]. Taking into account the degradation of mycotoxins during extrusion, it is possible that the concentrations of mycotoxins included in the investigated samples are lower than in the grain used for pet food production. Simultaneously, it can be stated that efficiency of these processes is not satisfactory in the light of fast progressing climate change. Forecast for Europe predicts climate warming and rising humidity, thus, better conditions for *Fusarium* sp. development and, eventually, extensive mycotoxin contamination and reduction of yield and quality are expected [22–25].

Mycotoxigenic fungal genera were identified in tested pet food samples. Both storage genera, like *Penicillium* sp. (38%) and *Aspergillus* sp. (12%), as well as plant-derived *Fusarium* sp. (33%) were identified. This result confirms the findings reported by previously published papers [26–29]. Interesting observation is the presence of *Fusarium* sp. responsible for biosynthesis of fumonisin B_1 (*F. proliferatum, F. verticillioides*) only, while zearalenone and trichothecenes produced mainly by *F. graminearum* and *F. culmorum* were also detected. Possible explanation of this phenomenon is the fact that *F. graminearum* and *F. culmorum* spores and hyphae are more sensitive to high temperatures and were inactivated during the manufacturing process. It is worth emphasizing that *Aspergillus* sp. was identified in collected material. This might result in the presence of aflatoxins, so the analysis of these highly harmful mycotoxins would bring the broader context into mycotoxicological research of pet food.

Most studies on pet food contamination with mycotoxins were focused on single group or a narrow range of mycotoxins [11,30–35], but currently the studies are focused on comprehensive approach, as demonstrated by Bohm et al. [11] and Gazzotti et al. [33], who analyzed *Fusarium* mycotoxins: Fumonisin B_1, deoxynivalenol, and zearalenone [34,35]. In our research FB_1, DON, and NIV, as well as ZEN, were quantified in both dogs' and cats' food. The presence of mycotoxins of interest in feed material was expected and the most frequent compounds were DON and ZEN. The surprising result is the higher concentrations of FB1 observed in both cat and dog food containing maize and wheat than in those with maize only, which is regarded a main vector for this mycotoxin. In general, wheat is considered free of fumonisins because it is not a host-plant for fumonisins-producing *Fusarium* species (e.g., *F. proliferatum, F. verticillioides*), but some recent reports revealed this mycotoxin in wheat and wheat by-products [36,37]. According to a study conducted by Chehri et al. [38] *F. verticillioides* was not only detected in wheat grain samples but it was also the second most frequent species after *F. graminearum*. Nevertheless, concentration values of investigated mycotoxins in pet food were below the acceptable levels recommended by the EU regulations. Our results were in accordance with the data obtained during a long-term international research on mycotoxins in feed [25]. According to these authors up to 88% of samples were contaminated by at least one of the mycotoxins but in Europe feed contamination did not exceed the permitted level. The same study indicates the *Fusarium* mycotoxins—DON and ZEN—as one of the most significant feed contaminants.

Tegzes et al. tested 60 samples of grain and grain-free dog food for aflatoxins, ochratoxin A as well as *Fusarium* mycotoxins contamination [34]. On the contrary to the previous studies, aflatoxins and ochratoxin A were not detected, which has been explained by the effectiveness of quality management regulations. Furthermore, it might be concluded that cereal additives are the main source of mycotoxins contamination because grain-free samples were not contaminated by these compounds. Our study reveals that dog and cat foods without cereals are generally free of mycotoxins; however, very low levels of DON (21.0–42.2 ng/g) were detected.

Recent studies increasingly indicated that high resolution mass spectrometry is a powerful tool in a large scale of mycotoxins screening as well as food and feed comprehensive analyses [35,39]. Castaldo et al. examined 89 pet food samples, which then were subjected to a targeted analysis of 28 most common mycotoxins, i.e., *Fusarium* mycotoxins, aflatoxins and ochratoxin A [35]. Then, a retrospective analysis was performed resulting in the detection of 245 different toxic bacterial and fungal metabolites, among which still *Fusarium* mycotoxins were the most common. Another problem which has been highlighted by this paper, is the presence of about 16 toxins per sample and their possible implication on animal health [25]. The interactions between mycotoxins can be synergistic, additive, less than additive and antagonistic. The possible implications of mycotoxins' synergism were summarized in extensive review article by Greiner and Oswald [40]. Results of a large-scale research by Gruber-Dorninger et al. proved that the presence of few different mycotoxins is a general rule so the risk of affection of human and animals' health's through synergistic action of different mycotoxins seems to be high [25]. Our studies revealed the co-occurrence of *Fusarium* mycotoxins as well, and it seems to be natural because of the ability of species to produce more than one mycotoxin (for example *F. proliferatum* is able to produce fumonisins, moniliformin, and beauvericin) [41]. Moreover, food and its by-product might be contaminated by different fungi during product lifetime (on the field, during storage) and fungal secondary metabolites can be accumulated in it. At this moment, a global list of feed and food contaminants as well as cumulative effect of these toxins is unknown, hence this is the future perspective in feed and food research. Both studies show the results similar to those obtained in the present study, thereby confirming the hypothesis that *Fusarium* mycotoxins are present in pet food and threaten pets' health. Moreover, it has been proven that the topic of food contamination by these mycotoxins is still alive and no crucial solution has been applied so far. Despite the fact that the concentrations of chosen mycotoxins did not exceed the permitted levels established by EU, it does not mean that pets' health is not in danger. Dry food is mostly a basis of pet diet and even extremely low doses of mycotoxins might result in chronic diseases when consumed daily. Mycotoxins could be ingested by the animal with contaminated food and accumulate in their tissues or excrete with body fluid, hence, meat, milk, and eggs might be also potentially an origin of mycotoxins in diet [42–45]. It might be concluded that multi-exposure to mycotoxins is an inevitable phenomenon.

Our previous study was dedicated to veterinary diets for dogs and cats. It has shown that these products contain molds as well as *Fusarium* mycotoxins [46]. Detected levels of DON, ZEN, and FBs were much below FDA recommendations but in this particular case it does not guarantee safety. Veterinary diet is a special purpose food which is applied during the treatment of sick animal, possibly more susceptible to harmful mycotoxins. This suggests that the issue covers also other pet food products and, hence, research should be extended to include other types of pet food.

4. Conclusions

After years of research on *Fusarium* mycotoxins in food and feed—especially pet food—the knowledge is still insufficient. It is necessary to find the effective solutions on how to reduce mycotoxins in feed product to protect human and animal lives. Limitation of grain used for pet food production might significantly reduce *Fusarium* mycotoxins presence. On the other hand, introduction of new standards in quality assurance based on high resolution mass spectrometry techniques gives a possibility of comprehensive mycotoxin detection as well as their by-products, and effective monitoring and management of mycotoxin contamination.

5. Materials and Methods

Studied material was collected from hermetically sealed packages of dry food for dogs and cats of 16 producers. They were obtained from pet shops and veterinary clinics in Poland. In total, the collected research material was represented by 38 samples including 12 foods for cats and 26 for dogs. The weights of the samples ranged from 60 to 150 g and the moisture was from 3.8% to 7.1%. Cereals composition was used as division criterion and the following groups were made: (i) Samples containing only maize, (ii) wheat, (iii) maize and wheat, and (iv) none of these cereals. According to this criterion we obtained 3 groups of cat food and 5 groups of dog food (Table 1). Samples of food formulated in differently-sized croquettes were homogenized before the analysis using liquid nitrogen and ceramic mortar, and then milled. Samples were stored at −20 °C until analysis (Figure 4).

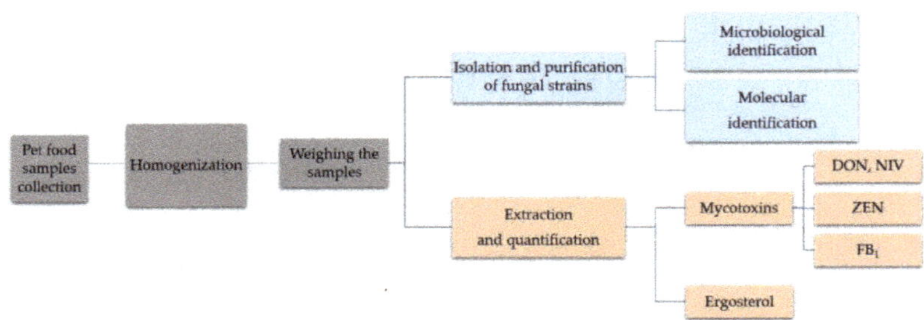

Figure 4. Scheme of sampling of cat and dog food and design of the experiment.

5.1. *Identification of Fungal Isolates*

Identification of fungi was performed using microscopic and molecular methods. Genera of fungi were determined on morphological basis and isolates belonging to *Fusarium* genus were then subjected to molecular analyses to obtain species identification.

Two microbiological media: Synthetic nutrient agar (SNA) and potato dextrose agar (PDA) were used for microbiological analysis. First, about 1 g of milled food samples was incubated for one week at room temperature on the PDA medium. Pure fungal isolates were obtained through multiple passaging onto fresh PDA media and then transferred to the SNA medium. The plates were incubated at room temperature for 4 weeks until the appearance of spores. Macrospores were viewed microscopically at 200× magnification and their morphological features were used for genus identification.

Molecular methods were applied only for *Fusarium* isolates. At first, the mycelia were frozen in liquid nitrogen and homogenized using plastic mortar. Then, genomic DNA was extracted according to the CTAB method [47].

The next step was setting up the polymerase chain reaction (PCR) which was performed using Tef1R and Ef728M primers and Taq DNA polymerase (Thermo, Waltham, MA, USA). The quality of PCR products was evaluated based on electrophoretic separation. Subsequently, PCR products were purified with exonuclease I (Epicentre, Madison, WI, USA) and alkaline phosphatase SAP (Promega, Madison, WI, USA). Fluorescent labeling was performed using Ef728M primer and BigDyeTerminator v 3.1 Cycle Sequencing Kit (Applied Biosystems, Foster City, CA, USA).

Sequence reading was done by Laboratory of DNA Sequencing and Oligonucleotide Synthesis at Institute of Biochemistry and Biophysics, Polish Academy of Sciences, Warsaw, Poland. Molecular identification of fungal species was done by comparing the sequences to the reference sequences deposited in the NCBI GenBank database using the BLASTn algorithm. The method was described in detail by Witaszak et al. [46].

5.2. *Standards and Chemical Reagents*

Ergosterol (ERG) and mycotoxins (fumonisin B_1—FB_1, zearalenone—ZEN, deoxynivalenol—DON, and nivalenol—NIV) standards were purchased with a standard grade certificate (purity above 98%) from Sigma-Aldrich (Steinheim, Germany). Stock solutions of standards were prepared in acetonitrile (except ERG in methanol) at 1.0 mg/mL concentrations and stored at −20 °C. Sodium dihydrophosphate, potassium hydroxide, sodium hydroxide, potassium chloride, acetic acid, hydrochloric acid, and o-phosphoric acid were purchased from POCh (Gliwice, Poland). Organic solvents (HPLC grade), disodium tetraborate, n-pentane, 2-mercaptoethanol, sodium acetate, and all the other chemicals were also purchased from Sigma-Aldrich (Steinheim, Germany). Water for the HPLC mobile phase was purified using a Milli-Q system (Millipore, Bedford, MA, USA).

5.3. Extraction and Purification Procedure

5.3.1. Ergosterol

Samples (each in triplicate) containing 100 mg of ground material were suspended in 2 mL methanol in a culture tube and treated with 0.5 mL of 2 M aqueous sodium hydroxide. Samples were irradiated twice in a microwave oven (370 W) for 20 s according to Waśkiewicz et al. [48]. After 15 min contents of cultures tubes were neutralized with 1 M aqueous hydrochloric acid, then 2 mL of methanol were added and samples were extracted with pentane (3 × 4 mL). The combined pentane extracts were evaporated to dryness in a stream of nitrogen, before analysis dissolved in 1 mL of methanol and 10 μL of thus prepared mixture were analyzed by HPLC.

5.3.2. Mycotoxins

Ground samples (10 g) were soaked with 30 mL of acetonitrile:water (80:20, *v*/*v*) and mycotoxins (DON, NIV, ZEN) were extracted overnight. After filtration (Whatman No. 5 paper, Whatman International Ltd, Maidstone, UK), 10 mL of the extract was collected and diluted with 40 mL of water. Half of the extract was used for DON and NIV analysis, while the second part was used for ZEN determination. In both cases, the extract was applied on top of an immunoaffinity DON-NIV column and the Zearala Test column (IAC, Vicam, Milford, MA, USA) for DON/NIV and ZEN analysis, respectively in accordance with the manufacturer's procedure.

Fumonisin B_1 was extracted from ground samples of veterinary diet (10 g) by homogenization for 3 min in 30 mL of methanol-water (3:1, *v*/*v*) and filtration through Whatman no. 4 filter paper (Whatman International Ltd, Maidstone, UK) according to Stępień et al. [49]. The filtered extract (pH = 5.8–6.3) was applied at the top of the conditioned cartridge (a SAX cartridge, Bond Elut, Agilent Santa Clara, CA, USA) with flow rate of 2 mL/min. Fumonisin B_1 was eluted from the column with 10 mL of 1% acetic acid in methanol. The elute was evaporated to dryness at 40 °C under a stream of nitrogen. Dry residue was stored at −20 °C until HPLC analyses.

5.4. HPLC Analysis

The chromatographic system consisted of Waters 2695 high-performance liquid chromatograph (Waters, Milford, USA) with detectors: (i) Waters 2996 Photodiode Array Detector with Nova Pak C−18 column (150 × 3.9 mm) and methanol:acetonitrile (90:10, *v*/*v*) as a mobile phase for ERG (λ_{max} = 282 nm) analysis and with Nova Pak C−18 column (300 × 3.9 mm) and methanol:water (25:75, *v*/*v*) as a mobile phase for DON and NIV analysis (λ_{max} = 224 nm); (ii) Waters 2475 Multi λ Fluorescence Detector (λ_{ex} = 274 nm, λ_{em} = 440 nm) and Waters 2996 Photodiode Array Detector with Nova Pak C−18 column (150 × 3.9 mm) and acetonitrile:water:methanol (46:46:8, *v*/*v*/*v*) as a mobile phase for ZEN analysis; (iii) Waters 2475 Multi λ Fluorescence Detector (λ_{ex} = 335 nm, λ_{em} = 440 nm) with an XBridge column (3.0 × 100 mm) and methanol:sodium dihydrogen phosphate (0.1 M in water) solution (77:23, *v*/*v*) adjusted to pH 3.35 with o-phosphoric acid as a mobile phase for FB_1 analysis after pre-column derivatization with *o*-phthaldialdehyde (OPA) reagent.

The linearity of the standard curves was 0.9897, 0.9991, 0.9990, 0.9989, and 0.9982 for FB_1, ZEN, DON, NIV, and ERG, respectively. The limits of quantification (LOQ) were (in ng/g): 1.0 for ZEN, 2.0 for FB_1 and 10.0 for DON, NIV, and ERG. The recovery experiment was performed on mycotoxin-free veterinary diet samples, spiked with three different levels of each mycotoxin separately at a concentration of 5.0, 8.0, and 10.0 ng/g for FB_1 and ZEN and 20.0, 40.0, and 80 ng/g for DON and NIV. On the basis of these experiments, recovery rates and standard deviations (RSD) were calculated. For the majority of the tested mycotoxins, the recovery rates ranged between 88% and 109% (depending on the matrix), whereas the RSD values did not exceed 13%.

5.5. Statistical Analysis

Firstly, the normality of distributions for studied traits was tested using the Shapiro–Wilk normality test [50]. One-way analysis of variance (ANOVA) was carried out to determine the effects of cereal composition on the variability of ergosterol (ERG), and mycotoxins: zearalenone (ZEN), deoxynivalenol (DON), nivalenol (NIV), and fumonisin B_1 (FB_1), independently for cat and dog food samples. Minimal, maximal and mean values of individual traits were calculated. The Fisher's least significant differences (LSDs) were calculated for individual characteristics and on this basis homogeneous groups were determined. The relationships between the ERG, FB_1, ZEN, DON, and NIV were determined using the simple correlation analysis. Relationships of five observed traits were presented in Table 4. Data analysis was performed with the GenStat 18 package software for cat and dog food samples independently.

Author Contributions: N.W., Ł.S. and A.W. conceived and designed the experiments; N.W. performed the experiments; N.W., Ł.S. and A.W. analyzed the data; J.B. performed statistical analyses; N.W., Ł.S., and A.W. wrote the paper. All authors have read and agreed to the published version of the manuscript.

Funding: This research received no external funding.

Conflicts of Interest: The authors declare no conflict of interest.

References

1. Arahori, M.; Huroshima, H.; Hori, Y.; Takagi, S.; Chijiiwa, H.; Fujita, K. Owners' view of their pets; emotions, intellect and mutual relationship: Cats and dogs compared. *Behav. Process.* **2017**, *141*, 316–321. [CrossRef] [PubMed]
2. Zilcha-Mano, S.; Mikulincer, M.; Shaver, P.R. Pet in the therapy room: An attachment perspective on Animal-Assisted Therapy. *Attach Hum. Dev.* **2011**, *13*, 541–546. [CrossRef] [PubMed]
3. Edenburg, N.; van Lith, H.A. The influence of animals on the development of children. *Vet. J.* **2011**, *190*, 208–214. [CrossRef] [PubMed]
4. Statista. Statista GmbH, Johannes-Brahms-Platz 1, 20355 Hamburg, Germany. 2020. Available online: https://www.statista.com (accessed on 20 January 2020).
5. Plantinga, E.A.; Bosch, G.; Hendriks, W.H. Estimation of the dietary nutrient profile of free roaming feral cats: Possible implications for nutrition of domestic cats. *Br. J. Nutr.* **2011**, *106*, 835–848. [CrossRef]
6. Axelsson, E.; Ratnakumar, A.; Arendt, M.L.; Magbool, K.; Webster, M.T.; Perloski, M.; Liberg, O.; Arnemo, J.M.; Hedhammar, A.; Lidblad-Toh, K. The genomic signature of dog domestication reveals adaptation to a starch-rich diet. *Nature* **2013**, *495*, 360–364. [CrossRef]
7. Di Cerbo, A.; Morales-Medina, J.C.; Palmieri, B.; Pezzuto, F.; Cocco, R.; Flores, G.; Iannitti, T. Functional foods in pet nutrition: Focus on dogs and cats. *Res. Vet. Sci.* **2017**, *112*, 161–166. [CrossRef]
8. Buffington, C.A.T. Dry foods and risk of disease in cats. *Can. Vet. J.* **2008**, *49*, 561–563. Available online: https://www.ncbi.nlm.nih.gov/pmc/articles/PMC2387258/ (accessed on 20 January 2020).
9. Desjardins, A.E.; Proctor, R.H. Molecular biology of *Fusarium* mycotoxins. *Int. J. Food Microbiol.* **2007**, *119*, 47–50. [CrossRef]
10. European Food Safety Authority (EFSA) Scientific opinion. Risk to human and animal health related to the presence of deoxynivalenol and its acetylated and modified forms in food and feed. *EFSA J.* **2017**, *15*, 4718. [CrossRef]
11. Böhm, J.; Koinig, L.; Razzazi-Fazeli, E.; Błajet-Kosicka, A.; Twarużek, M.; Grajewski, J.; Lang, C. Survey and risk assessment of the mycotoxins deoxynivalenol, zearalenone, fumonisins, ochratoxin A, and aflatoxins in commercial dry dog food. *Mycotoxin Res.* **2010**, *26*, 147–153. [CrossRef]
12. Hughes, D.M.; Gahl, M.J.; Graham, C.H.; Grieb, S.L. Overt signs of toxicity to dogs and cats of dietary deoxynivalenol. *J. Anim. Sci.* **1999**, *77*, 693–700. Available online: https://www.ncbi.nlm.nih.gov/pubmed/10229366 (accessed on 20 January 2020).
13. Boermans, H.J.; Leung, M.C.K. Mycotoxins and the pet food industry: Toxicological evidence and risk assessment. *Inter. J. Food Microbiol.* **2007**, *119*, 95–102. [CrossRef]

14. Zicker, S.C. Evaluating pet foods: How confident are you when you recommend a commercial pet food? *Top. Companion Anim. Med.* **2008**, *23*, 121–126. [CrossRef]
15. Bullerman, L.B.; Bianchini, A. Stability of mycotoxins during food processing. *Int. J. Food Microbiol.* **2007**, *119*, 140–146. [CrossRef]
16. Richard, J.L.; Payne, G.A. Mycotoxins: Risk in plant, animal, and human systems. In *Council for Agricultural Science and Technology Task Force Report No. 139*; Council for Agricultural: Ames, IA, USA, 2003.
17. *Risk Assessment for Mycotoxins in Human and Animal Food Chains*; Summary Report; French Food Safety Agency: Maisons-Alfort, France, 2006.
18. Milićević, D.R.; Skrinjar, M.; Baltić, T. Real and perceived risks for mycotoxin contamination in foods and feeds: Challenges for food safety control. *Toxins* **2010**, *2*, 572–592. [CrossRef]
19. Commission Recommendation of 17 August 2006 on the Prevention and Reduction of Fusarium Toxins in Cereals and Cereal Products (2006/583/EC). Available online: https://eur-lex.europa.eu/LexUriServ/LexUriServ.do?uri=OJ:L:2006:234:0035:0040:EN:PDF (accessed on 20 January 2020).
20. Commission Regulation (EC) No 1126/2007 of 28 September 2007 Amending Regulation (EC) No 1881/2006 Setting Maximum Levels for Certain Contaminants in Foodstuffs as Regards Fusarium Toxins in Maize and Maize Products. Available online: https://www.fsai.ie/uploadedFiles/Commission_Regulation_EC_No_1126_2007.pdf (accessed on 20 January 2020).
21. Gutarowska, B.; Zakowska, Z. Estimation of fungal contamination of various plant materials with UV-determination of fungal ergosterol. *Ann. Microbiol.* **2010**, *60*, 415–422. [CrossRef]
22. Milani, J.; Maleki, G. Effects of processing on mycotoxin stability in cereals. *J. Sci. Food Agric.* **2014**, *94*, 2372–2375. [CrossRef]
23. Moretti, A.; Pascale, M.; Logrieco, A.F. Mycotoxin risks under a climate change scenario in Europe. *Trends Food Sci. Tech.* **2019**, *84*, 38–40. [CrossRef]
24. Vaughan, M.M.; Backhouse, D.; Del Ponte, E.M. Climate change impacts on the ecology of *Fusarium graminearum* species complex and ausceptibility of wheat to Fusarium head blight: A review. *World Mycotoxin J.* **2016**, *9*, 1–16. [CrossRef]
25. Gruber-Dorninger, C.; Jenkins, T.; Schatzmayr, G. Global Mycotoxin Occurrence in Feed: A Ten-Year Survey. *Toxins* **2019**, *11*, 375. [CrossRef]
26. Martins, M.L.; Martins, H.M.; Bernardo, F. Fungal flora and mycotoxins detection in commercial pet food. *Rev. Port. Cienc. Vet.* **2003**, *98*, 179–183. Available online: http://www.fmv.ulisboa.pt/spcv/PDF/pdf12_2003/548_179_183.pdf (accessed on 20 January 2020).
27. Błajet-Kosicka, A.; Kosicki, R.; Twarużek, M.; Grajewski, J. Determination of moulds and mycotoxins in dry dog and cat food using liquid chromatography with mass spectrometry and fluorescence detection. *Food Addit. Contam. Part B Surveill.* **2014**, *7*, 302–308. Available online: https://www.ncbi.nlm.nih.gov/pubmed/24919718 (accessed on 20 January 2020). [CrossRef]
28. Bueno, D.J.; Silva, J.O.; Oliver, G. Mycoflora in commercial pet foods. *J. Food Prot.* **2001**, *64*, 741–743. Available online: https://www.ncbi.nlm.nih.gov/pubmed/11348013 (accessed on 20 January 2020). [CrossRef]
29. Campos, S.G.; Cavaglieri, L.R.; Fernández Juri, M.G.; Dalcero, A.M.; Krüger, C.; Keller, L.A.M.; Rosa, C.A.R. Mycobiota and aflatoxins in raw materials and pet food in Brazil. *J. Anim. Physiol. Anim. Nutr.* **2008**, *92*, 377–383. Available online: https://www.ncbi.nlm.nih.gov/pubmed/18477320 (accessed on 20 January 2020). [CrossRef]
30. Hołda, K.; Głogowski, R. A survey of deoxynivalenol and zearalenone content in commercial dry foods for growing dogs. *Ann. Wars. Univ. Life Sci. SGGW Anim. Sci.* **2014**, *53*, 111–117. Available online: http://annals-wuls.sggw.pl/?q=node/729 (accessed on 20 January 2020).
31. Songsermsakul, P.; Razzazi-Fazeli, E.; Böhm, J.; Zentek, J. Occurrence of deoxynivalenol (DON) and ochratoxin A (OTA) in dog foods. *Mycotoxin Res.* **2007**, *23*, 65–67. [CrossRef]
32. Zwierzchowski, W.; Gajęcki, M.; Obremski, K.; Zielonka, Ł.; Baranowski, M. The occurrence of zearalenone and its derivatives in standard and therapeutic feeds for companion animals. *Pol. J. Vet. Sci.* **2004**, *7*, 289–293. Available online: https://www.ncbi.nlm.nih.gov/pubmed/15633789 (accessed on 20 January 2020).
33. Gazzotti, T.; Biagi, G.; Pagliuca, G.; Pinna, C.; Scardilli, M.; Grandi, M.; Zaghini, G. Occurrence of mycotoxins in extruded commercial dog food. *Anim. Feed Sci. Technol.* **2015**, *202*, 81–89. [CrossRef]
34. Tegzes, J.H.; Oakley, B.B.; Brennan, G. Comparison of mycotoxin concentrations in grain versus grain-free dry and wet commercial dog food. *Toxicol. Commun.* **2019**, *3*, 61–66. [CrossRef]

35. Castaldo, L.; Graziani, G.; Gaspari, A.; Izzo, L.; Tolosa, J.; Rodriguez-Carrasco, Y.; Ritieni, A. Target Analysis and Retrospective Screening of Multiple Mycotoxins in Pet Food Using UHPLC-Q-Orbitrap HRMS. *Toxins* **2019**, *11*, 434. [CrossRef]
36. Shephard, G.S.; van der Westhuizen, L.; Gatyeni, P.M.; Katerere, D.R.; Marasas, W.F.O. Do fumonisin mycotoxins occur in wheat? *J. Agric. Food Chem.* **2005**, *53*, 9293–9296. [CrossRef]
37. Cendoya, E.; Chiotta, M.L.; Zachetti, V.; Chulze, S.N.; Ramirez, M.L. Fumonisins and fumonisin-producing *Fusarium* occurrence in wheat and wheat by products: A review. *J. Cereal Sci.* **2018**, *80*, 158–166. [CrossRef]
38. Chehri, K.; Jahromi, S.T.; Reddy, K.R.N.; Abbasi, S.; Salleh, B. Occurrence of *Fusarium* spp. And fumonisins in stored wheat grains marketed in Iran. *Toxins* **2010**, *2*, 2816–2823. [CrossRef]
39. Stadler, D.; Berthiller, F.; Suman, M.; Schuhmacher, R.; Krska, R. Novel analytical methods to study the fate of mycotoxins during thermal processing. *Anal. Bioanal. Chem.* **2019**. [CrossRef]
40. Grainer, B.; Oswald, I. Mycotoxin co-contamination of food and feed: Meta-analysis of publications describing toxicological interactions. *World Mycotoxin J.* **2011**, *4*, 285–313. [CrossRef]
41. Abbas, H.K.; Cartwright, R.D.; Xie, W.; Mirocha, C.J.; Richard, J.L.; Dvorak, T.J.; Sciumbato, G.L.; Shier, W.T. Mycotoxin production by *Fusarium proliferatum* isolates from rice with Fusarium sheath rot disease. *Mycopathologia* **1999**, *147*, 97–104. [CrossRef]
42. Hof, H. Mycotoxins in milk for human nutrition: Cow, sheep and human breast milk. *GMS Infect. Dis.* **2016**, *4*. [CrossRef]
43. Wang, L.; Zhang, Q.; Yan, Z.; Tan, Y.; Zhu, R.; Yu, D.; Yang, H.; Wu, A. Occurrence and quantitative risk assessment of twelve mycotoxins in eggs and chicken tissues in China. *Toxins* **2018**, *10*, 477. [CrossRef]
44. Bailly, J.D.; Guerre, P. Mycotoxins in meat and processed meat products. In *Safety of Meat and Processed Meat. Food Microbiology and Food Safety*; Toldrá, F., Ed.; Springer: New York, NY, USA, 2009; pp. 83–124.
45. Tomczyk, Ł.; Stępień, Ł.; Urbaniak, M.; Szablewski, T.; Cegielska-Radziejewska, R.; Stuper-Szablewska, K. Characterisation of the Mycobiota on the Shell Surface of Table Eggs Acquired from Different Egg-Laying Hen Breeding Systems. *Toxins* **2018**, *10*, 293. [CrossRef]
46. Witaszak, N.; Stępień, Ł.; Bocianowski, J.; Waśkiewicz, A. *Fusarium* Species and Mycotoxins Containing Veterinary Diets for Dogs and Cats. *Microorganisms* **2019**, *7*, 26. [CrossRef]
47. Stępień, Ł.; Jestoi, M.; Chełkowski, J. Cyclic hexadepsipeptides in wheat field samples and *esyn1* gene divergence among enniatin producing *Fusarium avenaceum* strains. *World Mycotoxin J.* **2013**, *6*, 399–409. [CrossRef]
48. Waśkiewicz, A.; Morkunas, I.; Bednarski, W.; Mai, V.C.; Formela, M.; Beszterda, M.; Wiśniewska, H.; Goliński, P. Deoxynivalenol and oxidative stress indicators in winter wheat inoculated with *Fusarium graminearum*. *Toxins* **2014**, *6*, 575–591. [CrossRef] [PubMed]
49. Stępień, Ł.; Koczyk, G.; Waśkiewicz, A. FUM cluster divergence in fumonisins-producing *Fusarium* species. *Fungal Biol.* **2011**, *115*, 112–123. [CrossRef] [PubMed]
50. Shapiro, S.S.; Wilk, M.B. An analysis of variance test for normality (complete samples). *Biometrika* **1965**, *52*, 591–611. [CrossRef]

© 2020 by the authors. Licensee MDPI, Basel, Switzerland. This article is an open access article distributed under the terms and conditions of the Creative Commons Attribution (CC BY) license (http://creativecommons.org/licenses/by/4.0/).

Article

The Effect of Using New Synbiotics on the Turkey Performance, the Intestinal Microbiota and the Fecal Enzymes Activity in Turkeys Fed Ochratoxin A Contaminated Feed

Katarzyna Śliżewska [1,*], Paulina Markowiak-Kopeć [1,*], Anna Sip [2], Krzysztof Lipiński [3] and Magdalena Mazur-Kuśnirek [3]

[1] Institute of Fermentation Technology and Microbiology, Department of Biotechnology and Food Sciences, Lodz University of Technology, Wólczańska 171/173, 90-924 Łódź, Poland

[2] Department of Biotechnology and Food Microbiology, Poznan University of Life Sciences, Wojska Polskiego 48, 60-627 Poznań, Poland; anna.sip@up.poznan.pl

[3] Department of Animal Nutrition and Feed Science, University of Warmia and Mazury, Oczapowskiego 5/248, 10-719 Olsztyn, Poland; krzysztof.lipinski@uwm.edu.pl (K.L.); magdalena.mazur@uwm.edu.pl (M.M.-K.)

* Correspondence: katarzyna.slizewska@p.lodz.pl (K.Ś.); paulina.markowiak-kopec@edu.p.lodz.pl (P.M.-K.)

Received: 24 August 2020; Accepted: 7 September 2020; Published: 9 September 2020

Abstract: The feed supplementation of probiotic microorganisms is a promising method for detoxification of ochratoxin A (OTA) in poultry. The aim of the study was to investigate the effect of newly elaborated synbiotics on the turkey performance, the intestinal microbiota and its enzymatic activity in turkeys (0–15 weeks) fed OTA contaminated feed (198.6–462.0 μg/kg) compared to control group (OTA-free feed). The studies determined the composition of intestinal microorganisms by the culture method and the activity of fecal enzymes by spectrophotometry. It was found that OTA had an adverse effect on the body weight, the intestinal microbiota and the fecal enzymes activity in turkeys. On the other hand, synbiotics resulted in an increase in the count of beneficial bacteria while reducing the number of potential pathogens in the digestive tract. Moreover, synbiotics caused an increase in the activity of α-glucosidase and α-galactosidase, while decreasing the activity of potentially harmful fecal enzymes (β-glucosidase, β-galactosidase, β-glucuronidase) in the turkey's excreta. Results indicate a beneficial effect of elaborated synbiotics on the health of turkeys and a reduction of the negative impact of OTA contaminated feed. These synbiotics can be successfully used as feed additives for turkeys.

Keywords: synbiotics; turkeys; intestinal microbiota; fecal enzymes; ochratoxin A

Key Contribution: New elaborated synbiotics have beneficial effects on turkey health and reduce the negative impact of OTA contaminated feed. These synbiotics can be successfully used as feed additives for turkeys.

1. Introduction

The task of rational nutrition of farm animals is not only to obtain the maximum production result but also to care for animal welfare through a beneficial effect on the digestive tract, metabolism and stimulation of the immune system. An important factor responsible for the deterioration of production results, damage to the liver and other organs and the weakening of the immune system are mycotoxins [1,2]. Over 500 mycotoxins are known, and only about 50 are relatively well characterized [3]. The best feed components are used in mixtures for poultry, rarely containing an increased number of mycotoxins. However, the results of many studies indicate that even relatively low levels of mycotoxins in feed for young animals can cause a weakening of the immune system or reduce feed intake, and an increased content of mycotoxins in feed, especially containing corn grain and wheat bran, occurs in practice [4]. It is assumed that about 25% of the world's harvest may be contaminated with mycotoxins [3]. Mycotoxins are dangerous, and low doses lead to a weakening of the immune system and a decrease in animal performance [1,5,6]. Aflatoxins (AFB1, AFM1) and ochratoxin A (OTA) are commonly present in poultry feeds. Losses in poultry farming caused by these substances are mainly due to a decrease in production results and slaughter efficiency. In addition, there is a reduction in the quality of animal products and increased immunosuppression [3,7]. Penetration of mycotoxins is possible through the respiratory and digestive system of birds, and their metabolites accumulating in meat and eggs can negatively affect human health [2]. OTA belongs to the carcinogenic mycotoxins and can be produced by many species of *Aspergillus* or *Penicillium*. The European Commission Recommendation 2006/576/EC suggests that the maximum level of OTA in poultry feeds should be set to 0.1mg OTA kg^{-1} [8]. One of the effects of its action on the digestive system is damage to the intestinal mucosa. Damage to the intestinal barrier, epithelial necrosis, impaired glucose absorption and intense diarrhea expose animals to colonization by harmful microorganisms, e.g., *Salmonella* Typhimurium [9]. Adverse effects of OTA in poultry are manifested by increased mortality, limited feed intake and nephrotoxicity [10]. The negative effects of OTA on performance traits of poultry have been confirmed by many authors [11–14]. Symptoms of poisoning poultry due to the presence of OTA may include weight loss, depression and muscle paralysis. In addition, there is a decrease in mobility and energy in poultry, as well as rapid breathing; catarrhal inflammation of the gastrointestinal mucosa; focal hyperemia and especially local ecchymoses on the surface of the liver, kidneys and less often the heart [15].

The use of feed additives such as probiotics, prebiotics and synbiotics is beneficial for the balance of the intestinal microbiota of poultry and may minimize the adverse effects of mycotoxins to which the animals are exposed [16–18]. According to the definition formulated in 2002 by World Health Organization (WHO) and Food and Agriculture Organization of the United Nations (FAO), probiotics are "live microorganisms which, when administered in sufficient amounts, confer a health benefit on the host" [19]. This definition was maintained in 2013 by the International Scientific Association for Probiotics and Prebiotics (ISAPP) and is still presently used [20]. In other side, prebiotics are defined as "a nonviable food component that confers a health benefit on the host associated with modulation of the microbiota" [21]. Finally, synbiotics are formulas containing both synergistically acting prebiotics and probiotics. Detoxification by probiotic microorganisms may occur through adsorption or as a result of microbial metabolism of the mycotoxin [17,18]. Not all microorganisms have detoxifying properties, these properties are mainly related to the strain and not the species. Microorganisms with confirmed detoxification properties are *Acinetobacter calcoaceticus*, *Bifidobacterium* spp. and *Enterococcus faecium* [22,23]. However, particular interest should be paid to *Saccharomyces cerevisiae* yeast and lactic acid bacteria (LAB) strains. Scientific evidence shows that detoxification by lactic acid bacteria and yeast is the result of adhesion to cell wall structures, confirming the fact that even dead cells retain this activity [24].

The beneficial effect of the new elaborated synbiotic preparations on the intestinal microbiota of poultry has so far been confirmed in studies on SPF (specific pathogen free) and broiler chickens [25,26]. In chickens, the effect of these synbiotics on dominant intestinal microorganisms was tested, and it was found that synbiotic preparations increase the number of beneficial microorganisms such as *Bifidobacterium* spp. and *Lactobacillus* spp., while reducing the number of potentially pathogenic bacteria such as *Clostridium* spp. and *Escherichia coli*. In addition, promising results have been obtained in in vitro as well as in vivo studies on OTA detoxification [16]. Due to the beneficial effect of synbiotics on the health of chickens, we also decided to examine the effect in turkeys.

Confirmation of the effectiveness of the developed synbiotics in the feeding of turkeys is an important area of research feed additives that can potentially replace antibiotic growth promoters. The aim of the research was to evaluate of effects newly elaborated synbiotic preparations on the turkey performance, the intestinal microbiota and activity of fecal enzymes in turkeys Big-6 fed ochratoxin A contaminated feed. The analyzes were carried out for animals fed with the administration of synbiotic preparations (A, B or C) and commercial probiotic preparations (BioPlus 2B or Cylactin) and also comparative control animals (no feed additives).

2. Results and Discussion

2.1. The Effect of Tested Formulas on Turkey Performance

Determining the performance of farm animals is very important in testing the effectiveness of feed additives such as probiotics, prebiotics or synbiotics. It was found that after 6 weeks the average body weight of turkeys was significantly higher (by approx. 8%) for turkeys consuming feed supplemented with synbiotic A and C in comparison to OTA group and turkeys consuming the feed with probiotics (BioPlus 2B, Cylactin) (Table 1). After 15 weeks, the highest weight of animals was found in the control group receiving OTA-free feed and was higher by an average of 2 kg compared to animals receiving OTA-contaminated feed without additives. FCR after 15 weeks of bird's life was lower for turkeys consuming feed with synbiotic A compared to other groups consuming feed supplemented synbiotics or probiotics, but the observed differences were not statistically significant. The OTA contamination of feed did not significantly affect the mortality rate of the turkeys. After 6 weeks, EPEF was significantly higher in the control group (248.18) and in groups of turkeys consuming feed supplemented with synbiotic A (189.60) and synbiotic C (182.02) in comparison to OTA group (169.66) and animals consuming feed with probiotics (165.06 and 165.17). On the other hand, after 15 weeks, EPEF was the highest in the control group of animals (393.65) (Table 1).

Table 1. Rearing parameters of turkeys after 6 and 15 weeks of life. One-way ANOVA with post-hoc Tukey's test ([a,b]—values in rows with different letters are significantly different for $p < 0.05$).

The Age of Birds (Weeks)	Feed Additives	The Body Weight (Mean ± SD) (kg)	Daily Cumulative Mortality Rate (%)	FCR	EPEF
6	Control	1.85 ± 0.07	0.83	1.76 [b]	248.18 [a]
	OTA	1.65 ± 0.17	0.24	2.31 [a]	169.66 [b]
	Synbiotic A + OTA	1.78 ± 0.15	0.24	2.23 [a]	189.60 [ab]
	Synbiotic B + OTA	1.67 ± 0.14	0.24	2.34 [a]	169.52 [b]
	Synbiotic C + OTA	1.78 ± 0.15	0.36	2.32 [a]	182.02 [ab]
	BioPlus 2B + OTA	1.64 ± 0.19	0.24	2.36 [a]	165.06 [b]
	Cylactin + OTA	1.65 ± 0.16	0.36	2.37 [a]	165.17 [b]
15	Control	9.67 ± 0.13 [a]	0.83	2.32	393.65 [a]
	OTA	7.60 ± 0,40 [b]	0.48	2.79	257.27 [b]
	Synbiotic A + OTA	7.80 ± 0.47 [b]	0.24	2.71	273.46 [b]
	Synbiotic B + OTA	7.63 ± 0.48 [b]	0.24	2.80	258.91 [b]
	Synbiotic C + OTA	7.72 ± 0.67 [b]	0.36	2.80	261.65 [b]
	BioPlus 2B + OTA	7.63 ± 0.51 [b]	0.24	2.79	258.91 [b]
	Cylactin + OTA	7.79 ± 0.52 [b]	0.36	2.80	264.97 [b]

The addition of probiotic, prebiotic and synbiotic preparations to poultry feed mixtures can reduce the negative effects of mycotoxins, which has been confirmed in many studies [16,27–30]. The ability of probiotic microorganisms (e.g., LAB, *S. cerevisiae*) to detoxify OTA leads to stabilization of the intestinal microbiota and improvement of rearing rates [17,18,23]. Santin et al. have shown that the use of *S. cerevisiae* cell walls under exposure to OTA in compound feed for broiler chickens can improve FCR [31]. The authors found no effect of the additive on other growth parameters. On the other hand, the use of esterified glucomannan in compound feed contaminated with OTA and other mycotoxins has increased the weight gain of broiler chickens [11]. The use of synbiotic preparations tested in this research improved the body weight of chickens exposed to OTA [16]. However, in this study, feed supplementation with synbiotics caused a slight increase in body weight of turkeys; however, it was statistically significantly lower compared to the control group.

2.2. The Effects of Synbiotics on the Ochratoxin A Content in Tissues and Intestinal Contents of Turkeys

Turkeys consuming feed supplemented with probiotics and synbiotics were characterized by lower OTA concentrations in the liver relative to birds from OTA group after 6 and 15 weeks of life (Table 2). However, it was only after 15 weeks that the differences were statistically significant. Synbiotic A and synbiotic B caused the greatest reduction in the OTA concentration in the liver after 15 weeks.

Table 2. The content of OTA in the liver, the kidneys, and the jejunum and the caecum content of turkeys after 6 and 15 weeks of life. One-way ANOVA with post-hoc Tukey's test ([a,b,c,d]—values in rows with different letters are significantly different for $p < 0.05$).

The Age of Birds (weeks)	Feed Additives	The Content of OTA (Mean ± SD) (μg/kg)			
		The Liver	The Kidneys	The Jejunum	The Caecum
6	Control	0.00 ± 0.00	0.00 ± 0.00 [d]	0.00 ± 0.00 [d]	0.00 ± 0.00 [d]
	OTA	4.51 ± 1.55	5.78 ± 0.61 [a]	151.82 ± 46.88 [a]	100.36 ± 4.85 [a]
	Synbiotic A + OTA	3.40 ± 0.27	2.93 ± 0.52 [c]	57.04 ± 18.87 [c]	78.79 ± 7.49 [b]
	Synbiotic B + OTA	2.46 ± 0.61	3.62 ± 0.56 [bc]	105.97 ± 18.31 [b]	46.87 ± 1.73 [c]
	Synbiotic C + OTA	2.95 ± 0.36	3.66 ± 1.50 [bc]	106.83 ± 15.22 [b]	111.82 ± 11.11 [a]
	BioPlus 2B + OTA	3.60 ± 0.96	4.46 ± 1.02 [ab]	102.81 ± 16.83 [b]	53.73 ± 1.84 [c]
	Cylactin + OTA	3.52 ± 0.41	3.67 ± 1.11 [bc]	126.07 ± 29.28 [ab]	106.43 ± 7.65 [a]
15	Control	0.00 ± 0.00 [c]	0.00 ± 0.00 [b]	0.00 ± 0.00 [d]	0.00 ± 0.00 [c]
	OTA	3.80 ± 0.68 [a]	4.54 ± 0.74 [a]	177.62 ± 33.04 [a]	73.34 ± 50.21 [a]
	Synbiotic A + OTA	2.16 ± 0.24 [b]	3.17 ± 0.73 [a]	97.40 ± 33.36 [bc]	34.63 ± 7.37 [bc]
	Synbiotic B + OTA	2.14 ± 0.28 [b]	3.23 ± 1.24 [a]	96.21 ± 43.98 [c]	54.93 ± 2.16 [ab]
	Synbiotic C + OTA	2.74 ± 0.42 [b]	3.52 ± 0.54 [a]	147.26 ± 24.38 [ab]	45.98 ± 3.92 [ab]
	BioPlus 2B + OTA	2.36 ± 0.57 [b]	3.82 ± 1.77 [a]	163.31 ± 19.80 [a]	60.16 ± 34.54 [ab]
	Cylactin + OTA	2.83 ± 1.18 [b]	4.50 ± 0.59 [a]	116.66 ± 33.23 [bc]	40.26 ± 11.43 [ab]

After 6 weeks of experiment, the lowest OTA concentration was noted in the kidneys of turkeys consuming feed supplemented with synbiotic A in comparison to birds from control-positive group (Table 2). Dietary supplementation with synbiotics (A, B and C) decreased the content of OTA in kidneys of turkeys in 15 weeks of experiment compared to control-positive group and birds consuming feed with Cylactin and BioPlus 2B.

After 6 weeks of birds' life, the concentration of OTA in the jejunum content of turkeys consuming feed supplemented with synbiotic A (57.04 μg/kg) was the lowest (Table 2). Moreover, the concentration of OTA in the jejunum content in animals fed with feed with the addition of synbiotic B and C (150.97 and 106.83 μg/kg, respectively), and also BioPlus 2B (102.81μg/kg) and Cylactin (126.07 μg/kg) was lower compared to control-positive group. After 15 weeks of experiment, the supplementation of synbiotic A and B was the most effective, and the concentration of OTA was, respectively, 97.40 and 96.21 μg/kg (Table 2).

Turkeys consuming feed supplemented with synbiotics A, B and probiotic BioPlus 2B had lower OTA concentration in the content of caecum (78.79, 46.87 and 53.73 μg/kg, respectively) in comparison to other groups of animals after 6 weeks rearing (Table 2). After 15 weeks of experiment, the concentration of OTA in the caecum content of birds consuming feed supplemented with synbiotics A, B and C (34.63, 54.93, 45.98 μg/kg respectively) was significantly lower in comparison to control-positive group of turkeys (Table 2).

The main site of OTA absorption is the small intestine, especially the proximal jejunum. Absorption occurs through both active transport and passive diffusion. After being absorbed with the blood of the portal vein, mycotoxins enter the liver and other organs [32]. Subsequently, mycotoxins can be partially metabolized in the body, excreted in urine, bile and feces, or accumulated in the liver, kidneys and muscles [32].

In the study of Pozzo et al., OTA residues were detected in kidneys (3.58 ng/g) and liver (1.92 ng/g) of broiler chickens receiving OTA contaminated mixtures in the amount of 0.1 mg/kg [33]. Similar results were obtained in other studies [34,35]. Zaghini et al. found that esterified glucomannan and *Saccharomyces cerevisiae* do not reduce the concentration of mycotoxins in kidneys of chickens fed with OTA contaminated feed at 2 mg/kg [36]. The administration of synbiotics tested in this research caused a decrease in concentration of OTA and the genotoxicity of fecal water in broiler chickens contaminated OTA [16]. On the other hand, Kozaczyński found that broiler chickens fed OTA contaminated mixtures (0.2 mg/kg) for 20 weeks did not cause OTA accumulation in the kidneys or liver [37]. Differences in OTA concentrations between organs result from the length of exposure and the concentration of mycotoxins in bird's feed doses [38]. The particularly strong affinity of OTA for accumulation in the liver and kidneys may be due to the function of these organs responsible for detoxification and excretion of toxins [39,40]. The use of probiotics, prebiotics or synbiotics can be a solution that limits the negative impact of OTA on monogastric animals. Preparations of this type can bind mycotoxins in the gastrointestinal tract (GIT), thus reducing their absorption and systemic toxicity, which was confirmed by our own research [24,41].

2.3. The Effect of Synbiotics on the Dominant Intestinal Microbiota of Turkeys

In the study, contents of the jejunum and the caecum were examined after 6 and 15 weeks, while the excreta were examined after 3, 9 and 15 weeks of animal rearing. The number of anaerobic bacteria in the content of the jejunum, the caecum and the excreta did not change significantly during 15 weeks of turkeys rearing (Table S1). This is a beneficial effect, as the balance of intestinal microbiota in turkeys is maintained. The number of anaerobic bacteria in the excreta was similar in each group after 3 weeks. In a further period of turkeys rearing, no differences were statistically significant between groups of birds, and after 6 weeks, the numbers of anaerobic bacteria were on average 1.95×10^8–1.37×10^9 CFU/g (the jejunum) and 1.25–6.10×10^9 CFU/g (the caecum). After 15 weeks of rearing, it was found that the total numbers of anaerobic bacteria were, respectively, 2.60×10^8–2.97×10^9 CFU/g (the jejunum), 1.05–5.67×10^9 CFU/g (the caecum) and 6.12×10^8–2.19×10^9 CFU/g (the excreta) (Table S1). These results were higher on average by one order of magnitude in comparison to the results reported by Dibaji, Seidavi, Asadpour and da Silva, who tested the effect of the Biomin IMBO synbiotic, containing the probiotic bacteria *Enterococcus faecium* (5×10^{11} CFU/kg) and fructooligosaccharides (a prebiotic), administered for 42 days, on the intestinal microbiota of chickens [42].

During the experiment, in samples of the intestinal content and the excreta, the average number of *Enterobacteriaceae* family bacteria was similar in turkeys consuming fed supplemented with probiotics, synbiotics and also control groups (positive and negative) of animals. After 15 weeks of bird's life, the total counts of *Enterobacteriaceae* family bacteria were on average 1.27–5.50×10^8 CFU/g (the jejunum), 1.84–5.50×10^8 CFU/g (the caecum) and 2.07–9.25×10^8 CFU/g (the excreta) (Table S1). Our results are comparable to the study by Biernasiak et al., when the number of *Enterobacteriaceae* family bacteria in the excreta of chickens receiving probiotics was on average 10^8 CFU/g [43].

The number of *Bifidobacterium* spp. and *Lactobacillus* spp. in samples of the intestinal content and the excreta of animals fed with OTA contaminated feed without additives decreased after 15 weeks of rearing (Table S1). The administration of feed supplemented with probiotics and synbiotics caused an increased count of these bacteria in the content of the jejunum, the caecum and the excreta of turkeys. In the case of new elaborated synbiotic preparations, the average numbers of *Bifidobacterium* spp. were 1.79–4.00 × 10^7 CFU/g (the jejunum), 1.11–2.40 × 10^7 CFU/g (the caecum) and 1.50–4.00 × 10^8 CFU/g (the excreta) after 15 weeks of rearing. Furthermore, the average counts of *Lactobacillus* spp. in these groups of animals were 2.03–4.90 × 10^6 CFU/g (the jejunum), 1.60–3.15 × 10^6 CFU/g (the caecum) and 1.74–2.91 × 10^7 CFU/g (the excreta) after 15 weeks of the experiment. It was found that the highest growth in number of these bacteria was in turkeys consuming feed supplemented with synbiotic C. In the content of intestines and the excreta, the numbers of *Bifidobacterium* spp. and *Lactobacillus* spp. in turkeys fed with feed supplemented with the synbiotic C were respectively 2 and 3 orders of magnitude higher compared to the control-positive group. Feeding turkeys with feed supplemented with probiotics resulted in an increase in the number of *Bifidobacterium* spp. and *Lactobacillus* spp. in the intestinal contents (1 order of magnitude) and the excreta of birds (2 and 1 order of magnitude, respectively) compared to control-positive group (Table S1).

In the jejunum content of control-positive turkeys, the count of *Clostridium* spp. and *Escherichia coli* increased to 1.20 × 10^8 CFU/g and 2.20 × 10^8 CFU/g, respectively, after 15 weeks of animals rearing (Table S1). In the case of turkeys consuming fed supplemented with synbiotics, the numbers of *Clostridium* spp. and *Escherichia coli* in the intestinal content after 6 weeks were significantly lower in comparison to control-positive group (2 and 1 orders of magnitude). The administration of turkeys with synbiotic supplemented feed for 15 weeks reduced the numbers of *Clostridium* spp. to 4.67 × 10^3–1.67 × 10^4 CFU/g (the jejunum), 1.05–2.77 × 10^5 CFU/g (the caecum) and 2.25–8.85 × 10^4 CFU/g (the jejunum). Moreover, the synbiotic supplementation of fodder resulted in a decrease in the numbers of *Escherichia coli* in the intestinal contents and the excreta of turkeys (3 and 4 orders of magnitude, respectively) compared to control-positive group. After 15 weeks of rearing, the numbers of *Escherichia coli* in these groups were 9.57 × 10^3–2.47 × 10^4 CFU/g (the jejunum), 4.00 × 10^4–4.25 × 10^5 CFU/g (the caecum) and 1.25–4.00 × 10^5 CFU/g (the excreta). The best of results was found after the feed supplementation with the synbiotic C. Furthermore, probiotic preparations (BioPlus 2B and Cylactin) caused the reduction in the number *Clostridium* spp. and *Escherichia coli* by an average of 1–2 orders of magnitude compared to control-positive group (Table S1). Thus, all elaborated synbiotic preparations proved to be more effective in reducing the number of potentially pathogenic microorganisms in the intestinal content and the excreta of turkeys compared to tested commercial probiotic preparations.

The administration of probiotics and synbiotics to animals by 15 weeks did not have a statistically significant effect on the number of *Enterococcus* spp. and *Bacteroides* spp. in the intestinal content and the excreta of turkeys Big-6 (Table S1). In another study, no effect of two probiotics on the number of *Enterococcus* spp. in the intestinal microbiota of chickens was observed [44]. In previous studies in chickens regarding the developed synbiotic preparations, no change in the number of *Bacteroides* spp. as a result of treatment was observed either [25,26].

During the rearing of turkeys, the numbers of yeast in the intestinal content and the excreta of animals fed with synbiotics were significantly higher (the average of 2–3 orders the magnitude after 15 weeks) compared to birds consuming feed supplemented with probiotics or control-positive and control-negative animals (Table S1). Yeast are not commensal microorganisms in the intestinal of poultry. These results are very promising, because synbiotic preparations administrated to turkeys contained *Saccharomyces cerevisiae*. Hence, a high total count of yeast in samples obtained from these animals may indicate the survival of those microorganisms in the gastrointestinal tract of turkeys.

In order to better visualize the results, principal component analysis (PCA) was used. The biplot for PC1 and PC2 showed the influence of tested synbiotic formulas and commercial preparations on dominant microorganisms in the jejunum content (Figure 1), the caecum content (Figure 2) and the excreta (Figure 3) in tested groups of animals. The visualization using the scatter plot showed a distinct clustering of individuals in each group of turkeys. In addition, from agglomerative hierarchical clustering analysis (AHC), it was determined that animals were divided into 3 clusters (Figures 1–3). The clusters visible on presented dendrograms clearly show that the intestinal microbiota is an individual feature.

This study confirmed the beneficial effects of synbiotics A, B and C on the intestinal microbiota of turkeys. An important factor in these studies is the additional exposure of animals to contaminated OTA feed. Despite contamination of the feed, the synbiotic preparations had a very beneficial effect on the composition of the intestinal microbiota of animals and minimized the adverse effects of OTA. In addition, the effect of synbiotic preparations was better compared to commercial probiotic preparations BioPlus 2B and Cylactin.

2.4. The Effect of Synbiotics on the Activity of Fecal Enzymes in the Excreta of Turkeys

After 15 weeks of rearing, the lowest α-glucosidase activity was found in the excreta of turkeys fed with no additives—control-positive group (69.46 μMh/g) (Table 3). The activity of this enzyme was higher in the excreta of animal consuming feed with probiotics (BioPlus 2B and Cylactin) and was 75.91 μMh/g and 78.85 μMh/g, respectively. Furthermore, the highest α-glucosidase activity was found in groups of consuming feed supplemented with synbiotics A, B and C. Results obtained in these groups were 13–39% higher compared to the control-positive group and were 90.32 μMh/g, 78.77 μMh/g and 86.47 μMh/g, respectively. In the control-negative group, the activity of this enzyme after 15 weeks was 88.33 μMh/g (Table 3).

In the case of β-glucosidase, the highest activity of this enzyme was found in the excreta of control-positive animals and was on average 11.23 μMh/g after 15 weeks of treatment (Table 3). The administration of probiotics BioPlus 2B and Cylactin to animals reduced the activity of this enzyme to 8.31 μMh/g and 9.41 μMh/g, respectively. Furthermore, the average β-glucosidase activity in the excreta of animals consuming feed with the addition of synbiotics was on average 18–27% lower compared to the control-positive group and was, respectively, 8.50 μMh/g (Synbiotic A), 8.22 μMh/g (Synbiotic B) and 9.47 μMh/g (synbiotic C). In the control-negative group, the activity of this enzyme after 15 weeks was 8.09 μMh/g (Table 3).

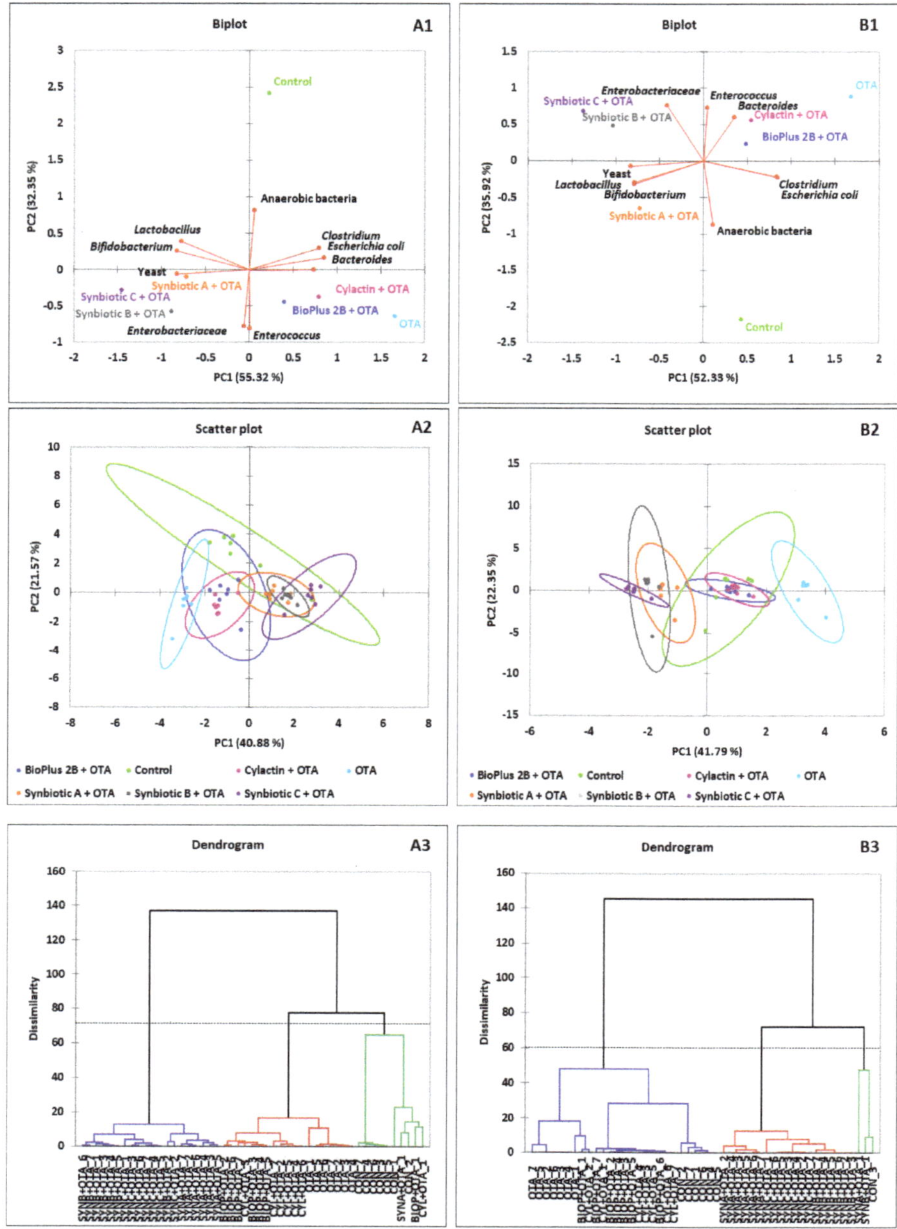

Figure 1. PCA plot of counts of microorganisms that dominate in the jejunum of turkeys. The scatter plot showing a clustering of individuals in groups of animals. AHC of 42 birds by the Ward's method. The dotted line on the dendrogram is located at the node before the largest relative increase in dissimilarity level, (**A1–A3**) after 6 and (**B1–B3**) after 15 weeks of animals rearing.

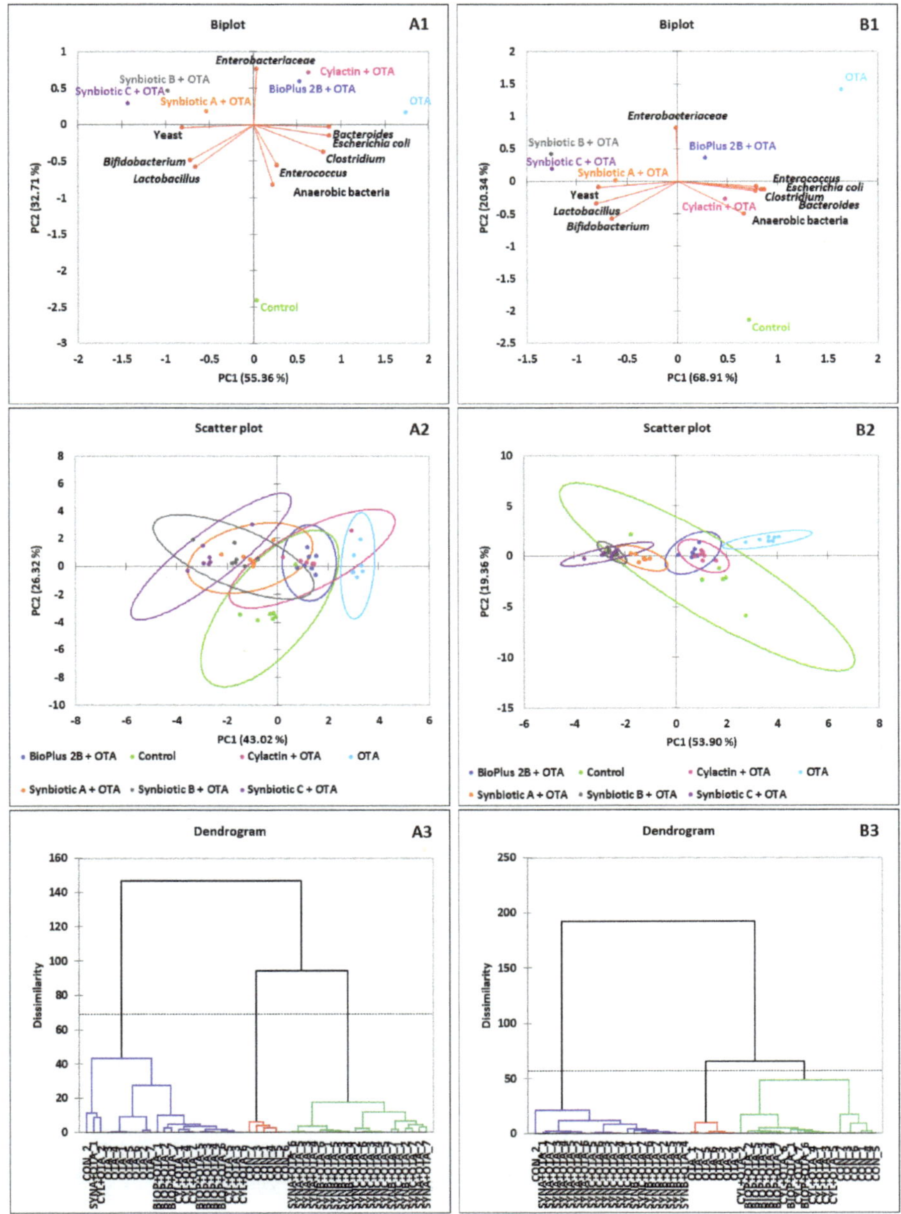

Figure 2. PCA plot of counts of microorganisms which dominate in the caecum of turkeys. The scatter plot showing a clustering of individuals in groups of animals. AHC of 42 birds by the Ward's method. The dotted line on the dendrogram is located at the node before the largest relative increase in dissimilarity level; (**A1–A3**) after 6, and (**B1–B3**) after 15 weeks of animals rearing.

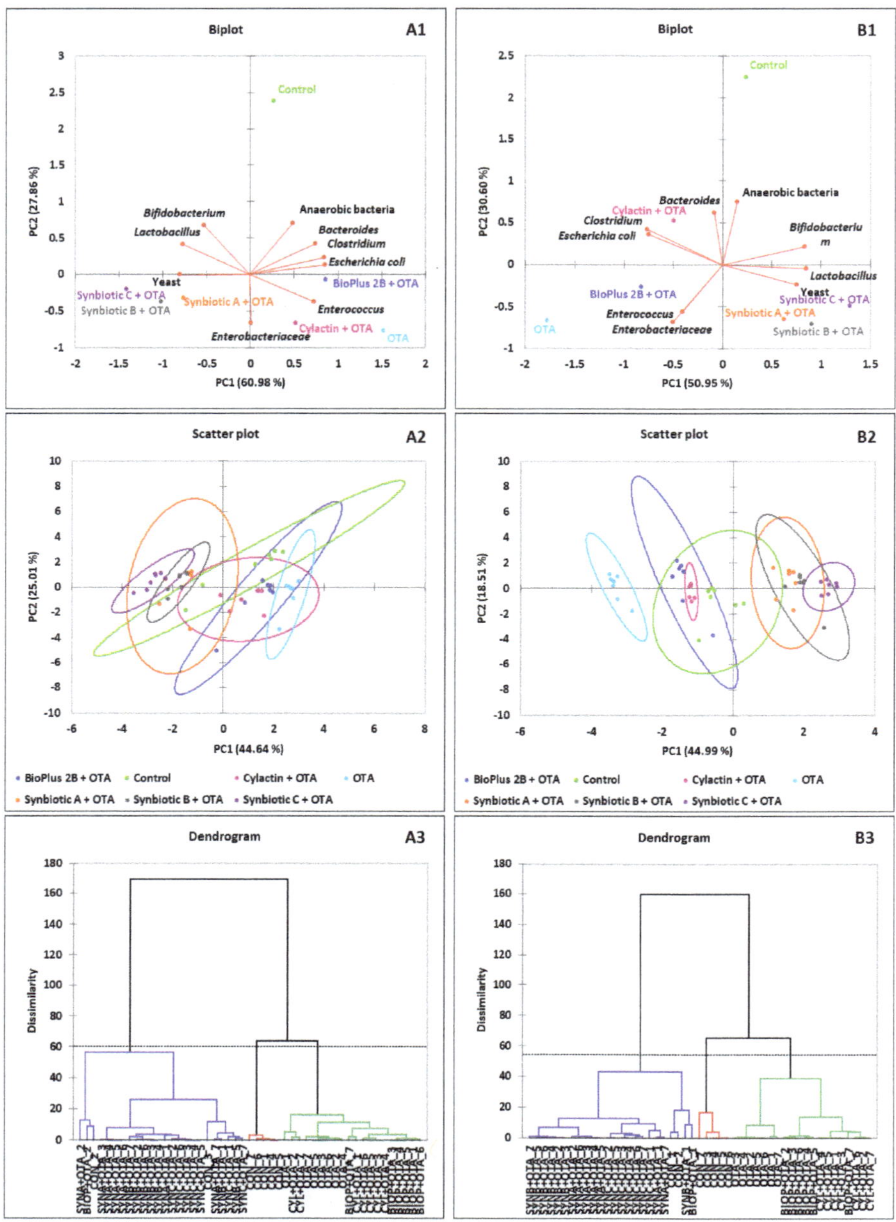

Figure 3. *Cont.*

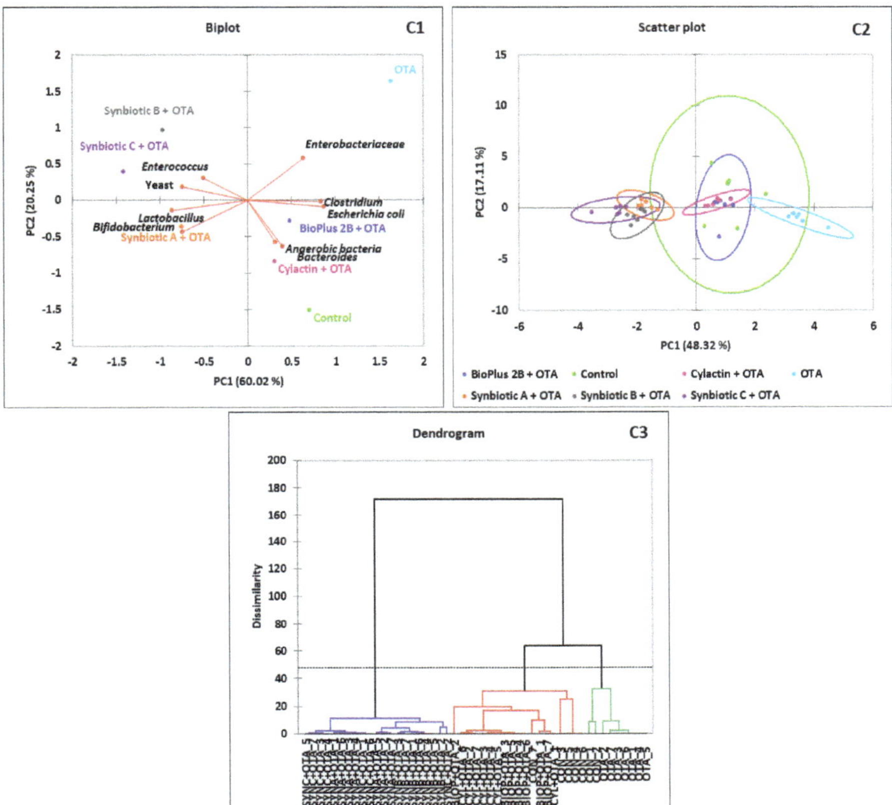

Figure 3. PCA plot of counts of microorganisms that dominate in the excreta of turkeys. The scatter plot showing a clustering of individuals in groups of animals in each time point of rearing, respectively. AHC of 42 birds by the Ward's method. The dotted line on the dendrogram is located at the node before the largest relative increase in dissimilarity level, (**A1–A3**) after 3, (**B1–B3**) after 9 and (**C1–C3**) after 15 weeks of animals rearing.

Table 3. The activity of fecal enzymes in excreta of turkeys after 2 days and 15 weeks of rearing. One-way ANOVA with post-hoc Tukey's test ([a,b,c,d] values in rows with different letters are significantly different for $p < 0.05$).

Age	Feed Additives	The Activity of Enzyme (Mean ± SD) [μMh/g]				
		α-glucosidase	β-glucosidase	α-galactosidase	β-galactosidase	β-glucuronidase
2 days	Control	86.37 ± 1.32	7.01 ± 1.51	81.55 ± 2.88	65.36 ± 2.07	39.05 ± 2.14
15 weeks	Control	88.33 ± 2.77 [a]	8.09 ± 1.97 [b]	94.00 ± 2.89 [a]	89.33 ± 2.38 [cd]	42.31 ± 2.10 [c]
	OTA	69.46 ± 0.35 [c]	11.23 ± 1.14 [a]	81.03 ± 1.89 [c]	97.77 ± 1.57 [a]	51.92 ± 2.53 [a]
	Synbiotic A + OTA	90.32 ± 1.66 [a]	8.50 ± 1.04 [b]	93.84 ± 0.98 [a]	94.11 ± 1.52 [b]	50.50 ± 1.69 [a]
	Synbiotic B + OTA	78.77 ± 4.55 [b]	8.22 ± 1.11 [b]	96.85 ± 3.59 [a]	91.04 ± 1.04 [c]	47.99 ± 2.90 [ab]
	Synbiotic C + OTA	86.47 ± 2.48 [a]	9.47 ± 1.14 [ab]	98.03 ± 3.47 [a]	87.08 ± 0.89 [d]	42.51 ± 1.00 [c]
	BioPlus 2B + OTA	75.91 ± 2.99 [bc]	8.31 ± 0.67 [b]	87.17 ± 1.80 [b]	94.32 ± 0.87 [b]	45.00 ± 2.36 [bc]
	Cylactin + OTA	78.85 ± 2.86 [b]	9.41 ± 0.67 [ab]	84.77 ± 1.88 [a]	93.59 ± 1.67 [b]	49.42 ± 2.17 [ab]

After 15 weeks bird's life, the average α-galactosidase activity in the excreta of turkeys consuming feed contaminated OTA was 81.03 μMh/g (Table 3). In the excreta of turkeys fed with fodder supplemented probiotics BioPlus 2B and Cylactin, the activity of this enzyme after 15 weeks was higher compared to control-positive group and was 87.17 μMh/g and 84.77 μMh/g, respectively. In turkeys fed with synbiotic supplemented feed, the highest α-galactosidase activity (9–17% higher compared to the control-positive group) was found, which was 93.84 μMh/g (synbiotic A), 96.85 μMh/g (synbiotic B) and 98.03 μMh/g (synbiotic C). The average α-galactosidase activity in the excreta of control turkeys was 94.00 μMh/g after 15 weeks of rearing (Table 3).

The administration of the OTA contaminated feed without supplements for 15 weeks increased the activity of β-galactosidase in the excreta of turkeys to an average of 97.77 μMh/g (Table 3). The addition of probiotic preparations to the feed caused a decrease in the activity of this enzyme to an average of 94.32 μMh/g (BioPlus 2B) and 93.59 μMh/g (Cylactin) after 15 weeks of turkeys rearing. The most effective turned out to be synbiotics B and C, which caused a decrease in the activity of β-galactosidase in the excreta of turkeys to 91.04 μMh/g and 87.08 μMh/g, respectively. In the control-negative group, the activity of this enzyme after 15 weeks was 89.33 μMh/g (Table 3).

The highest β-glucuronidase activity after 15 weeks of rearing was found in control-positive animals and averaged 51.92 μMh/g (Table 3). In the case of turkeys consuming feed with the addition of probiotics, the activity of this enzyme was reduced to 45.00 μMh/g (BioPlus 2B) and 49.42 μMh/g (Cylactin). Furthermore, the group of turkeys fed with synbiotic supplemented feed had the lowest β-glucuronidase activity in the excreta and was 50.50 μMh/g (synbiotic A), 47.99 μMh/g (synbiotic B) and 42.51 μMh/g (synbiotic C), respectively. The average β-glucuronidase activity in the excreta of control turkeys was 42.31 μMh/g after 15 weeks of rearing (Table 3).

Some fecal enzymes (e.g., β-glucuronidase) can catalyze the release reaction of potentially harmful compounds, which can lead to carcinogenesis. Controlling the enzymatic activity of the intestinal microbiota by administering probiotics, prebiotics or synbiotics allows to reduce the activity of potentially harmful enzymes [45]. In the other research, it was found that 40-day lactic acid bacteria feed supplementation resulted in a decrease in β-glucuronidase and β-glucosidase activity in the excreta of broiler chickens [46]. In our own studies, administration of elaborated synbiotics also reduced the activity of these enzymes in the excreta of turkeys which additionally received the OTA contaminated feed. Shokryazdan et al. showed that the administration of a probiotic to broiler chickens significantly affects the activity of fecal enzymes [47]. The addition of three strains of *L. salivarius* to the feed resulted in a decrease in β-glucuronidase and β-glucosidase activity compared to control groups [47]. The activity of β-glucosidase and β-glucuronidase in 42-day-old chickens was reduced by 35% and 37% (dose 0.5 g/kg) and by 39% and 42% (dose 1.0 g/kg). Other studies tested the effects of Thepax® probiotic (*S. cerevisiae*, *Lb. acidophilus*) and yogurt as a probiotic (*Lb. delbrueckii*, *Lb. thermophilus*) [48]. Broiler chickens received yogurt along with drinking water or feed with the addition of Thepax. Both supplements have significantly reduced the number of *E. coli* and *Clostridium perfringens*, which show high activity of potentially harmful enzymes. However, in the other studies, the effect of fructooligosaccharides (FOS) on the turkeys' health was studied [49]. A diet containing 0.5%, 1% and 2% FOS resulted in reduction of the activity of β-glucuronidase and β-glucosidase. The research of Čokášová et al. studied the probiotic properties of *Lb. plantarum* and its effect on rats in which carcinogenesis has been chemically initiated [50]. In the feces of animals treated with the probiotic, an increased number of *Lactobacillus* spp. and a reduction in the number of coliforms were found. In addition, there was a decrease in β-glucuronidase and β-galactosidase activity and an increase in α-glucosidase and α-galactosidase activity. Śliżewska et al. investigated the effect of *Fusarium* mycotoxins, deoxynivalenol and zearalenone on β-glucosidase and β-glucuronidase activity in gilts [51]. After 6 weeks, a statistically significant increase in the activity of both enzymes was observed compared to the control group fed with uncontaminated feed. The reason for the increase in enzyme activity was the increase in the population of mesophilic microorganisms (e.g., *Escherichia coli*) due to the presence of mycotoxins in feed.

The results obtained in our study indicate a beneficial effect of the developed synbiotic preparations in the modulation of the intestinal microbiota and the activity of fecal enzymes in the excreta of turkeys fed with the OTA contaminated feed.

3. Conclusions

Results of this research showed that all elaborated synbiotic preparations have beneficial effect of on the turkey performance, the balance of the intestinal microbiota and the fecal enzymes activity in turkeys. Furthermore, the administration of synbiotics to animals fed OTA contaminated fodder produces a significant decrease in OTA concentration in the liver, the kidneys, the jejunum and the caecum of turkeys. All synbiotic formulas cause an increase in the count of beneficial intestinal microorganisms (*Lactobacillus* spp. and *Bifidobacterium* spp.) while a significant reduction in potentially pathogenic bacteria (*Escherichia coli* and *Clostridium* spp.) in the intestinal contents and the excreta of animals. Moreover, synbiotics produce beneficial changes in the fecal enzymes activity, causing a significant increase of α-glucosidase and α-galactosidase activity (beneficial enzymes) and a significant decrease of β-glucosidase, β-galactosidase and β-glucuronidase activity (potentially harmful enzymes) in comparison to the control group. Synbiotic C (the most complex preparation) proved to be the most effective one. The use of synbiotic preparations effectively minimizes the negative impact of the OTA contaminated feed on the turkey's health. Hence, elaborated synbiotics can be successfully used as feed additives for turkeys for fattening.

4. Materials and Methods

4.1. Probiotic and Synbiotic Preparations

Synbiotic preparations (A, B and C) were elaborated in the Institute of Fermentation Technology and Microbiology of the Lodz University of Technology (Poland). The final formula of synbiotics was developed by the Department of Biotechnology and Food Microbiology of the Poznan University of Life Sciences. Each synbiotic preparation contained: 2.0×10^9 CFU/g of *Lactobacillus* spp., 2.0×10^7 CFU/g of *Saccharomyces cerevisiae* yeast and 2% inulin (prebiotic). The content of each microorganism in synbiotics A, B and C was equal. These probiotic strains were isolated from various sources (*Lb. paracasei* ŁOCK 1091—caecal content of sow; *Lb. pentosus* ŁOCK 1094—broiler chicken dung; *Lb. plantarum* ŁOCK 0860—plant silage; *Lb. reuteri* ŁOCK 1092—piglet caecal content; *Lb. rhamnosus* ŁOCK 1087—turkey dung and *S. cerevisiae* ŁOCK 0119—distillers' yeast, grain) [26]. Commercial probiotic preparations contained 3.2×10^{10} CFU/g *Bacillus* spp. (BioPlus 2B; Biochem), and 1.0×10^{10} CFU/g *Enterococcus faecium* (Cylactin; DSM), respectively (Table 4). Probiotic microorganisms and prebiotic included in the synbiotic formulas have been tested for their probiotic and prebiotic properties according to the procedures recommended by FAO/WHO and EFSA [52,53]. Furthermore, the effectiveness of the developed synbiotic preparations in the feeding of monogastric animals has been confirmed so far in many in vitro and in vivo studies [16,25,26,54].

Table 4. The composition of new elaborated synbiotic formulas and commercial probiotic preparations.

Type of Preparation	Name of Preparation	Probiotic Microorganisms	Prebiotics
Synbiotics	A	*Lb. plantarum* ŁOCK 0860 *Lb. reuteri* ŁOCK 1092 *Lb. pentosus* ŁOCK 1094 *S. cerevisiae* ŁOCK 0119	inulin
	B	*Lb. plantarum* ŁOCK 0860 *Lb. reuteri* ŁOCK 1092 *Lb. pentosus* ŁOCK 1094 *Lb. rhamnosus* ŁOCK 1087 *S. cerevisiae* ŁOCK 0119	
	C	*Lb. plantarum* ŁOCK 0860 *Lb. reuteri* ŁOCK 1092 *Lb. pentosus* ŁOCK 1094 *Lb. rhamnosus* ŁOCK 1087 *Lb. paracasei* ŁOCK 1091 *S. cerevisiae* ŁOCK 0119	
Probiotics	BioPlus 2B	*B. licheniformis* DSM 5749 *B. subtilis* DSM 5750	without
	Cylactin	*E. faecium* NCIMB 10415 (SF68)	

4.2. Animal Treatment

The study was conducted on 140 turkeys Big-6 (females), randomly divided into seven groups. Turkeys were housed in standard environmental conditions in separate cages (20 animals per cage) in one room of animal experimental laboratory. The animals were reared for 15 weeks from the first day of their life. The feed was natural contamination by the addition of OTA-infected wheat. The OTA concentration in feed ranged from 198.6 to 462.0 µg/kg depending on the type of feed (Table 5). For the examination period, during 15 weeks of life, in each group, birds were administrated synbiotic preparation (A, B or C) in a dose of 0.50 g/kg of feed or commercial probiotic preparation (BioPlus 2B or Cylactin) in a dose of 0.4 g/kg of feed ad libitum, respectively. The positive control was the group of birds that were administrated ad libitum the contaminated feed without additives. The negative control was the group of birds where OTA-free feed was administrated ad libitum without additives.

For the nutrition of animals, standard mixtures (nutritional value was adapted to the requirements of intensively growing birds) were used (Table 5). Mixtures were made in powder form. In their composition, there were cereal components (wheat), high-protein ingredients (post-extraction soya meal) and also soybean oil as an additional source of energy. In order to obtain the required level of exogenous amino acids, pure amino acids such as methionine, lysine and threonine were used. Mixtures contained mineral–vitamin premixes involving enzyme preparations (xylanase and phytase) and coccidiostat Clinacox (Starter 1,2 and Grower 1). Detailed parameters of the applied feed are presented in Table 5.

Birds were grown in the Department of Animal Nutrition and Fodder Knowledge, University of Warmia and Mazury (Olsztyn, Poland). Experiments were conducted after obtaining the approval of the Local Ethical Commission for Animal Testing in Olsztyn (Poland) according to resolution No. 14/2012 (from 29 February 2012). All studies were performed in accordance with relevant guidelines and regulations. The scheme of the conducted experiment is presented in Figure 4.

Table 5. The composition and nutritional value of turkey compound feeds.

The Composition of Feed (g/kg) \ Type of Feed	Starter 1 0–3 Weeks	Starter 2 4–6 Weeks	Grower 1 7–9 Weeks	Grower 2 10–12 Weeks	Finisher 13–15 Weeks
Wheat	261.9	310.3	416.7	515.8	590.4
Corn	200.0	200.0	150.0	100.0	100.0
Soybean meal	358.2	360.9	347.4	300.0	225.1
Full-fat soybeans	100.0	50.0	-	-	-
Blood meal	20.0	10.0	-	-	-
Soybean oil	5.2	19.2	39.1	45.1	47.8
L-lysine HCl	3.1	3.6	3.7	3.2	4.0
DL-methionine	3.5	2.6	2.4	2.6	2.5
L-threonine	0.7	0.7	1.1	0.7	1.0
Limestone	18.8	14.5	13.4	11.1	9.7
Monocalcium phosphate	22.1	19.9	17.7	13.1	11.1
Sodium bicarbonate	0.1	1.3	1.2	0.7	0.7
NaCl	2.0	1.9	2.2	2.6	2.7
Feed enzymes	0.1	0.1	0.1	0.1	0.1
Premix *	5.0	5.0	5.0	5.0	5.0
The nutritional value of feed					
Metabolizable energy (kcal/kg)	2800	2880	3000	3100	3200
Crude protein (g)	27.00	25.20	23.00	21.50	19.00
Lysine (%)	1.77	1.65	1.45	1.30	1.17
Methionine and Cysteine (%)	1.15	1.02	0.95	0.93	0.85
Ca (g)	1.40	1.20	1.15	1.00	0.90
Available P (g)	0.70	0.65	0.60	0.50	0.45
Na (g)	0.13	0.15	0.15	0.15	0.15
The mycotoxin content in feed					
Ochratoxin A (µg/kg) **	198.6	251.6	331.0	397.2	462.0

* The composition of the premix for: Starter—12 500 IU Vitamin A, 4 500 IU Vitamin D_3, 87.5 mg Vitamin E, 3.75 mg Vitamin K_3, 3.5 mg Vitamin B_1, 10 mg Vitamin B_2, 75 mg Niacin, 22.5 mg Pantothenic acid, 6.0 mg Vitamin B_6, 30 µg Vitamin $B_{12,}$ 2.5 mg Folic acid, 400 µg Biotin, 800 mg Choline Chloride, 92.5 mg Fe, 130 mg Mn, 20 mg Cu, 105 mg Zn, 2.5 mg J, 0.3 mg Se; Grower—11 500 IU Vitamin A, 4 140 IU Vitamin D_3, 80.5 mg Vitamin E, 3.45 mg Vitamin K_3, 3.22 mg Vitamin B_1, 9.2 mg Vitamin B_2, 69 mg Niacin, 20.7 mg Pantothenic acid, 5.52 mg Vitamin B_6, 37.6 µg Vitamin $B_{12,}$ 2.3 mg Folid Acid, 368 µg Biotin, 600 mg Choline Chloride, 85.1 mg Fe, 120 mg Mn, 18.4 mg Cu, 96.6 mg Zn, 2.3 mg J, 0.26 mg Se; Finisher—10 500 IU Vitamin A, 3 780 IU Vitamin D_3, 66.5 mg Vitamin E, 2.85 mg Vitamin K_3, 2.66 mg Vitamin B_1, 7.6 mg Vitamin B_2, 57 mg Niacin, 17.1 mg Pantothenic acid, 4.6 mg Vitamin B_6, 22.8 µg Vitamin $B_{12,}$ 1.9 mg Folid Acid, 304 µg Biotin, 400 mg Choline Chloride, 70.3 mg Fe, 98.8 mg Mn, 15.2 mg Cu, 79.8 mg Zn, 1.9 mg J, 0.23 mg Se. ** It does not apply to the negative control group.

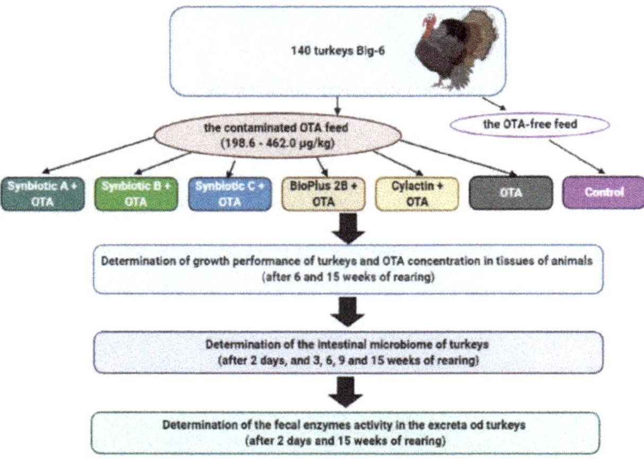

Figure 4. The scheme of the experiment.

4.3. Determination of the Turkey Performance

The turkeys' rearing time was 15 weeks. Turkeys were observed daily in order to detect potential undesirable effects or deaths of birds. The body weight of animals in individual study groups was determined after 6 and 15 weeks of life. Final production parameters such as daily cumulative mortality rate (%), feed conversion ratio (FCR), and European Production Efficiency Factor (EPEF) were determined according to the formulas [16]:

$$\text{daily cumulative mortality rate }(\%) = \frac{\text{the number of death turkeys (pc.)}\times 100\%}{\text{the number of turkeys in research (pc.)}}$$

$$\text{FCR} = \frac{\text{the feed consumption (kg)}}{\text{the body weight gain (kg)}}$$

$$\text{EPEF} = \frac{\text{the liveability }(\%)\times\text{the body weight (kg)}\times 100}{\text{the age (days)}\times\text{the feed conversion ratio (kg)}}$$

4.4. Determination of the Ochratoxin A Content in Feed, Tissues and the Intestinal Content

The concentrations of OTA in diets, wheat grain, tissue and intestine content were determined by immunoaffinity column clean-up and HPLC-fluorescence detection in accordance with Polish Standard PN-EN 16007-2012-U [55]. Deep-frozen tissue samples and intestinal contents were lyophilized at −70 °C using a lyophilizer (Alpha 1–2 LDplus, CHRIST, Germany). The extracts concentrations were measured by high-performance liquid chromatography (HPLC) (LC–20AD, Shimadzu, Japan). The HPLC apparatus consisted of Jupiter 5u C18 300A column chromatography (Phenomenex) 250 × 4.60 mm, column operating temperature: 24 °C; mobile phase—acetonitrile/methanol/aqueous acetic acid (35; 35; 30 v/v/v); flow rate: 0.5 mL/min; dosing volume: 100 µL; fluorescence detector: λex = 333 nm, λem = 467 nm. The qualitative interpretation of the obtained chromatograms was carried out by comparing the retention time of the analyzed toxin in the appropriate standard solution with the retention time of a given analyte in the actual sample. Quantitative analysis was carried out by reading the content of mycotoxin in the sample from the standard curve and making appropriate calculations. Each sample was analyzed in two parallel repetitions.

4.5. Determination of Intestinal Microbiota of Turkeys

The composition of dominant microorganisms in turkeys was determined after 3, 9 and 15 weeks (the excreta), after 6 and 15 weeks (the jejunum as part of the small intestine and the caecum) of rearing in seven randomly selected turkeys (three repetitions) in each experimental group. The count of analyzed microorganisms was determined using the culture method in accordance with the PN-ISO standards in triplicate, using selective microbial media [39]. The total anaerobic bacterial count (PCA, Merck), *Enterobacteriaceae* family bacteria count (VRBD, Merck), *Escherichia coli* count (TBX, Merck) and the count of bacteria belonging to the genes *Lactobacillus* (MRS, Merck), *Bifidobacterium* (RCA), *Clostridium* (TSC with D-cycloserine, Merck), *Enterococcus* (BAA, Merck), *Bacteroides* (VL, Merck) were determined. Considering the presence of yeast in the composition of synbiotic preparations, the yeast count was also determined on SDA (Merck). Plates were incubated in conditions appropriate for a given group of microorganisms: unlimited oxygen at 37 °C for 48 h (*Lactobacillus*, *Enterococcus*, *Enterobacteriaceae*), 44 °C for 48 h (*Escherichia coli*), 30 °C for five days (total yeast count) and in limited oxygen at 37 °C for 48 h (total anaerobic count, *Bifidobacterium*, *Clostridium* and *Bacteroides*) [25].

4.6. Determination of the Activity of Fecal Enzymes in the Excreta of Turkeys

The fecal enzymes activity in the excreta of turkeys was determined after 2 days and 15 weeks of the animals rearing in seven randomly selected turkeys (three repetitions) in each experimental group. The activity of α-glucosidase, β-glucosidase, α-galactosidase, β-galactosidase and β-glucuronidase in the excreta of turkeys was determined using the spectrophotometric method using the multi-plate microplate reader TriStar2 S LB 942 (Berthold Technologies, Germany). In order to prepare samples for analysis, an ultrasonic disintegrator (Sonificator Cole-Parmer Instrument Co., USA) was used with appropriate parameters (amplitude 60 Hz, pulse 6 s, break 2 s, total time 5 min). The method is based on a color reaction between the appropriate substrate and the determined enzyme. Absorbance was measured at $\lambda = 400$ nm (α-glucosidase, α-galactosidase, β-galactosidase), $\lambda = 450$ nm (β-glucosidase) and $\lambda = 540$ nm (β-glucuronidase). The activity of the determined enzyme [μMh/g] was expressed as the amount of p-nitrophenol [μM] (α-glucosidase, β-glucosidase, α-galactosidase, β-galactosidase) or phenolphthalein (β-glucuronidase) released under specific conditions within 1 h on 1 g of the excreta.

5. Statistical Analysis

The limit of quantification (the LOQ) for the OTA content in feed, tissues and the intestinal content was 0.025–0.03 μg/kg. The normality of the distribution of variables was examined with Shapiro–Wilk's test, and the homogeneity of variances was tested with Bartlett's test [16]. Following the confirmation of normality and equal variance, results were analyzed using analysis of variance with a one-way ANOVA test and Tukey's post hoc test. Differences between samples with normal distribution were also evaluated by Student's t-test. In addition, Principal components' analysis (PCA) and Agglomerative hierarchical cluster analysis (AHC) of overall diversity in the intestinal microbiota was performed to compare all groups of turkeys at the time of probiotics and synbiotics supplementation. Statistical analysis was performed using XLSTAT Software (Addinsoft, SARL, Paris, France) at the significance level of $p < 0.05$. The results were presented as mean ± standard deviation (SD).

6. Patents

The strains from the new elaborated synbiotic preparations possess full probiotic documentation described in patent applications and the patent description [56–61].

Supplementary Materials: The following are available online at http://www.mdpi.com/2072-6651/12/9/578/s1. Table S1: Counts of microorganisms which dominate in the content of the jejunum and the caecum and also the excreta of turkeys. One-way ANOVA with post-hoc Tukey's test ($p < 0.05$).

Author Contributions: K.Ś. initiated the research concept, supervised experiments and wrote the manuscript. P.M.-K. conducted microbiology experiments, analyzed results and wrote the manuscript. A.S. took part in the development of the composition and final formulas of synbiotic preparations. K.L. and M.M.-K. bred turkeys, determined the performance of animals, analyzed results and wrote the part of manuscript. All authors have read and agreed to the published version of the manuscript.

Funding: This research was supported by a grant from the Program of Applied Research funded by the National Centre for Research and Development no. PBS3/A8/32/2015 "Synbiotic preparation for health prophylaxis of monogastric animals, prevention of bacterial diseases and poisonings caused by toxins, and enhancing nutrition safety and breeding productivity of animals".

Acknowledgments: We would like to thank our students Milena Perek and Aleksanda Rybicka for help in determination of the intestinal microbiota and the fecal enzymes activity of turkeys.

Conflicts of Interest: The authors declare that they have no competing interests.

References

1. Yiannikouris, A.; Jouany, J.P. Mycotoxins in feeds and their fate in animals: A review. *Anim. Res.* **2002**, *51*, 81–99. [CrossRef]
2. Zain, M.E. Impact of mycotoxins on humans and animals. *J. Saudi Chem. Soc.* **2011**, *15*, 129–144. [CrossRef]
3. Patil, R.D.; Sharma, R.; Asrani, R.K. Mycotoxicosis and its control in poultry: A review. *J. Poult. Sci. Technol.* **2014**, *2*, 1–10.
4. Santos Pereira, C.; C Cunha, S.; Fernandes, J.O. Prevalent Mycotoxins in Animal Feed: Occurrence and Analytical Methods. *Toxins* **2019**, *11*, 290. [CrossRef]
5. Hassen, W.; Ayed-Boussema, I.; Bouslimi, A.; Bacha, H. Heat shock proteins (Hsp 70) response is not systematic to cell stress: Case of the mycotoxin ochratoxin A. *Toxicology* **2007**, *242*, 63–70. [CrossRef]
6. Mishra, S.; Dwivedi, P.D.; Pandey, H.P.; Das, M. Role of oxidative stress in Deoxynivalenol induced toxicity. *Food Chem. Toxicol.* **2014**, *72*, 20–29. [CrossRef]

7. Akande, K.E.; Abubakar, M.M.; Adegbola, T.A.; Bogoro, S.E. Nutritional and health implications of mycotoxins in animal feeds: A review. *Pak. J. Nutr.* **2006**, *5*, 398–403.
8. Official Journal of the European Union 2006/576/EU. Commission Recommendation of 17 August 2006 on the presence of deoxynivalenol, zearalenone, ochratoxin A, T-2 and HT-2 and fumonisins in products intended for animal feeding. *Off. J. Eur. Comm.* **2006**, *229*, 7–9.
9. Maresca, M.; Mahfoud, R.; Pfohl-Leszkowicz, A.; Fantini, J. The mycotoxin ochratoxin A alters intestinal barrier and absorption functions but has no effect on chloride secretion. *Toxicol. Appl. Pharmacol.* **2001**, *176*, 54–63. [CrossRef]
10. Hamilton, P.B.; Huff, W.E.; Harris, J.R.; Wyatt, R.D. Natural occurrences of ochratoxicosis in poultry. *Poult. Sci.* **1982**, *51*, 1832–1841. [CrossRef]
11. Raju, M.V.L.N.; Devegowda, G. Influence of esterified-glucomannan on performance and organ morphology, serum biochemistry and haematology in broilers exposed to individual and combined mycotoxicosis (aflatoxin, ochratoxin and T-2 toxin). *Br. Poult. Sci.* **2000**, *41*, 640–650. [CrossRef] [PubMed]
12. Verma, J.; Johri, T.S.; Swain, B.K.; Ameena, S. Effect of graded levels of aflatoxin, ochratoxin and their combinations on the performance and immune response of broilers. *Br. Poult. Sci.* **2004**, *45*, 512–518. [CrossRef] [PubMed]
13. Elaroussi, M.A.; Mohamed, F.R.; El Barkouky, E.M.; Atta, A.M.; Abdou, A.M.; Hatab, M.H. Experimental ochratoxicosis in broiler chickens. *Avian Pathol.* **2006**, *35*, 263–269. [CrossRef] [PubMed]
14. Abidin, Z.; Khan, M.Z.; Khatoon, A.; Saleemi, M.K.; Khan, A. Protective effects of lcarnitine upon toxicopathological alterations induced by ochratoxin A in white Leghorn cockerels. *Toxin Rev.* **2016**, *35*, 157–164. [CrossRef]
15. Grajewski, J. Mikotoksyny I Mikotoksykozy Zagrożeniem dla Człowieka i Zwierząt. In *Mikotoksyny I Grzyby Pleśniowe—Zagrożenia dla Człowieka I Zwierząt*; Grajewski, J., Ed.; Wydawnictwo UKW: Bydgoszcz, Poland, 2006; pp. 57–58. (In Polish)
16. Markowiak, P.; Śliżewska, K.; Nowak, A.; Chlebicz, A.; Żbikowski, A.; Pawłowski, K.; Szeleszczuk, P. Probiotic microorganisms detoxify ochratoxin A in both a chicken liver cell line and chickens. *J. Sci. Food Agric.* **2019**, *99*, 4309–4318. [CrossRef] [PubMed]
17. Schatzmayr, G.; Zehner, F.; Täubel, M.; Schatzmayr, D.; Klimitsch, A.; Loibner, A.P.; Binder, E.M. Microbiologicals for deactivating mycotoxins. *Mol. Nutr. Food Res.* **2006**, *50*, 543–551. [CrossRef]
18. McCormick, S.P. Microbial detoxification of mycotoxins. *J. Chem. Ecol.* **2013**, *39*, 907–918. [CrossRef]
19. Food and Agriculture Organization (FAO). *Guidelines for the Evaluation of Probiotics in Food*; Report of a Joint FAO/WHO Working Group on Drafting Guidelines for the Evaluation of Probiotics in Food; FAO: London, UK; Quebec City, ON, Canada, 2002.
20. Hill, C.; Guarner, F.; Reid, G.; Gibson, G.R.; Merenstein, D.J.; Pot, B.; Morelli, L.; Canani, R.B.; Flint, H.J.; Salminen, S.; et al. Expert consensus document: The International Scientific Association for Probiotics and Prebiotics consensus statement on the scope and appropriate use of the term probiotic. *Nat. Rev. Gastroenterol. Hepatol.* **2014**, *11*, 506–514. [CrossRef]
21. Food and Agriculture Organization (FAO). *Technical Meeting on Prebiotics: Food Quality and Standards Service (AGNS)*; FAO Technical Meeting Report; FAO: Rome, Italy, 2007.
22. Topcu, A.; Bulat, T.; Wishah, R.; Boyaci, I.H. Detoxification of aflatoxin B1 and patulin by *Enterococcus faecium* strains. *Int. J. Food Microbiol.* **2010**, *139*, 202–205. [CrossRef]
23. Vinderola, G.; Ritieni, A. Role of probiotics against mycotoxins and their deleterious effects. *J. Food Res.* **2015**, *4*, 10–21. [CrossRef]
24. Armando, M.R.; Pizzolitto, R.P.; Dogi, C.A.; Cristofolini, A.; Merkis, C.; Poloni, V.; Dalcero, A.M.; Cavaglieri, L.R. Adsorption of ochratoxin A and zearalenone by potential probiotic *Saccharomyces cerevisiae* strains and its relation with cell wall thickness. *J. Appl. Microbiol.* **2012**, *113*, 256–264. [CrossRef] [PubMed]
25. Śliżewska, K.; Markowiak, P.; Żbikowski, A.; Szeleszczuk, P. Effects of synbiotics on the gut microbiota, blood and rearing parameters of chickens. *FEMS Microbiol. Lett.* **2019**, *366*, fnz116. [CrossRef] [PubMed]
26. Śliżewska, K.; Markowiak-Kopeć, P.; Żbikowski, A.; Szeleszczuk, P. The effect of synbiotic preparations on the intestinal microbiota and her metabolism in broiler chickens. *Sci. Rep.* **2020**, *10*, 4281. [CrossRef]
27. Biernasiak, J.; Piotrowska, M.; Libudzisz, Z. Detoxification of mycotoxins by probiotic preparation for broiler chickens. *Mycotoxin Res.* **2006**, *22*, 230–235. [CrossRef] [PubMed]

28. Kabak, B.; Brandon, E.F.; Var, I.; Blokland, M.; Sips, A.J. Effects of probiotic bacteria on the bioaccessibility of aflatoxin B1 and ochratoxin A using an in vitro digestion model under fed conditions. *J. Environ. Sci. Health* **2009**, *44*, 472–480. [CrossRef]
29. Śliżewska, K.; Piotrowska, M. Reduction of ochratoxin A in chicken feed using probiotic. *Ann. Agric. Environ. Med.* **2014**, *21*, 676–680. [CrossRef]
30. Mazur-Kuśnirek, M.; Antoszkiewicz, Z.; Lipiński, K.; Fijałkowska, M.; Purwin, C.; Kotlarczyk, S. The effect of polyphenols and vitamin E on the antioxidant status and meat quality of broiler chickens fed diets naturally contaminated with ochratoxin A. *Arch. Anim. Nutr.* **2019**, *73*, 431–444. [CrossRef]
31. Santin, E.; Paulillo, A.C.; Nakagui, L.S.O.; Alessi, A.C.; Polveiro, W.J.C.; Maiorka, A. Evaluation of cell wall yeast as adsorbent of ochratoxin in broiler diets. *Int. J. Poult. Sci.* **2003**, *2*, 465–468.
32. Duarte, S.C.; Lino, C.M.; Pena, A. Ochratoxin A in feed of food-producing animals: An undesirable mycotixn with health and performance effects. *Vet. Microbiol.* **2011**, *154*, 1–13. [CrossRef]
33. Pozzo, L.; Cavallarin, L.; Antoniazzi, S.; Guerre, P.; Biasibetti, E.; Capucchio, M.T.; Schiavone, A. Feeding a diet contaminated with ochratoxin A for broiler chickens at the maximum level recommended by the EU for poultry feeds (0.1 mg/kg). 2. Effects on meat quality, oxidative stress, residues and histological traits. *Anim. Physiol. Anim. Nutr.* **2013**, *97*, 23–31. [CrossRef]
34. Bozzo, G.; Ceci, E.; Bonerba, E.; Desantis, S.; Tantillo, G. Ochratoxin A in laying hens: High-performance liquid chromatography detection and cytological and histological analysis of target tissues. *J. Appl. Poult. Res.* **2008**, *17*, 151–156. [CrossRef]
35. Denli, M.; Blandon, J.C.; Guynot, M.E.; Salado, S.; Perez, J.F. Efficacy of a new ochratoxin-binding agent (OcraTox) to counteract the deleterious effects of ochratoxin A in laying hens. *Poult. Sci.* **2008**, *87*, 2266–2272. [CrossRef] [PubMed]
36. Zaghini, A.; Simioli, M.; Roncada, P.; Rizzi, L. Effect of *Saccharomyces cerevisiae* and esterified glucomannan on residues of Ochratoxin A in kidney, muscle and blood of laying hens. *Ital. J. Anim. Sci.* **2007**, *6* (Suppl. S1), 737–739. [CrossRef]
37. Kozaczynski, W. Experimental ochratoxicosis A in chickens. Histopathological and histochemical study. *Arch. Vet. Pol.* **1994**, *34*, 205–219.
38. Khatoon, A.; Abidin, Z. An extensive review of experimental ochratoxicosis in poultry: I. Growth and production parameters along with histopathological alterations. *World Poult. Sci. J.* **2018**, *74*, 627–646. [CrossRef]
39. Petzinger, E.; Ziegler, K. Ochratoxin A from a toxicological perspective. *J. Vet. Pharmacol. Ther.* **2000**, *23*, 91–98. [CrossRef]
40. Ringot, D.; Chango, A.; Schneider, Y.J.; Larondelle, Y. Toxicokinetics and toxicodynamics of ochratoxin A, an update. *Chem. Biol. Interact.* **2006**, *159*, 18–46. [CrossRef]
41. Huwig, A.; Freimund, S.; Ka¨ppeli, O.; Dutler, H. Mycotoxin detoxication of animal feed by different adsorbents. *Toxicol. Lett.* **2001**, *122*, 179–188. [CrossRef]
42. Dibaji, S.M.; Seidavi, A.; Asadpour, L.; da Silva, F.M. Effect of a synbiotic on the intestinal microflora of chickens. *J. Appl. Poult. Res.* **2014**, *23*, 1–6. [CrossRef]
43. Biernasiak, J.; Śliżewska, K.; Libudzisz, Z.; Smulikowska, S. Effects of dietary probiotic containing *Lactobacillus* bacteria, yeasts and yucca extract on the performance and faecal microflora of broiler chickens. *Pol. J. Food. Nutr. Sci.* **2007**, *57*, 19–25.
44. Lan, P.T.N.; Binh, L.T.; Banno, Y. Impact of two probiotic *Lactobacillus* strains feeding on fecal lactobacilli and weigth gains in chickens. *J. Gen. Appl. Microbiol.* **2003**, *49*, 29–36. [PubMed]
45. Uccello, M.; Malaguarnera, G.; Basile, F.; D'agata, V.; Malaguarnera, M.; Bertino, G.; Vacante, M.; Drago, F.; Biondi, A. Potential role of probiotics on colorectal cancer prevention. *BMC Surg.* **2012**, *12*, S35. [CrossRef] [PubMed]
46. Jin, L.Z.; Ho, Y.W.; Abdullah, N.; Jalaludin, S. Digestive and bacterial enzyme activities in broilers fed diets supplemented with *Lactobacillus* cultures. *Poult. Sci.* **2000**, *79*, 886–891. [CrossRef] [PubMed]
47. Shokryazdan, P.; Jahromi, M.F.; Liang, J.B.; Ramasamy, K.; Sieo, C.C.; Ho, Y.W. Effects of a *Lactobacillus salivarius* mixture on performance, intestinal health and serum lipids of broiler chickens (online). *PLoS ONE* **2017**, *12*, e0175959. [CrossRef] [PubMed]

48. Boostani, A.; Mahmoodian Fard, H.R.; Ashayerizadeh, A.; Aminafshar, M. Growth performance, carcass yield and intestinal microflora populations of broilers fed diets containing Thepax and yogurt. *Braz. J. Poult. Sci.* **2013**, *15*, 1–6. [CrossRef]
49. Juśkiewicz, J.; Jankowski, J.; Zduńczyk, Z.; Mikulski, D. Performance and gastrointestinal tract metabolism of turkeys fed diets with different contents of fructooligosaccharides. *Poult. Sci.* **2006**, *85*, 886–891. [CrossRef]
50. Čokášová, D.; Bomba, A.; Strojný, L.; Pramuková, B.; Szabadosová, V.; Salaj, R.; Žofčáková, J.; Brandeburová, A.; Supuková, A.; Šoltésová, A.; et al. The effect of new probiotic strain *Lactobacillus plantarum* on counts of coliforms, lactobacilli and bacterial enzyme activities in rats exposed to N,N-dimethylhydrazine (chemical carcinogen). *Acta Vet. Brno* **2012**, *81*, 189–194. [CrossRef]
51. Śliżewska, K.; Nowak, A.; Piotrowska, M.; Żakowska, Z.; Gajęcka, M.; Zielonka, Ł.; Gajęcki, M. Cecal enzyme activity in gilts following experimentally induced *Fusarium mycotoxicosis*. *Pol. J. Vet. Sci.* **2015**, *18*, 191–197. [CrossRef]
52. Śliżewska, K.; Chlebicz-Wójcik, A. The In Vitro Analysis of Prebiotics to Be Used as a Component of a Synbiotic Preparation. *Nutrients* **2020**, *12*, 1272. [CrossRef]
53. Śliżewska, K.; Chlebicz-Wójcik, A.; Nowak, A. Probiotic Properties of New *Lactobacillus* Strains Intended to Be Used as Feed Additives for Monogastric Animals. *Probiotics Antimicrob. Prot.* **2020**, 1–17. [CrossRef]
54. Chlebicz, A.; Śliżewska, K. In Vitro Detoxification of Aflatoxin B1, Deoxynivalenol, Fumonisins, T-2 Toxin and Zearalenone by Probiotic Bacteria from Genus *Lactobacillus* and *Saccharomyces cerevisiae* Yeast. *Probiotics Antimicrob. Proteins* **2020**, *12*, 289–301. [CrossRef] [PubMed]
55. PN-EN 16007-2012-U. *Animal Feed—Determination of Ochratoxin A in Animal Feed by Immunoaffinity Column Clean-Up and High-Performance Liquid Chromatography with Fluorescence Detection*; PKN: Warszawa, Poland, 2012.
56. Śliżewska, K.; Chlebicz, A. Lactic Bacterial Strain of *Lactobacillus pentosus*. U.S. Patent 422589, Patent Description no. PL 233 261, 21 August 2017.
57. Śliżewska, K.; Chlebicz, A. Lactic Bacterial Strain of *Lactobacillus reuteri*. U.S. Patent 422593, Patent Description no. PL 233 263, 21 August 2017.
58. Śliżewska, K.; Chlebicz, A. Lactic Bacterial Strain of *Lactobacillus rhamnosus*. U.S. Patent 422602, Patent Description no. PL 233 582, 21 August 2017.
59. Śliżewska, K.; Chlebicz, A. Strain of Yeast *Saccharomyces cerevisiae*. U.S. Patent 422709, Patent Description no. PL 233 581, 31 August 2017.
60. Śliżewska, K.; Chlebicz, A. Lactic Bacterial Strain of *Lactobacillus paracasei*. U.S. Patent 422603, Patent Description no. PL 233 262, 21 August 2017.
61. Śliżewska, K.; Motyl, I.; Libudzisz, Z.; Otlewska, A.; Burchardt, H.; Klecha, J.; Henzler, J. *Lactobacillus plantarum* Lactic Bacteria Strain. U.S. Patent 401554, Patent Description no. PL 221 959 B1, 12 November 2012.

© 2020 by the authors. Licensee MDPI, Basel, Switzerland. This article is an open access article distributed under the terms and conditions of the Creative Commons Attribution (CC BY) license (http://creativecommons.org/licenses/by/4.0/).

Review

Bioactive Metabolites and Potential Mycotoxins Produced by *Cordyceps* Fungi: A Review of Safety

Bo Chen [1,2], Yanlei Sun [1,2], Feifei Luo [1] and Chengshu Wang [1,2,3,*]

1 Key Laboratory of Insect Developmental and Evolutionary Biology, CAS Center for Excellence in Molecular Plant Sciences, Shanghai Institute of Plant Physiology and Ecology, Chinese Academy of Sciences, Shanghai 200032, China; chenbo@sippe.ac.cn (B.C.); sunyanlei@sippe.ac.cn (Y.S.); ffluo@sibs.ac.cn (F.L.)
2 CAS Center for Excellence in Biotic Interactions, University of Chinese Academy of Sciences, Beijing 100049, China
3 School of Life Science and Technology, ShanghaiTech University, Shanghai 201210, China
* Correspondence: wangcs@sippe.ac.cn; Tel.: +86-21-549242157

Received: 30 May 2020; Accepted: 15 June 2020; Published: 19 June 2020

Abstract: Ascomycete *Cordyceps* fungi such as *C. militaris*, *C. cicadae*, and *C. guangdongensis* have been mass produced on artificial media either as food supplements or health additives while the byproducts of culture substrates are largely used as animal feed. The safety concerns associated with the daily consumption of *Cordyceps* fungi or related products are still being debated. On the one hand, the known compounds from these fungi such as adenosine analogs cordycepin and pentostatin have demonstrated different beneficial or pharmaceutical activities but also dose-dependent cytotoxicities, neurological toxicities and or toxicological effects in humans and animals. On the other hand, the possibility of mycotoxin production by *Cordyceps* fungi has not been completely ruled out. In contrast to a few metabolites identified, an array of biosynthetic gene clusters (BGCs) are encoded in each genome of these fungi with the potential to produce a plethora of as yet unknown secondary metabolites. Conservation analysis of BGCs suggests that mycotoxin analogs of PR-toxin and trichothecenes might be produced by *Cordyceps* fungi. Future elucidation of the compounds produced by these functionally unknown BGCs, and in-depth assessments of metabolite bioactivity and chemical safety, will not only facilitate the safe use of *Cordyceps* fungi as human food or alternative medicine, but will also benefit the use of mass production byproducts as animal feed. To corroborate the long record of use as a traditional medicine, future efforts will also benefit the exploration of *Cordyceps* fungi for pharmaceutical purposes.

Keywords: *Cordyceps* fungi; mass production; mycotoxins; biosynthetic gene cluster; toxicity; safety

Key Contribution: A few species of *Cordyceps* fungi have been used as traditional medicines for a long time. These fungi can now be mass-produced at magnitudes from dozens to thousands of tons a year in Asian countries. Based on the known compounds and putative mycotoxins that might be produced by these fungi; safety assessments and future efforts are still required to alleviate concerns when consuming these fungi as food supplements or health additives and using the related byproducts as animal feed.

---

1. Introduction

Filamentous fungi are rich producers of bioactive secondary metabolites, some of which either have been developed as commercial drugs to save lives or are carcinogenic or neurotoxic mycotoxins that threaten human health [1]. The production of these compounds has long been considered to be dispensable for fungal biology since the disruption of the biosynthetic gene clusters (BGCs) could barely, if at all, impair fungal growth and development under experimental conditions [2].

However, many studies have shown that the small molecules produced by fungi play essential roles in fungus–environment interactions [3,4]. For example, metabolites with antibiotic and antifungal activities are used by the producing fungi to outcompete other microbes [5,6]. The phytotoxins produced by plant pathogens are required in mediating fungus–plant interactions [7], and insecticidal toxins biosynthesized by insect pathogens facilitate fungal infection of insect hosts [8,9]. In addition, the mycotoxins produced by plant pathogens or endophytes can frequently result in toxic effects at different levels in herbivorous animals [10], which has been considered as a strategy employed by plants against grazers [11]. Thus, small molecules produced by fungi are of biological importance to producers and beyond.

In nature, there are more than 1000 species of fungi that can infect and kill insects, most of which are ascomycete entomopathogenic fungi [12]. In particular, three families of Hypocrealean fungi, i.e., Cordycipitaceae, Ophiocordycipitaceae, and Clavicipitaceae, contain species that have been used either for biocontrol of insect pests or as Traditional Chinese Medicine (TCM) in Asian countries and beyond [13,14]. For example, the species of *Ophiocordyceps sinensis*, *Cordyceps militaris*, and *C. cicadae* (syn. *Isaria cicadae*) have been used as TCM with recorded antibiotic, anti-inflammatory, anti-aging, anti-cancer, anti-proliferative, anti-metastatic, anti-fatigue, and immunomodulatory activities or effects (Table 1) for a long time in history [15,16]. On the other hand, similar to other traditional herbs, the safety of consuming these fungi has long been a concern [16]. A plethora of bioactive metabolites have been identified from Hypocrealean entomopathogens [14,17,18], including those which have been developed as commercial drugs such as the immunosuppressant drug cyclosporin A, a cyclodepsipeptide isolated from the Ophiocordycipitaceae fungus *Tolypocladium inflatum* [6] and the anti-leukemia drug pentostatin, first isolated from *Streptomyces*, found in *C. militaris* [19]. Apart from the medicinal or beneficial effects reported for the chemicals identified from *Cordyceps* fungi, side effects of cytotoxicity and or neurological toxicity have also been reported for these compounds (Table 1). Anecdotal records of nausea, diarrhea and even excessive post-extraction bleeding have been reported after daily consumption of *Cordyceps* fruiting bodies or related products [20,21]. Thus, in-depth safety evaluation is still required before consuming these fungi.

Table 1. Summary of the reported bioactivity and toxicity of the compounds identified from *Cordyceps* fungi.

Compound	Producing Fungus	Bioactivities	Toxic Effect
Cordycepin	*C. militaris*; *C. kyusyuensis*	Anticancer, anti- inflammatory, antioxidant, inhibition of RNA synthesis, insecticidal, antibiotic, antifungal, antivirus	Gastrointestinal toxicity, bone marrow toxicity, decrease in toxicity
Pentostatin	*C. militaris*	Immunosuppressive, inhibitor of adenosine deaminase, antineoplastic	Nausea, diarrhea, renal and neurological toxicities, pulmonary toxicity, gastrointestinal toxicity
N6-(2-Hydroxy-ethyl)-adenosine	*C. militaris*; *C. cicadae*	Renal protection, anti-cancer, insecticidal	Induction of oxidative stress
Tenellin	*C. bassiana*	Iron chelation, inhibitor of membrane ATPase	Toxic towards erythrocytes
Militarinones	*C. militaris*	Antimicrobial	Cytotoxicity
Fumosorinone	*C. fumosorosea*	Inhibitor of tyrosine phosphatase 1B, activation of insulin pathway, anti-diabetic	/
Farinosones	*C. farinosa*	Neuritotrophic activity	Cytotoxicity
Oosporein	*C. cicadae*; *C. bassiana*	Immunosuppressive, antimicrobial, metal detoxification	Cytotoxicity

Table 1. *Cont.*

Compound	Producing Fungus	Bioactivities	Toxic Effect
Beauveriolides	*C. militaris*; *C. bassiana*	Anti-aging, beta-amyloid lowering, anti-atherogenic	Cytotoxicity
Beauvericin	*C. cicadae*; *C. bassiana*	Insecticidal, nematicidal, induction of cell apoptosis, ionophoric property	Cytotoxicity
Cordyceamides	*O. sinensis*	/	Cytotoxicity
Cordycedipeptide	*O. sinensis*	/	Cytotoxicity
Cordysinins	*O. sinensis*	Anti-inflammatory	/

Cordyceps fungi are being mass produced for harvesting the fruiting bodies for food and health additives while the byproducts (largely culture substrates) of mass production are then mostly used as animal feed [22]. To alleviate the safety concerns for both purposes, this paper reviews the production and biological activity/toxicity of known metabolites identified from *Cordyceps* fungi and unknown metabolites deduced from the conserved biosynthetic gene clusters (BGCs) by comparative analysis with those BGCs involved in producing known mycotoxins based on the obtained genome information of these fungi [23–25]. The content of literature reviews conducted in the paper may benefit the future exploration and safety assessment of *Cordyceps* fungi and their related byproducts used for food, traditional medicine or animal feed.

2. Mass Production of *Cordyceps* Fungi

Ascomycete entomopathogenic fungi are facultative with saprophytic growth abilities. However, it is still technically challenging, often difficult, to induce fungal sexual fruiting bodies on artificial media or on insect hosts under laboratory conditions [26]. Until recently, induction of the fruiting-body formation of the caterpillar fungus *O. sinensis* (best known as *C. sinensis*, one of the most expensive traditional medicines) was not successful [27,28]. After inoculation of the ghost moth (*Hepialus* spp.) larvae, it is an energy-intensive process for fungal infection and development within the insect body cavity at a relatively low temperature (less than 18 °C) and the formation of fruiting bodies for a total period of more than half a year (Figure 1A). Dozens of tons of the fruiting bodies (attached with insect cadavers) can now be produced annually in China [28]. In contrast, mass production of *C. militaris* has long been successful by inoculation of fungal propagates on artificial media (e.g., rice medium) (Figure 1B). Different from the homothallic nature of *O. sinensis* [26], *C. militaris* is sexually heterothallic but its single mating-type can also fruit but without mating and meiosis to produce sexual perithecia [29]. Without considering the mass production of *C. militaris* in Japan, South Korea and Vietnam, the annual yield of the dried fruiting bodies reaches up to 10,000 tons per year in China [22]. The fruiting bodies of *C. cicadae* together with the mycosed cicada pupae have also been used as a TCM in renal protection or for the treatment of chronic kidney disease [23]. Mass production of this fungus has also been successful with a yield of hundreds of tons each year (Figure 1C). In contrast to *C. militaris*, asexual fruiting bodies, i.e., synnema-like structures, are largely produced by *C. cicadae* [23]. A few other species such as *C. guangdongensis* have also been mass produced at different magnitudes in Asian countries. Without consideration of the liquid fermentations of *Cordyceps* fungi [22], careful evaluation and safety assessment are still required regarding the (daily) consumption of enormous amounts of fruiting bodies and their related products as foods or health-promoting additives, and utilization of the leftover substrates as animal feed.

Figure 1. Mass production of *Cordyceps* fungi. (**A**) Successful induction of the fruiting bodies (arrowed) of *O. sinensis* after inoculation of the ghost moth larvae for more than 150 days (image taken from the Sunshine Lake LLC, Yichang, China). Bar, 1 cm. (**B**) Mass production of *C. militaris* in plastic bottles (image taken from the Honghao Biotech Company, Jiangmeng, China). Insert, the fruiting bodies formed in a bottle 45 days post-inoculation. Bar, 1 cm. (**C**) Mass production of *C. cicadae* in plastic boxes (image taken from the BioAisa Pharmaceuticals, Pinghu, China). Insert, the fruiting bodies formed in a box 20 days post-inoculation. Bar, 1 cm.

3. Known Metabolites Produced by *Cordyceps* Fungi

Different compounds have been isolated from entomopathogenic fungi including *Cordyceps* species [17]. One of the most important but arguable metabolites produced by *C. militaris* is cordycepin, i.e., the adenosine analog 3′-deoxyadenosine (Figure 2). This compound is not produced in *O. sinensis* and *C. cicadae;* elucidation of its biosynthetic mechanism showed that the responsible BGC is only present in *C. militaris* and *C. kyusyuensis* as well as the evolutionarily distant mold fungus *Aspergillus nidulans* [19]. It was later found that the field-collected samples of *O. sinensis*, i.e., the complex of fruiting body and insect cadaver, were frequently contaminated with *A. nidulans* and even *C. militaris* that might contribute to the detection of trace amounts of cordycepin in caterpillar fungus [30]. Cordycepin can inhibit RNA synthesis and has demonstrated immense medicinal potential (Table 1), including anti-cancer, anti-inflammatory, antibiotic, anti-virus, and antioxidant activities [31]. However, in addition to its dosage-dependent toxicity to different cells, cordycepin has shown an effect in stimulating testosterone production in the models of both mouse Ledydig cells and mice, which may alter male fertility [32]. Recently, it has been shown that the anti-leukemia drug pentostatin (Figure 2), i.e., the 2′-deoxycoformycin originally isolated from *Streptomyces antibioticus* being an irreversible inhibitor of adenosine deaminase, can also be produced by *C. militaris* through the same BGC for cordycepin production via a protector–protégé strategy [19]. Similar to other chemotherapeutic drugs/agents, dosage- and schedule-dependent side effects have also been observed for pentostatin that include nausea, diarrhea, and renal and neurological toxicities [33]. The test of cordycepin in combination with pentostatin can trigger severe gastrointestinal toxicity and bone marrow toxicity in dogs [34].

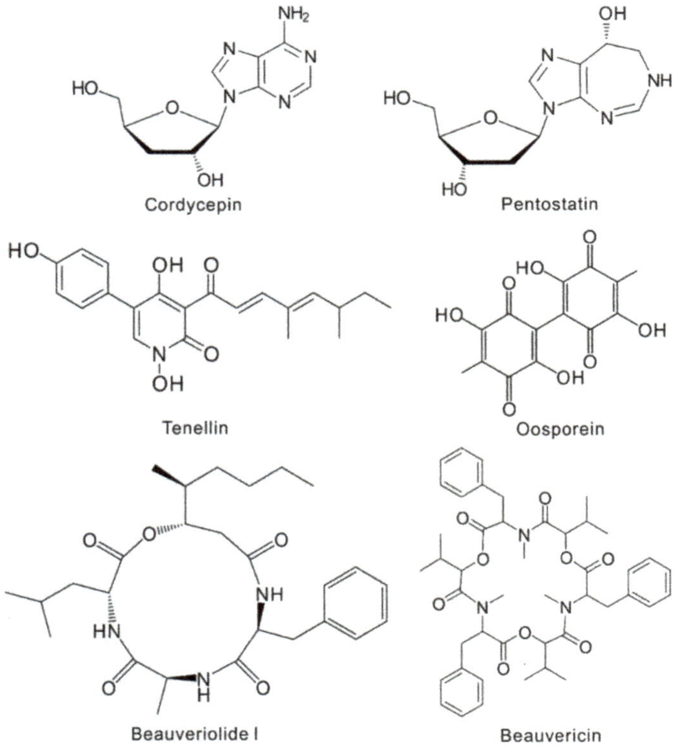

Cordycepin

Pentostatin

Tenellin

Oosporein

Beauveriolide I

Beauvericin

Figure 2. Structure of the selected metabolites identified from *Cordyceps* fungi.

Apart from cordycepin and pentostatin, another adenosine analog N^6-(2-Hydroxyethyl)-adenosine (HEA) has been identified in *C. cicadae*, *C. militaris*, and other species of *Cordyceps* with renal protection and anti-cancer activities [35,36]. Insecticidal activity of HEA has also been demonstrated by targeting the adenosine receptor (AdoR) of insects [37], suggesting that adenosine analogs can be recognized by AdoR(s) [38]. A different family of AdoRs, the G-protein coupled receptors with seven transmembrane domains, has been identified in humans as potential drug targets [39]. Since adenosine is multifunctional in the physiology of different organisms, the functions of cordycepin and HEA are still unclear, notably whether they act as agonists or antagonists of AdoRs in mammals. Deletion of *AdoRa1* in mice has led to decreased fertility and an increased risk of seizures [38]. The long-term effect of *Cordyceps* consumption and the effects of cordycepin and HEA administration on activation or inactivation of AdoRs require further investigation.

Other known metabolites produced by either *C. militaris* or *C. cicadae* include 2-pyridone alkaloid tenellin-like compounds and the bibenzoquinone oosporein (Figure 2). Different structures of Tenellin-like pyridones have been identified in different Cordycipitaceae fungi including fumosorinone produced by *C. (Isaria) fumosorosea* [40], farinosones by *I. farinosus* [41], and militarinones by *C. militaris* [42]. These 2-pyridones (Table 1) can maintain iron hemostasis and have also shown profound neuritogenic activity and cell cytotoxicities [42]. Oosporein, originally identified from the insect pathogen *Beauveria bassiana* (Syn. *Cordyceps bassiana*), shows insecticidal and antibiotic activities that promote fungal infection of insect hosts [9,43]. The conserved gene cluster and production of oosporein has been detected in *C. cicadae* [23]. This compound can also cause gout in avian species including chickens, turkeys and other birds, and can therefore threaten the safety of the poultry industry if contaminated substrates are used as feed [44].

A few cyclodepsipeptides (Figure 2) have also been identified from *Cordyceps* fungi, e.g., beauveriolide I and III from *C. militaris* [45] and beauvericin from *C. cicadae* [23,46]. Beauveriolides have demonstrated anti-aging [47], beta-amyloid-lowering [48], and anti-atherogenic activity by inhibition of lipid droplet accumulation in macrophages without any obvious side effects [49]. However, insecticidal and nematicidal beauvericin, first isolated from *B. bassiana*, can induce cytotoxicity and cell apoptosis in a dose-dependent manner due to its ionophoric property that can increase ion permeability in membranes [50]. Overall, along with the beneficial medicinal and or biological activities, different negative effects are also evident with the compounds identified from *Cordyceps* fungi that raise safety concerns about its consumption.

4. Unknown Mycotoxins May be Produced by *Cordyceps* Fungi

Apart from the metabolites described above, many more can be expected after genomic analysis of *Cordyceps* fungi that identifies an array of BGCs encoded in each fungus (Figure 3A). For example, eight non-ribosomal peptide synthetase (NRPS), seven polyketide synthase (PKS), five NRPS-PKS and four terpene synthase (TS) BGCs are encoded in the genome of *C. militaris* [25,51]. Likewise, the BGCs of eight NRPSs, eight PKSs, six NRPS-PKSs and six TSs are encoded in *C. cicadae* [23]. Consistently, metabolomic analysis also suggested that diverse compounds can be produced by these fungi [23,52]. Due to the frequent occurrence of gene silence of fungal BGCs under experimental conditions [2], many unknown metabolites remain to be determined. Genome-wide phylogenetic, gene cluster content, and core enzyme structure analyses of NRPS, PKS, and NRPS-PKS BGCs suggest that the *Cordyceps* fungi may not produce carcinogenic mycotoxins such as aflatoxins [23,25]. However, our further analysis indicated that the putative aristolochen synthase CCM_03050 is highly similar (65% identity at amino acid level) to PRX2, and the clustered CCM_03051 is similar to the short chain dehydrogenase (SDR) PRX4 of *Penicillium roqueforti* involved in PR-toxin production (Figure 3B). The bicyclic sesquiterpene PR-toxin can cause significant damage to liver and kidney, induce abortions and reduce fertility in cattle. However, its derivatives such as PR-acid and PR-imine are largely non-toxic and unstable [53]. The biosynthesis of PR-toxin requires at least four genes, i.e., *Prx1-Prx4* in *P. roqueforti* [54]. Considering that the tailoring enzyme genes *Prx1* (for a SDR) and *Prx3* (for a quinone oxidoreductase) are missing in *C. militaris* whereas the clustered *CCM_03049* encodes a putative dioxygenase/oxygenase and *CCM_03052* encodes a cytochrome P450 enzyme, this TS BGC may not biosynthesize PR-toxin but its analog(s) requiring further investigation. Intriguingly, these conserved genes are absent in the genomes of the closely-related *C. cicadae* and other Cordycipitaceae fungi, based on the survey of their genome contents.

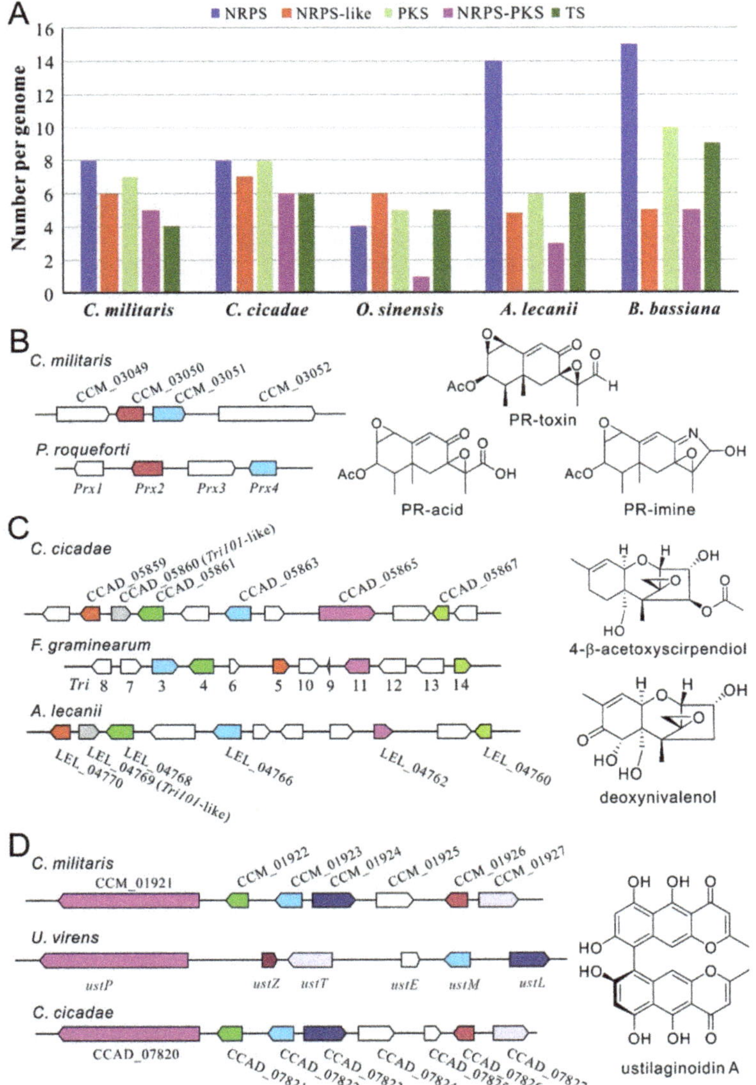

Figure 3. Conservation analysis of the gene clusters between *Cordyceps* and other fungi involved in toxin production. (**A**) Comparative analysis of the biosynthetic gene clusters (BGCs) encoded in the genomes of the selected Cordycipitaceae fungi. NRPS, non-ribosomal peptide synthetase; PKS, polyketide synthase; TS, terpene synthase. (**B**) Conservation between the gene clusters of *C. militaris* and those of *P. roqueforti* involved in PR-toxin production. (**C**) Conservation between the gene clusters of the Cordycipitaceae fungi *C. cicadae* and *A. lecanii* and that of *F. graminearum* involved in the biosynthesis of trichothecenes. (**D**) Conservation between the gene clusters of *C. militaris* and *C. cicadae* and that of *U. virens* involved in the production of ustilaginoidins. The genes labeled in the same color within the panels (**B–D**) represent orthologous relationships. The compound(s) shown in the right of the panels (**B–D**) indicates the structures of the representative metabolites produced by the known BGCs in the related or reference fungal species.

We also found that the Cordycipitaceae fungi *C. cicadae* and *Akanthomyces lecanii* (anamorph: *Lecanicillium lecanii*) each encode a gene cluster that shows conservation with the BGC involved in the production of trichothecene (Tri) mycotoxins in wheat head blight *Fusarium graminearum* (Figure 3C). Phytotoxic trichothecenes are a class of sesquiterpenes that contain more than 150 chemically-related compounds such as type A mycotoxins T-2, HT-2, and NX-2, and type B deoxynivalenol (DON), acetylated DON, and nivalenol toxins that can inhibit protein synthesis and are neurotoxic, immunosuppressive, and nephrotoxic in mammals [55,56]. The biosynthesis of trichothecenes has been clarified with the involvement of 15 *Tri* genes located at three different loci including a 12-gene core cluster, a single gene *Tri101* and two genes *Tri1* and *Tri16* loci on different chromosomes of *F. graminearum* [57,58]. We found that the homologs of the core gene *Tri5* for trichodiene synthase are conservedly present in *C. cicadae* (CCAD_05859, 70% identity) and *A. lecanii* (LEL_0770, 79% identity). The homologs of the essential biosynthetic genes *Tri3*, *Tri4*, *Tri11* and *Tri14* are also present in the genomes of these two fungi. In addition, interestingly, the homologs of the isolated gene *Tri101* (encoding an acetyltransferase) in *F. graminearum* are present in the core cluster of *Cordyceps* fungi, i.e., *LEL_04769* (40% identity) and *CCAD_05860* (37% identity) (Figure 3C). Likewise, the homologs of the isolated gene *Tri1* are also present in the separated loci of two fungi (*CCAD_02575*, 42% identity; *LEL_06734*, 41% identity). Taken together, it is considerably likely that *C. cicadae* and *A. lecanii* may produce trichothecene-like mycotoxin(s) that requires clarification and toxicity tests. In support, the trichothecene derivative 4-β-acetoxyscirpendiol (4-acetyl-12,13-epoxyl-9-trichothecene-3,15-diol) was once isolated from the fruit-body samples of *Isaria japonica* (syn. *Cordyceps tenuipes*) and the compound could induce apoptosis of human leukemia cells [59]. This *Tri*-like BGC is not present in *C. militaris* but is present in the closely-related biocontrol fungus *B. bassiana* (TRI5 vs. BBA_08696, 69% identity; TRI101 vs. BBA_08697, 41%) after genome survey [56]. A *Tri*-like BGC has also been identified in diverse fungal species including *Trichoderma* spp., which is responsible for the production of the trichothecene derivatives harzianums A and B with antimicrobial activities [60,61].

The bis-naphthopyrone-type pigments are widely produced by different fungi that can protect filamentous fungi from fungivores [62] or abiotic stress factors like UV radiation and high temperatures [63,64]. Genome survey indicated that both *C. militaris* and *C. cicadae* contain a conserved PKS BGC like that of rice false smut *Ustilaginoidea virens* (Figure 3D), which is responsible for the production of ustilaginoidins [65]. Different analogs of ustilaginoidins have varied level of antibacterial, cytotoxic, and phytotoxic activities [66]. The exact compound(s) produced by this PKS gene cluster remains to be determined in *Cordyceps* fungi.

In contrast to other *Cordyceps* fungi, the caterpillar fungus *O. sinensis* has a highly repetitive genome with a limited number of genes and BGCs (Figure 3A) [24]. Besides the reported bioactive constituents like nucleosides, sterols and polysaccharides [18,67–69], a few compounds have been identified from this fungus with different activities such as the aurantiamides cordyceamides A and B with unclear activity/toxicity [70], the cyclodipeptide cordycedipeptide A with cytotoxicity [71], and the cyclodipeptide cordysinins with antioxidant activity [69,72]. Overall, the reputed health benefits of this fungus are still unclear [73]. Despite the mass production of sexual fruiting bodies being recently successful [74], the slow growing nature of this fungus makes it technically problematic to perform transgenic and chemical genetic investigations [73]. With the help of heterologous expression systems, future efforts can be taken to explore the production, if any, of mycotoxin as well as the pharmaceutical potential of this fungus.

5. Requirement of Safety Assessments

As indicated above, toxicological concerns about consuming *Cordyceps* fungi as food supplements or using their byproducts as animal feed are raised based on the evaluations of either the known compounds they produce or those previously unreported toxins being putatively produced by the conserved BGCs of *Cordyceps* fungi. Obviously, the production of the avian-gout toxin oosporein by *C. cicadae* suggests that mass-production byproducts cannot be used as feed for poultry. In particular, except for the unknown metabolites biosynthesized by other BGCs, the potential of the production of analogous PR-toxin in *C. militaris* and Tri-like toxin(s) in *C. cicadae* raises substantial concerns that require further investigation and safety assessments. Traditionally, instead of using the purified compounds, the fruiting bodies or mycelium samples of *Cordyceps* fungi were used for direct safety assessments, including bacterial Ames tests, different cell- and/or animal-model tests for mutagenic, clastogenic, genotoxic, and (sub-)accurate toxic effects, which suggested that consumption of *Cordyceps* fungi might be safe [75,76]. However, considering the magnitude of the mass production of *Cordyceps* fungi in China and other countries, in-depth monitoring and assessment are still required, given the daily consumption of *Cordyceps* fruiting bodies or related products as tonics or food/health additives and the use of culture substrates as feed. In particular, aside from the feature of culture stability, high titer of cordycepin and pentostatin production is being used as the key standard for screening of *C. militaris* strains for industrial mass production. Consistent with the side effects of pentostatin and adenosine analogs [33,38], the negative effects of nausea and diarrhea have been anecdotally reported by enthusiasts after consuming products with enriched contents of cordycepin/pentostatin [14]. Indeed, it is common and typical that mycotoxin production and accumulation are dependent on the fungal culturing media and stage, and the side effects of daily consumptions are dose- and even consumer-dependent [20]. For example, trichothecene production was not detected in *A. lecanii* and *B. bassiana* after inductions in multiple artificial media used for the successful induction of toxin formations in other fungi [56]. It is critical at least that toxin production is clarified for the *Cordyceps* fungi under mass production conditions.

6. Conclusions and Prospective

Along with the increasing level of the mass production of *Cordyceps* fungi such as *C. militaris* and *C. cicadae*, there is increasing consumption of the fruiting bodies or related products as food supplements or health additives, and use of the byproducts as animal feed. To alleviate safety concerns, full elucidation of the BGCs' capacities in production of different compounds is critically needed. In addition to promoting toxicological tests with the known and newly-identified compounds, in-depth investigations may also benefit the exploration of these fungi for pharmaceutical potential.

Author Contributions: Conceptualization, C.W.; methodology, B.C. and C.W.; data curation, B.C., Y.S., F.L., and C.W.; manuscript writing, C.W.; supervision, C.W.; funding acquisition, C.W. All authors have read and agreed to the published version of the manuscript.

Funding: This work was supported by the National Natural Science Foundation of China (31530001) and the Shanghai Academic/Technology Research Leader Program (grant 18XD1404500).

Acknowledgments: The authors thank Chunru Li from BioAsia Pharmaceuticals for provision of the mass production pictures of *C. cicadae*.

Conflicts of Interest: The authors declare no conflict of interest.

References

1. Marin, S.; Ramos, A.J.; Cano-Sancho, G.; Sanchis, V. Mycotoxins: Occurrence, toxicology, and exposure assessment. *Food Chem. Toxicol.* **2013**, *60*, 218–237. [CrossRef] [PubMed]
2. Keller, N.P. Fungal secondary metabolism: Regulation, function and drug discovery. *Nat. Rev. Microbiol.* **2019**, *17*, 167–180. [CrossRef] [PubMed]
3. Spiteller, P. Chemical ecology of fungi. *Nat. Prod. Rep.* **2015**, *32*, 971–993. [CrossRef] [PubMed]
4. Macheleidt, J.; Mattern, D.J.; Fischer, J.; Netzker, T.; Weber, J.; Schroeckh, V.; Valiante, V.; Brakhage, A.A. Regulation and role of fungal secondary metabolites. *Annu. Rev. Genet.* **2016**, *50*, 371–392. [CrossRef] [PubMed]
5. Cray, J.A.; Bell, A.N.; Bhaganna, P.; Mswaka, A.Y.; Timson, D.J.; Hallsworth, J.E. The biology of habitat dominance; can microbes behave as weeds? *Microb. Biotechnol.* **2013**, *6*, 453–492. [CrossRef] [PubMed]
6. Yang, X.; Feng, P.; Yin, Y.; Bushley, K.; Spatafora, J.W.; Wang, C. Cyclosporine biosynthesis in *Tolypocladium inflatum* benefits fungal adaptation to the environment. *mBio* **2018**, *9*, e01211-18. [CrossRef]

7. Perincherry, L.; Lalak-Kańczugowska, J.; Stępień, Ł. *Fusarium*-produced mycotoxins in plant-pathogen interactions. *Toxins* **2019**, *11*, 664. [CrossRef]
8. Rohlfs, M.; Churchill, A.C. Fungal secondary metabolites as modulators of interactions with insects and other arthropods. *Fungal Genet. Biol.* **2011**, *48*, 23–34. [CrossRef]
9. Feng, P.; Shang, Y.; Cen, K.; Wang, C. Fungal biosynthesis of the bibenzoquinone oosporein to evade insect immunity. *Proc. Natl. Acad. Sci. USA* **2015**, *112*, 11365–11370. [CrossRef]
10. Philippe, G. Lolitrem B and indole diterpene alkaloids produced by endophytic fungi of the genus *Epichloë* and their toxic effects in livestock. *Toxins* **2016**, *8*, 47. [CrossRef]
11. Rohlfs, M. Fungal secondary metabolite dynamics in fungus-grazer interactions: Novel insights and unanswered questions. *Front. Microbiol.* **2014**, *5*, 788.
12. Shang, Y.; Feng, P.; Wang, C. Fungi that infect insects: Altering host behavior and beyond. *PLoS Pathog.* **2015**, *11*, e1005037. [CrossRef] [PubMed]
13. Wang, C.; Wang, S. Insect pathogenic fungi: Genomics, molecular interactions, and genetic improvements. *Annu. Rev. Entomol.* **2017**, *62*, 73–90. [CrossRef] [PubMed]
14. Olatunji, O.J.; Tang, J.; Tola, A.; Auberon, F.; Oluwaniyi, O.; Ouyang, Z. The genus *Cordyceps*: An extensive review of its traditional uses, phytochemistry and pharmacology. *Fitoterapia* **2018**, *129*, 293–316. [CrossRef] [PubMed]
15. Paterson, R.R. *Cordyceps*: A traditional Chinese medicine and another fungal therapeutic biofactory? *Phytochemistry* **2008**, *69*, 1469–1495. [CrossRef]
16. Das, S.K.; Masuda, M.; Sakurai, A.; Sakakibara, M. Medicinal uses of the mushroom *Cordyceps militaris*: Current state and prospects. *Fitoterapia* **2010**, *81*, 961–968. [CrossRef]
17. Zhang, L.; Fasoyin, O.E.; Molnar, I.; Xu, Y. Secondary metabolites from hypocrealean entomopathogenic fungi: Novel bioactive compounds. *Nat. Prod. Rep.* **2020**. [CrossRef]
18. Zhao, J.; Xie, J.; Wang, L.Y.; Li, S.P. Advanced development in chemical analysis of *Cordyceps*. *J. Pharm. Biomed. Anal.* **2014**, *87*, 271–289. [CrossRef]
19. Xia, Y.L.; Luo, F.F.; Shang, Y.F.; Chen, P.L.; Lu, Y.Z.; Wang, C.S. Fungal cordycepin biosynthesis is coupled with the production of the safeguard molecule pentostatin. *Cell Chem. Biol.* **2017**, *24*, 1479–1489. [CrossRef]
20. Hatton, M.N.; Desai, K.; Le, D.; Vu, A. Excessive postextraction bleeding associated with *Cordyceps sinensis*: A case report and review of select traditional medicines used by Vietnamese people living in the United States. *Oral Surg. Oral Med. Oral Pathol. Oral Radiol.* **2018**, *126*, 494–500. [CrossRef]
21. Tuli, H.S.; Sandhu, S.S.; Sharma, A.K. Pharmacological and therapeutic potential of *Cordyceps* with special reference to cordycepin. *3 Biotech.* **2014**, *4*, 1–12. [CrossRef] [PubMed]
22. Dong, C.; Guo, S.; Wang, W.; Liu, X. *Cordyceps* industry in China. *Mycology* **2015**, *6*, 121–129. [CrossRef] [PubMed]
23. Lu, Y.; Luo, F.; Cen, K.; Xiao, G.; Yin, Y.; Li, C.; Li, Z.; Zhan, S.; Zhang, H.; Wang, C. Omics data reveal the unusual asexual-fruiting nature and secondary metabolic potentials of the medicinal fungus *Cordyceps cicadae*. *BMC Genom.* **2017**, *18*, 668. [CrossRef] [PubMed]
24. Hu, X.; Zhang, Y.J.; Xiao, G.H.; Zheng, P.; Xia, Y.L.; Zhang, X.Y.; St Leger, R.J.; Liu, X.Z.; Wang, C.S. Genome survey uncovers the secrets of sex and lifestyle in caterpillar fungus. *Chin. Sci. Bull.* **2013**, *58*, 2846–2854. [CrossRef]
25. Zheng, P.; Xia, Y.; Xiao, G.; Xiong, C.; Hu, X.; Zhang, S.; Zheng, H.; Huang, Y.; Zhou, Y.; Wang, S.; et al. Genome sequence of the insect pathogenic fungus *Cordyceps militaris*, a valued traditional Chinese medicine. *Genome Biol.* **2011**, *12*, R116. [CrossRef] [PubMed]
26. Zheng, P.; Xia, Y.L.; Zhang, S.W.; Wang, C.S. Genetics of *Cordyceps* and related fungi. *Appl. Microbiol. Biotechnol.* **2013**, *97*, 2797–2804. [CrossRef]
27. Liu, G.; Han, R.; Cao, L. Artificial cultivation of the Chinese cordyceps from injected ghost moth larvae. *Environ. Entomol.* **2019**, *48*, 1088–1094. [CrossRef]
28. Li, X.; Liu, Q.; Li, W.; Li, Q.; Qian, Z.; Liu, X.; Dong, C. A breakthrough in the artificial cultivation of Chinese cordyceps on a large-scale and its impact on science, the economy, and industry. *Crit. Rev. Biotechnol.* **2019**, *39*, 181–191. [CrossRef]
29. Lu, Y.; Xia, Y.; Luo, F.; Dong, C.; Wang, C. Functional convergence and divergence of mating-type genes fulfilling in *Cordyceps militaris*. *Fungal Genet. Biol.* **2016**, *88*, 35–43. [CrossRef]

30. Zhang, S.W.; Cen, K.; Liu, Y.; Zhuo, X.W.; Wang, C.S. Metatranscriptomics analysis of the fruiting caterpillar fungus collected from the Qinghai-Tibetan plateau. *Scientia Sinica Vitae* **2018**, *48*, 562–570.
31. Liao, Y.; Ling, J.; Zhang, G.; Liu, F.; Tao, S.; Han, Z.; Chen, S.; Chen, Z.; Le, H. Cordycepin induces cell cycle arrest and apoptosis by inducing DNA damage and up-regulation of p53 in Leukemia cells. *Cell Cycle* **2015**, *14*, 761–771. [CrossRef] [PubMed]
32. Qin, P.; Li, X.; Yang, H.; Wang, Z.Y.; Lu, D. Therapeutic potential and biological applications of cordycepin and metabolic mechanisms in cordycepin-producing fungi. *Molecules* **2019**, *24*, 2231. [CrossRef] [PubMed]
33. Margolis, J.; Grever, M.R. Pentostatin (Nipent): A review of potential toxicity and its management. *Semin. Oncol.* **2000**, *27*, 9–14. [PubMed]
34. Rodman, L.E.; Farnell, D.R.; Coyne, J.M.; Allan, P.W.; Hill, D.L.; Duncan, K.L.; Tomaszewski, J.E.; Smith, A.C.; Page, J.G. Toxicity of cordycepin in combination with the adenosine deaminase inhibitor 2′-deoxycoformycin in beagle dogs. *Toxicol. Appl. Pharmacol.* **1997**, *147*, 39–45. [CrossRef] [PubMed]
35. Zheng, R.; Zhu, R.; Li, X.; Li, X.; Shen, L.; Chen, Y.; Zhong, Y.; Deng, Y. N6-(2-Hydroxyethyl) adenosine from *Cordyceps cicadae* ameliorates renal interstitial fibrosis and prevents inflammation via TGF-beta1/Smad and NF-kappaB signaling pathway. *Front. Physiol.* **2018**, *9*, 1229. [CrossRef] [PubMed]
36. Liu, K.; Wang, F.; Wang, W.; Dong, C. *Beauveria bassiana*: A new N(6)-(2-hydroxyethyl)-adenosine-producing fungus. *Mycology* **2017**, *8*, 259–266. [CrossRef]
37. Fang, M.; Chai, Y.; Chen, G.; Wang, H.; Huang, B. N6-(2-Hydroxyethyl)-Adenosine Exhibits Insecticidal Activity against *Plutella xylostella* via Adenosine Receptors. *PLoS ONE* **2016**, *11*, e0162859. [CrossRef]
38. Chen, J.F.; Eltzschig, H.K.; Fredholm, B.B. Adenosine receptors as drug targets—What are the challenges? *Nat. Rev. Drug Discov.* **2013**, *12*, 265–286. [CrossRef]
39. Effendi, W.I.; Nagano, T.; Kobayashi, K.; Nishimura, Y. Focusing on Adenosine Receptors as a Potential Targeted Therapy in Human Diseases. *Cells* **2020**, *9*, 785. [CrossRef]
40. Liu, L.X.; Zhang, J.; Chen, C.; Teng, J.T.; Wang, C.S.; Luo, D.Q. Structure and biosynthesis of fumosorinone, a new protein tyrosine phosphatase 1B inhibitor firstly isolated from the entomogenous fungus *Isaria fumosorosea*. *Fungal Genet. Biol.* **2015**, *81*, 191–200. [CrossRef]
41. Cheng, Y.; Schneider, B.; Riese, U.; Schubert, B.; Li, Z.; Hamburger, M. Farinosones A-C, neurotrophic alkaloidal metabolites from the entomogenous deuteromycete *Paecilomyces farinosus*. *J. Nat. Prod.* **2004**, *67*, 1854–1858. [CrossRef] [PubMed]
42. Schmidt, K.; Riese, U.; Li, Z.; Hamburger, M. Novel tetramic acids and pyridone alkaloids, militarinones B, C, and D, from the insect pathogenic fungus *Paecilomyces militaris*. *J. Nat. Prod.* **2003**, *66*, 378–383. [CrossRef] [PubMed]
43. Fan, Y.; Liu, X.; Keyhani, N.O.; Tang, G.; Pei, Y.; Zhang, W.; Tong, S. Regulatory cascade and biological activity of *Beauveria bassiana* oosporein that limits bacterial growth after host death. *Proc. Natl. Acad. Sci. USA* **2017**, *114*, E1578–E1586. [CrossRef] [PubMed]
44. Pegram, R.A.; Wyatt, R.D. Avian gout caused by oosporein, a mycotoxin produced by *Caetomium trilaterale*. *Poult. Sci.* **1981**, *60*, 2429–2440. [CrossRef] [PubMed]
45. Wang, X.; Gao, Y.L.; Zhang, M.L.; Zhang, H.D.; Huang, J.Z.; Li, L. Genome mining and biosynthesis of the Acyl-CoA: Cholesterol acyltransferase inhibitor beauveriolide I and III in *Cordyceps militaris*. *J. Biotechnol.* **2020**, *309*, 85–91. [CrossRef]
46. Wang, J.-H.; Zhang, Z.-L.; Wang, Y.-Q.; Yang, M.; Wang, C.-H.; Li, X.-W.; Guo, Y.-W. Chemical constituents from mycelia and spores of fungus *Cordyceps cicadae*. *Chin. Herb. Med.* **2017**, *9*, 188–192. [CrossRef]
47. Nakaya, S.; Mizuno, S.; Ishigami, H.; Yamakawa, Y.; Kawagishi, H.; Ushimaru, T. New rapid screening method for anti-aging compounds using budding yeast and identification of beauveriolide I as a potent active compound. *Biosci. Biotechnol. Biochem.* **2012**, *76*, 1226–1228. [CrossRef]
48. Witter, D.P.; Chen, Y.; Rogel, J.K.; Boldt, G.E.; Wentworth, P., Jr. The natural products beauveriolide I and III: A new class of beta-amyloid-lowering compounds. *ChemBioChem* **2009**, *10*, 1344–1347. [CrossRef]
49. Ohshiro, T.; Kobayashi, K.; Ohba, M.; Matsuda, D.; Rudel, L.L.; Takahashi, T.; Doi, T.; Tomoda, H. Selective inhibition of sterol O-acyltransferase 1 isozyme by beauveriolide III in intact cells. *Sci. Rep.* **2017**, *7*, 4163. [CrossRef]
50. Mallebrera, B.; Juan-Garcia, A.; Font, G.; Ruiz, M.J. Mechanisms of beauvericin toxicity and antioxidant cellular defense. *Toxicol. Lett.* **2016**, *246*, 28–34. [CrossRef]

51. Shang, Y.F.; Xiao, G.H.; Zheng, P.; Cen, K.; Zhan, S.; Wang, C.S. Divergent and convergent evolution of fungal pathogenicity. *Genome Biol. Evol.* **2016**, *8*, 1374–1387. [CrossRef] [PubMed]
52. Choi, J.N.; Kim, J.; Lee, M.Y.; Park, D.K.; Hong, Y.S.; Lee, C.H. Metabolomics revealed novel isoflavones and optimal cultivation time of *Cordyceps militaris* fermentation. *J. Agric. Food Chem.* **2010**, *58*, 4258–4267. [CrossRef] [PubMed]
53. Dubey, M.K.; Aamir, M.; Kaushik, M.S.; Khare, S.; Meena, M.; Singh, S.; Upadhyay, R.S. PR Toxin-Biosynthesis, Genetic Regulation, Toxicological Potential, Prevention and Control Measures: Overview and Challenges. *Front. Pharmacol.* **2018**, *9*, 288. [CrossRef]
54. Hidalgo, P.I.; Ullán, R.V.; Albillos, S.M.; Montero, O.; Fernández-Bodega, M.; García-Estrada, C.; Fernández-Aguado, M.; Martín, J.F. Molecular characterization of the PR-toxin gene cluster in *Penicillium roqueforti* and *Penicillium chrysogenum*: Cross talk of secondary metabolite pathways. *Fungal Genet. Biol.* **2014**, *62*, 11–24. [CrossRef] [PubMed]
55. Woloshuk, C.P.; Shim, W.B. Aflatoxins, fumonisins, and trichothecenes: A convergence of knowledge. *FEMS Microbiol. Rev.* **2013**, *37*, 94–109. [CrossRef]
56. Proctor, R.H.; McCormick, S.P.; Kim, H.S.; Cardoza, R.E.; Stanley, A.M.; Lindo, L.; Kelly, A.; Brown, D.W.; Lee, T.; Vaughan, M.M.; et al. Evolution of structural diversity of trichothecenes, a family of toxins produced by plant pathogenic and entomopathogenic fungi. *PLoS Pathog.* **2018**, *14*, e1006946. [CrossRef]
57. Chen, Y.; Kistler, H.C.; Ma, Z. *Fusarium graminearum* trichothecene mycotoxins: Biosynthesis, regulation, and management. *Annu. Rev. Phytopathol.* **2019**, *57*, 15–39. [CrossRef]
58. Alexander, N.J.; Proctor, R.H.; McCormick, S.P. Genes, gene clusters, and biosynthesis of trichothecenes and fumonisins in *Fusarium*. *Toxin Rev.* **2009**, *28*, 198–215. [CrossRef]
59. Oh, G.S.; Hong, K.H.; Oh, H.; Pae, H.O.; Kim, I.K.; Kim, N.Y.; Kwon, T.O.; Shin, M.K.; Chung, H.T. 4-Acetyl-12,13-epoxyl-9-trichothecene-3,15-diol isolated from the fruiting bodies of *Isaria japonica* Yasuda induces apoptosis of human leukemia cells (HL-60). *Biol. Pharm Bull.* **2001**, *24*, 785–789. [CrossRef]
60. Liu, H.; Wang, G.; Li, W.; Liu, X.; Li, E.; Yin, W.B. A highly efficient genetic system for the identification of a harzianum B biosynthetic gene cluster in *Trichoderma hypoxylon*. *Microbiology* **2018**, *164*, 769–778. [CrossRef]
61. Lindo, L.; McCormick, S.P.; Cardoza, R.E.; Brown, D.W.; Kim, H.S.; Alexander, N.J.; Proctor, R.H.; Gutiérrez, S. Effect of deletion of a trichothecene toxin regulatory gene on the secondary metabolism transcriptome of the saprotrophic fungus *Trichoderma arundinaceum*. *Fungal Genet. Biol.* **2018**, *119*, 29–46. [CrossRef] [PubMed]
62. Xu, Y.; Vinas, M.; Alsarrag, A.; Su, L.; Pfohl, K.; Rohlfs, M.; Schafer, W.; Chen, W.; Karlovsky, P. Bis-naphthopyrone pigments protect filamentous ascomycetes from a wide range of predators. *Nat. Commun.* **2019**, *10*, 3579. [CrossRef]
63. Chen, Y.X.; Feng, P.; Shang, Y.F.; Xu, Y.J.; Wang, C.S. Biosynthesis of non-melanin pigment by a divergent polyketide synthase in *Metarhizium robertsii*. *Fungal Genet. Biol.* **2015**, *81*, 142–149. [CrossRef] [PubMed]
64. Zeng, G.; Zhang, P.; Zhang, Q.; Zhao, H.; Li, Z.; Zhang, X.; Wang, C.; Yin, W.B.; Fang, W.G. Duplication of a *Pks* gene cluster and subsequent functional diversification facilitate environmental adaptation in *Metarhizium* species. *PLoS Genet.* **2018**, *14*, e1007472. [CrossRef] [PubMed]
65. Obermaier, S.; Thiele, W.; Furtges, L.; Muller, M. Enantioselective Phenol Coupling by Laccases in the Biosynthesis of Fungal Dimeric Naphthopyrones. *Angew. Chem. Int. Ed. Engl.* **2019**, *58*, 9125–9128. [CrossRef] [PubMed]
66. Lu, S.; Sun, W.; Meng, J.; Wang, A.; Wang, X.; Tian, J.; Fu, X.; Dai, J.; Liu, Y.; Lai, D.; et al. Bioactive bis-naphtho-γ-pyrones from rice false Smut pathogen *Ustilaginoidea virens*. *J. Agric. Food Chem.* **2015**, *63*, 3501–3508. [CrossRef]
67. Liu, Y.; Wang, J.; Wang, W.; Zhang, H.; Zhang, X.; Han, C. The chemical constituents and pharmacological actions of *Cordyceps sinensis*. *Evid. Based Complement. Alternat. Med.* **2015**, *2015*, 575063.
68. Chu, Z.B.; Chang, J.; Zhu, Y.; Sun, X. Chemical Constituents of *Cordyceps cicadae*. *Nat. Prod. Commun.* **2015**, *10*, 2145–2146. [CrossRef]
69. Chen, P.X.; Wang, S.; Nie, S.; Marcone, M. Properties of *Cordyceps sinensis*: A review. *J. Funct. Foods* **2013**, *5*, 550–569. [CrossRef]
70. Jia, J.M.; Tao, H.H.; Feng, B.M. Cordyceamides A and B from the culture liquid of *Cordyceps sinensis* (BERK.) SACC. *Chem. Pharm. Bull.* **2009**, *57*, 99–101. [CrossRef]
71. Jia, J.M.; Ma, X.C.; Wu, C.F.; Wu, L.J.; Hu, G.S. Cordycedipeptide A, a new cyclodipeptide from the culture liquid of *Cordyceps sinensis* (Berk.) Sacc. *Chem. Pharm. Bull.* **2005**, *53*, 582–583. [CrossRef] [PubMed]

72. Yang, M.L.; Kuo, P.C.; Hwang, T.L.; Wu, T.S. Anti-inflammatory principles from *Cordyceps sinensis*. *J. Nat. Prod.* **2011**, *74*, 1996–2000. [CrossRef] [PubMed]
73. Martel, J.; Ko, Y.F.; Liau, J.C.; Lee, C.S.; Ojcius, D.M.; Lai, H.C.; Young, J.D. Myths and realities surrounding the mysterious caterpillar fungus. *Trends Biotechnol.* **2017**, *35*, 1017–1021. [CrossRef] [PubMed]
74. Liu, G.; Cao, L.; Rao, Z.; Qiu, X.; Han, R. Identification of the genes involved in growth characters of medicinal fungus *Ophiocordyceps sinensis* based on *Agrobacterium tumefaciens*-mediated transformation. *Appl. Microbiol. Biotechnol.* **2020**, *104*, 2663–2674. [CrossRef]
75. Yan, W.J.; Li, T.H.; Lin, Q.Y.; Song, B.; Jiang, Z.D. Safety assessment of *Cordyceps guangdongensis*. *Food Chem. Toxicol.* **2010**, *48*, 3080–3084. [CrossRef]
76. Fung, S.Y.; Lee, S.S.; Tan, N.H.; Pailoor, J. Safety assessment of cultivated fruiting body of *Ophiocordyceps sinensis* evaluated through subacute toxicity in rats. *J. Ethnopharmacol.* **2017**, *206*, 236–244. [CrossRef]

© 2020 by the authors. Licensee MDPI, Basel, Switzerland. This article is an open access article distributed under the terms and conditions of the Creative Commons Attribution (CC BY) license (http://creativecommons.org/licenses/by/4.0/).

MDPI
St. Alban-Anlage 66
4052 Basel
Switzerland
Tel. +41 61 683 77 34
Fax +41 61 302 89 18
www.mdpi.com

Toxins Editorial Office
E-mail: toxins@mdpi.com
www.mdpi.com/journal/toxins

www.ingramcontent.com/pod-product-compliance
Lightning Source LLC
LaVergne TN
LVHW070100170726
843515LV00007B/1577

* 9 7 8 3 0 3 9 4 3 8 4 7 1 *